Driving Factors for Venture Creation and Success in Agricultural Entrepreneurship

Mohd Yasir Arafat
Aligarh Muslim University, Aligarh, India

Imran Saleem
Aligarh Muslim University, Aligarh, India

Jabir Ali
Indian Institute of Management, Jammu, India

Adil Khan
O.P. Jindal University, Raigarh, India

Hamad Hussain Balhareth
Saudi Electronic University, Saudi Arabia

IGI Global
PUBLISHER of TIMELY KNOWLEDGE

A volume in the Advances in Business Strategy and Competitive Advantage (ABSCA) Book Series

Published in the United States of America by
IGI Global
Business Science Reference (an imprint of IGI Global)
701 E. Chocolate Avenue
Hershey PA, USA 17033
Tel: 717-533-8845
Fax: 717-533-8661
E-mail: cust@igi-global.com
Web site: http://www.igi-global.com

Library of Congress Cataloging-in-Publication Data

Names: Arafat, Mohd Yasir, 1989- editor.
Title: Driving factors for venture creation and success in agricultural entrepreneurship / Mohd Yasir Arafat, Imran Saleem, Jabir Ali, Adil Khan, Hamad Hussain Balhareth, editor.
Description: Hershey, PA : Business Science Reference, [2022] | Includes bibliographical references and index. | Summary: "This book highlights the contextual dimensions of the agribusiness industry through which entrepreneurship researchers would be able to enhance their understanding of entrepreneurship by focusing on the following research question: "Why do individuals, farmers, agrarian, start a new business in the agricultural sector and how do they manage entrepreneurial performance, and what impact it has on the economy?""-- Provided by publisher.
Identifiers: LCCN 2021031882 (print) | LCCN 2021031883 (ebook) | ISBN 9781668423493 (hardcover) | ISBN 9781668423509 (paperback) | ISBN 9781668423516 (ebook)
Subjects: LCSH: Agricultural industries. | Agriculture--Economic aspects. | Entrepreneurship.
Classification: LCC HD9000.5 .D684 2022 (print) | LCC HD9000.5 (ebook) | DDC 338.1--dc23
LC record available at https://lccn.loc.gov/2021031882
LC ebook record available at https://lccn.loc.gov/2021031883

This book is published in the IGI Global book series Advances in Business Strategy and Competitive Advantage (ABSCA) (ISSN: 2327-3429; eISSN: 2327-3437)

British Cataloguing in Publication Data
A Cataloguing in Publication record for this book is available from the British Library.

EDITORIAL ADVISORY BOARD

Table of Contents

Detailed Table of Contents

This piece of research aims to explain the drivers of early-stage entrepreneurship in factor-driven economies and how these are affected by several cognitive factors. This study covers literature on several driving factors of entrepreneurial activity, trying to formulate a framework of determinants of early-stage agricultural entrepreneurial activity. For this purpose, the adult population survey (APS) data of factor-driven economies published by GEM has been used. The selected respondents (848) include those individuals who, alone or with other individuals, presently involved in venture creation, including any self-employment in the agricultural sector. The impact of cognitive and social capital factors on early-stage entrepreneurial activity is measured using logistic regression. The findings suggest that its opportunity perception and self-efficacy, which are the major motivators of early-stage entrepreneurship in developing nations. Also, there are gender biases and age-related negativity with respect to new agri-business creation in developing countries.

must alter from its largely sustenance status to becoming a competitive enterprise in the gauging of entrepreneurs, incorporation of business concepts in maneuvering issues of agriculture is the way to go. The present study is a review work engrossed in using agribusiness to boost job creation and raise productivity and the income bases of farmers.

One of the business models that has been developed in the palm oil plantations in Indonesia is the consolidation of the land of smallholders while professionally managed by cooperatives. While there are cooperatives that perform well, some are not so much resulting in the business cease. This chapter aims to analyze a smallholding palm oil business model that consolidates 367 smallholders with a land area of 734 ha in Srimulyo Village, Tungkal Jaya District, South Sumatra. The results of the analysis show that the cooperative in this village has performed well, and the palm oil land has had major productivity. Several factors are associated with the success of the palm oil smallholding business model, such as the trust from members, the cooperative initiatives, and the sustainability of the plantations. In the framework of developing sustainable plantations, this cooperative has received the RSPO certificate. At the same time, the smallholders have also been able to set up savings funds for replanting to be used during the time to rejuvenate plants.

The purpose of this chapter is to extend the research on determinants of entrepreneurial intentions in the agricultural industry by using the theoretical framework of determinants of entrepreneurial intentions and entrepreneurial event model. By employing the DEI and EEM, the researchers evaluate how perceived desirability, perceived feasibility, individual background, and triggering events can influence the attitude of an individual and in turn how entrepreneurial attitude can control entrepreneurial intentions. To achieve the objective, a questionnaire survey was held using the sample of 335 PhD, master, and bachelor students in commerce, business, and economics from an Indian central university. The data was analysed

The need for energy in the agriculture field is increasing as a result of increasing the productivity of these fields. The appearance of smart grid and IoT (internet of objects) enabled farmers to control, manage, and optimize the energy consumption. Agriculture will continue to rely on energy to increase its productivity in line with increasing population and great demand. In this chapter, the authors present an integrated model between SmartFarms, the smart-grid, and optimization methods. In this way, smart forms can participate actively and benefit from the energy market. In this chapter, they consider the electrical energy that is directly used in activities ranging from field processes such as irrigation of land. Energy is also indirectly consumed in synthetic additives notably fertilizers, pesticides, and herbicides. In addition, the authors can consider the electrical energy, which is used for powering some agricultural machines.

Numerous nations hail the agriculture sector as a critical source of wealth creation, and past researches have shown the importance of entrepreneurship in the agriculture industry. However, there is a substantial difference in men and women's rates of taking entrepreneurial initiatives. Prior research has overlooked the significance of entrepreneurial inclination in creating agricultural start-ups from the perspective of gender. The primary objective of this study is to investigate the factors that influence women entrepreneurs working in the business endeavours of the agriculture sector. Using logistic regression, the study looked at a representative interview of 581 samples with individuals (18–65 years of age) from GEM countries. This model demonstrates the connection between the variables' qualities reliant on the data and the determinants. The chapter suggests that policymakers consider the consequences of promoting women's entrepreneurship in the agricultural industry and evolve the policies accordingly.

Preface

Entrepreneurship has been one of the areas of concern across the world for enhancing economic growth and employment. Though agriculture is the mainstay for the majority of the developing countries, it has not attracted much entrepreneurial attention due to its high level of risk and uncertainty. With the recent focus of the government professionalization of agriculture and rural sectors, there has been increasing emphasis on changing the entrepreneurial ecosystem in rural and agriculture domains. The available literature on agricultural entrepreneurship is limited to address the current business challenges and opportunities in developing countries like India. Therefore, it became imperative to understand the theoretical underpinnings and practical sophistication in the process of agricultural venturing. This book tries to capture the key journey of agricultural entrepreneurship with coverage of a broad spectrum of agribusiness, agricultural start-ups, entrepreneurial ecosystem, emerging business models in agriculture, agripreneurship, social entrepreneurship, sustainable development. This book aims at providing a comprehensive reference guide to various stakeholders of agricultural entrepreneurship such as budding entrepreneurs, incubation centres, financing agencies, policymakers, and researchers.

The book is composed of 16 powerful chapters full of academic rigors for enhancing the knowledge base on agricultural entrepreneurship. Chapter 1, "Drivers of Agricultural Entrepreneurship in Factor-Driven Economies,'' highlights the drivers of early-stage entrepreneurship in factor-driven economies and how these are affected by several cognitive factors. Interestingly, Chapter 2, "Transformations of Urban Agro-Ecology Landscape in Territory Transition," analyses the transformation process of the urban agro-ecology landscape in territory transition. It is concluded that agro-ecology is the key element in the construction of food system sovereignty and security which requires the transition towards urban agroecology based on the transformation of social and political power structures moving away from corporate control towards community governance aimed to improve ecosystem services and environmental sustainability of the city.

Chapter 3 evaluated the business solution for migrants as "Scholastic Review of Food, Memory, and Identity: Classical Migrant Experiences" and reviewed several experiences of relevant migrants, using different approaches, creating a link between food, identity, and memory of migrants as well as looking at the sequential movement of food and its interactions, by reviewing extant literature in the global and African contexts. Migration and migrants are evident across the borders of countries around the universe. It was revealed that migrants are encumbered with different experiences-accepting and repelling in the course of migration, as it is glaring that there is a conglomerate between food, memory, and identity.

Chapter 4, " Is Agripreneurship a mitigating measure for agricultural issues in India?" argued about agripreneurship as a tool for mitigating agricultural issues in India. Agripreneurship may be defined as the amalgamation of agriculture and entrepreneurship. It is the choice to assimilate in the quest to make agriculture an enterprise of adjuration in contemporary business engagements. The chapter is a review of works that are engrossed in using agribusiness to boost job creation, raise productivity and the income bases of farmers, considering India as an agrarian economy.

Chapter 5, "Business Model of Palm Oil Smallholding in South Sumatra, Indonesia: Challenges and Future Prospects," showed that the cooperative in this village has performed well and the palm oil land has had major productivity. Several factors are associated with the success of the palm oil smallholding business model, such as the trust from members, the cooperative initiatives, and the sustainability of the plantations.

Chapter 6, "A Study on Cognitive Antecedents of Agricultural Entrepreneurship Intentions of Indian University Students," extended the research on determinants of entrepreneurial intentions in the agricultural industry by using the theoretical framework of determinants of entrepreneurial intentions and entrepreneurial event model. By employing the DEI and EEM, the researchers evaluated how perceived desirability, perceived feasibility, individual background, and triggering events can influence the attitude of an individual and in turn how entrepreneurial attitude can control entrepreneurial intentions.

Chapter 7, "The Implications of the New Geography Framework of Urban Agroecology on Urban Planning," explains Amongst various SDGs, one of the fundamental issue to survival of humanity is about food subsistence and sustenance. The changing dynamics of agro ecology calls for apt policy interventions. It is noted here that urban society has economic, social, political and environmental implications for sustaining urban agro ecology. Thus, urban planning need to be so dynamic so as to address not only the niche class but also to the marginalized sections of the society.

Chapter 8, "A Comprehensive Review of Agricultural Policies in India,'' discusses that despite the diminishing value added of agricultural activities to total domestic production or GDP, the agriculture sector remained the largest employer and thus proved to be the driver of growth and poverty reduction. Therefore, this chapter aims to provide an outline of Indian agriculture growth – reviewing its agricultural policy reforms and observing the fundamental concerns that have shaped the expansion of the agriculture sector.

Chapter 9 presented innovation in the farm sector by optimization of energy usage through "Optimization of Energy in Smart Farms Using a Genetic Algorithm." The appearance of Smart Grid and IoT (internet of objects) enabled farmers to control, manage and optimize energy consumption. This chapter presents an integrated model between Smart Farms, the Smart-grid, and optimization methods as energy are also indirectly consumed in synthetic additives notably fertilizers, pesticides, and herbicides. In addition, we can consider the electrical energy, which is used for powering some agricultural machines.

Chapter 10, "Women enterpreneurs and agricultural start-ups: Cognitive and Social Capital perspective," investigated the factors that influence women entrepreneurs working in the business endeavours of the agriculture sector. They argued that prior research has overlooked the significance of entrepreneurial inclination in creating agricultural start-ups from the perspective of gender. The study looked at a representative interview of 581 samples with individuals (18–65 years of age) from GEM countries using logistic regression. Using logistic regression, the model demonstrates the relationship between the characteristics of variables reliant on the data and the determinants. The paper suggested that the policymakers should consider the consequences of promoting women's entrepreneurship in the agricultural industry and evolve the policies accordingly.

Chapter 11, "The Role of Micro-Entrepreneurship and Self-Help Groups (SHGs) in Women Empowerment and Development: A Bottom of Pyramid Perspective," argued that women's entrepreneurship emerging through Self-Help Groups (SHGs) contributes to economic well-being, sustainability, and poverty reduction. In the recent decade micro-enterprises and SHGs have come up as a prominent solution to the entrepreneurial crisis existing in the country and their role in empowering women and their development needs to be checked. The research study results show that the contributory role of women entrepreneurs to society is considerably worth appreciation. The causal relationship has also surfaced demonstrating the connection between women empowerment and development brought about by SHG and micro-entrepreneurship from a bottom of the pyramid perspective.

Chapter 12, "Rural Innovation Using Technology in Progressive Agriculture: Empowering Women Towards Sustainable Rural Development," attempted to identify the use of sustainable technology with rural innovation, progressive agriculture, and women empowerment for rural development. They argued that sustainable technology is an individual and group effort through the experts and professionals in rural as well as urban people. Rural innovation involves efficient execution of ideas, information, imagination, and innovative initiative in fulfilling social requirements and new ideas are converted into something useful for rural development. The results of this study include the difficulties in progressive effort on rural innovation vis-a-vis planning and executing of advanced farming, protecting traditional arts, and crafting through maximum involvement of rural women.

Chapter 13 on "Social Capital and Entrepreneurship" presented a social capital point of view of entrepreneurship and how the assets and value embedded in entrepreneur social relations could support the success of the enterprise, especially small ones. Social connections are unique in nature, are personal and stable in the long run. These features make them very unique and difficult to imitate. Therefore, if used properly, based on the position of the entrepreneur in the social hierarchy, can generate or at least behave as a very unique and inimitable source of competitive advantage. Capitalizing on such resources could be of help to entrepreneurs especially in times of highly competitive rivalry.

Chapter 14, "Role and Prospects of Women Entrepreneurs in the Agriculture Sector of India," examined the trend and pattern of women agri-entrepreneurs in India and also highlights the challenges faced by them. They argued that despite this appreciable growth, the role of women entrepreneurs remains devitalized and underutilized which requires proper attention by the government and other stakeholders of the country. However, over the past many years, the government has taken various crucial initiatives to promote the role of women entrepreneurs, especially in the agriculture sector. As a result, there has been a noteworthy transformation in terms of the share and contribution of women entrepreneurs who are engaged in the agriculture sector.

Chapter 15 is "Identifying the Barriers and Drivers to Agriculture Entrepreneurship in India." In this paper authors have identified the Barriers and Drivers to Agriculture Entrepreneurship in their study and a comprehensive understanding of barriers and drivers to agri-entrepreneurship has become a necessity in current times. Agriculture entrepreneurship can raise the living standards and helps in the creation of wealth not only for the entrepreneurs but also for other farmers. Agriculture entrepreneurship is very helpful in the global food supply in least-developed and developing countries. The purpose of this paper is to familiarize the readers with the meaning, prospects, barriers, and key drivers of agri-entrepreneurship to engage in agri-entrepreneurship.

Chapter 16 presented the business opportunities in high-value agriculture entitled "An Insight into Research and Studies on Botrytis Cinerea-Control and Analysis." This chapter highlighted that agriculture and agriculture-based industries are the backbones of the Indian economy. Crop cultivation in rural areas has become a challenging task due to competition in terms of cost and quality. Increasing the yield of the crop by using different modern techniques has become a crucial part of research in this field.

Overall, this book provides a comprehensive collection of insights on process, challenges, opportunities and policies related to agricultural entrepreneurship for the world in general and developing nations in particular.

Acknowledgment

It is a matter of great pleasure for us to thank individuals and organizations, who contributed and facilitated the evolution of this project. At the outset we acknowledge authors who poured their wisdom, experience and efforts in enriching this project. Needless to say that it is their contribution and online discussions that this book could see the light of the day. Our whole heartedly thanks to the esteemed authors.

We are also thankful to the reviewers who tirelessly and selflessly shouldered the responsibility of further enriching the book well in time. Their critical evaluation of papers was indeed a learning experience for one and all involved in this project.

This project has taken shape largely at Aligarh Muslim University, Aligarh, India. We thank the authorities and more specially the support staff, who very willingly made their services available even at odd hours.

Each one of us is equally grateful to our respective families, as each one of us parried on the quality time that otherwise they deserved.

Chapter 1
Drivers of Agricultural Entrepreneurship in Factor-Driven Economies:
An Analysis Based on GEM Data

Mohd Saeem Khan
Aligarh Muslim University, Aligarh, India

Mohd Yasir Arafat
Aligarh Muslim University, Aligarh, India

Mohd Asif Khan
Aligarh Muslim University, Aligarh, India

Hashem Abdullah Al Nemer
University of Jeddah, Saudi Arabia

ABSTRACT

This piece of research aims to explain the drivers of early-stage entrepreneurship in factor-driven economies and how these are affected by several cognitive factors. This study covers literature on several driving factors of entrepreneurial activity, trying to formulate a framework of determinants of early-stage agricultural entrepreneurial activity. For this purpose, the adult population survey (APS) data of factor-driven economies published by GEM has been used. The selected respondents (848) include those individuals who, alone or with other individuals, presently involved in venture creation, including any self-employment in the agricultural sector. The impact of cognitive and social capital factors on early-stage entrepreneurial activity is measured using logistic regression. The findings suggest that its opportunity perception and self-efficacy, which are the major motivators of early-stage entrepreneurship in developing nations. Also, there are gender biases and age-related negativity with respect to new agri-business creation in developing countries.

DOI: 10.4018/978-1-6684-2349-3.ch001

INTRODUCTION

Entrepreneurship is regarded as the most important phenomenon resulting in employment generation along with the promotion of creativity and innovation, leading to socio-economic welfare (Acs et al. 2008). Taking this notion to the next level, Sieger et al. (2016) relate the future and wellbeing of the economies with the ongoing and future entrepreneurial activities. In an earlier study, Reynolds et al. (2004) define entrepreneurship as an economy adjuster, shaping the economy with the help of several courses of action. Since entrepreneurship is imperative for multi-dimensional growth of the economy, it is essential to take it to another level and for that, proper policy implementation is required taking into consideration all those factors which promote and restrict individuals from taking up entrepreneurial activities. If this issue is not dealt with appropriate understanding, there are chances of underutilization of human capital, resulting in the continuation of lower standards of living along with poor policy implementation (Langowitz and Minniti 2007).

During the 1900s and 2000s, when the focus of research was on developing a common research framework which could forecast and illustrate a unique set of empirical phenomena (Davidsson, 2005; Shane & Venkataraman, 2000), the entrepreneurship scholars also put their focus on advancing new theories and paradigms which could predict the individual behaviour (Massis et al. 2018). This attention towards theory-driven research helps the entrepreneurship researchers to observe that there exist differences with respect to entrepreneuship trends across sectors (McDougall, 1989; Zahra, 1996). Several scholars laid stress on sector specific entrepreneurial study as it can provide more sensible theoretical explanations, more accuracy while testing things empirically and a more robust implication of theory (Baker, Gedajlovic, &Lubatkin, 2005; Navis &Ozbek, 2016; Welter, 2011; Zahra & Wright, 2011). Even recently, considering the necessity of entrepreneurial process, several studies (Afandi et al. 2017; Arafat et al. 2018; Pindado et al. 2018; Pindado and Sánchez 2017; Arafat and Saleem 2017a) laid stress on the sector-specific study of entrepreneurship which may result in a more precise understanding of the subject.

Furthermore, many studies show that the models and methods applied to generic entrepreneurship research can also be used in studying the same for agriculture entrepreneurship (Carter, 1998; Carter and Rosa, 1998; Borsch and Forsman, 2001; McNally, 2001). Thus, this study concentrates on the agriculture sector's early-stage entrepreneurial activity within the factor-driven economies and on recognising the drivers of new agri-business firm.

There has been little unanimity among the researchers with respect to the meaning of agricultural entrepreneurship (Vik and McElwee 2011; Lans et al. 2013). Many researchers considered agricultural entrepreneurship as setting up a non-agribusiness by people involved in farming (Seuneke et al. 2013). A more acceptable approach is presented by Pindado and Sánchez (2017), who defined it as a decision of an individual to start a new agri-business. This approach goes hand in hand with the definition given by the GEM, which focuses on measuring an individuals' tendency to start a new agri-business. Therefore, this study also considers the same criteria for the measurement of agricultural entrepreneurship. Further, this study considers an agricultural entrepreneur to be a person who, alone or with other individuals, presently involved in venture creation, including any self-employment or selling any goods or services to others in the agricultural sector.

Since this study incorporates the data provided by the Global Entrepreneurship Monitor, the classification of factor-driven or developing economies is also derived from the same source. GEM uses the classification of economic development given by the World Economic Forum (WEF) in its Global Competitiveness Report. WEF identifies developed economies as countries where businesses are more knowledge-intensive, and the service sector expands and on the contrary, the factor-driven economies as the least developed, dominated by agriculture with heavy reliance on (unskilled) labour and natural resources. In the developed economies, agricultural entrepreneurship is characterised by a knowledge-based system, inclusion of new players through the concept of business angel and enhanced linkages of systems between the agencies while in factor-driven economies, these characteristics are yet to find a place (Rajalahti, Woelcke&Pehu, 2005). Table 1 presents a list of factor-driven economies given by the Global Entrepreneurship Monitor.

Table 1. Factor-driven economies as per GEM reports

Angola	Bangladesh	Bolivia
Botswana	BurkinaFaso	Cameroon
Egypt	ElSalvador	Ethiopia
Georgia	Ghana	India
Indonesia	Iran	Libya
Madagascar	Malawi	Morocco
Nigeria	Pakistan	Palestinian Territories
Philippines	Senegal	Sudan
Syria	Tonga	Tunisia
Uganda	Vanuatu	Vietnam
Zambia		

Source: Economy Profiles, GEMCONSORTIUM

However, for this research, data of 10 developing countries is used as only these 10 countries participated in the GEM survey in 2014. The total number of respondents in the forementioned survey is 848. The following graph represents the countries and the number of respondents from each country.

Figure 1. Factor- driven countries participating in GEM 2014 survey
Source: GEM report, 2014

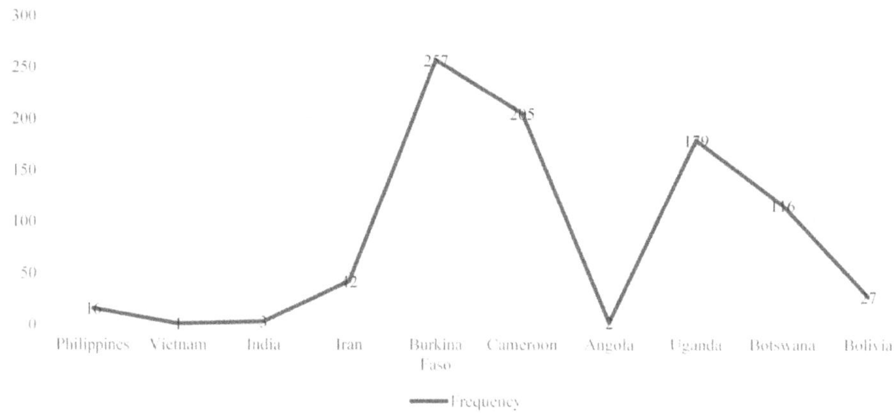

For exploring the basic research question on entrepreneurship (what factors drive new venture creation?) the agriculture sector provides a captivating platform as it holds a major position among the world's largest sectors, providing employment to over one billion people and accounting for 3% of global GDP (FAO, 2016). Observing the work of Alsos et al. (2011) it can be seen that the agricultural sector was never given due importance among the entrepreneurship researches. However, it consists of some specific characteristics distinguishing its entrepreneurial structure from other economic activities. The highlighting characteristics include heavy impact on environment in comparison to other sectors (Britz et al.2012), its basis on biological process with high longitudinal and historical unpredictability (Trnkaet al. 2011). And along with food production, agriculture also helps in conserving bio-diversity, determining the landscapes and creating a cultural legacy (Daugstad et al. 2006). Besides, a recent decline is observed in start-ups in the developing countries as evident from GEM reports (2015/16 and 2016/17). Therefore, in order to survive in this unstable situation it is of utmost importance to explain the configuration of resources for the venture creation process in this sector (Grande et al. 2011; Alsos et al. 2011; Deakins et al. 2016;). Also, the policymakers are of the view that the key to promoting entrepreneurship in agrarian industries is woven in an apt understanding of the agriculture industry start-up process. Hence, according to Vesala and Vesala (2010), there is a need to enrich the farmers' behaviour related to entrepreneurial activities.

Examining the literature related to entrepreneurial behaviour, it becomes clear that there exists a crunch concerning agriculture sector-specific studies (Brunjes and Revilla 2013; McElwee 2006) and the studies are tilted more towards generic entrepreneurial behaviour (Alsos et al. 2011). Most of the entrepreneurship-related studies are inclining towards portfolio expansion and corporate entrepreneurship (Barbieri and Mahoney 2009; Hansson et al. 2013).

It was in 2006 when McElwee published his research which triggered a slew of researches on agricultural entrepreneurship (Lans et al. 2017; Lansetal.2014; Barnes et al. 2015; Grande2011; Pindado and Sánchez 2017), but majority (Barnes et al. 2015; 2016; Ali 2016; Lans et al. 2014; Senger et al. 2017; Grande 2011) only measured existing agri-business's performance. So, what is evident from the available literature is that nearly all the studies focusing on the agricultural sector are targeting the already established business rather than focusing their attention on new venture creation. Also, as far as the researcher's knowledge is concerned, just a handful of researches are concentrated on a specific sector, dealing with the process of venture creation, rather than focusing interest on already established businesses. A study conducted by Ramos-Rodríguez et al. (2012) can be served as an example, though their focus area was hotel and restaurant entrepreneurship.

A recent contribution to the field of research has been made by Yaseen, et al. (2018) where the subject area is Pakistan's agriculture sector and the subject studied is the start-up propensity of that population, but what plagued that study is its small sample size (174). In one of his studies, McElwee (2008) put forth a classification of agriculture entrepreneurs, but what is missing is empirical evidence regarding venture creation motivator in the agriculture industry. Seuneke et al. (2013) also presented their work on agricultural entrepreneurship but rather than focusing on new venture creation, they studied the process through which entrepreneurial skills are learned. In their work, Yaseen et al. present perceived desirability, feasibility, and readiness as entrepreneurship motivators for farmers. However, again, the main problem is small size of sample which cannot be generalised across the globe. Another recently published work (Pindado and Sánchez, 2017) took up the study of new venture creation in agriculture sector but targeted only European countries along with using only a single dimension of social capital. In their more recent research Pindado, Sánchez, Verstegen and Lans (2018) studied European countries through the lenses of social and human capital. Though the authors contributed to the existing literature, the result of this cannot be generalised to the developing world.

Moreover, the work published recently by Arafat et al. (2018) has also studied agricultural entrepreneurship but overlooked developing countries. Another recent piece of entrepreneurship research (Cinar, Du and Heinkel 2018) based on GEM data, covering China, put forth some interesting findings, but their research focuses on generic entrepreneurship rather than industry-specific approach. Therefore, it now seems imperative to undergo a study focusing on new venture creation in the agriculture sector of the developing nations.

Most of the literature on the subject is inclined towards entrepreneurs' internal environment {cognition} (Mitchell et al. 2002; Baron 2004; Baron 2007;) and their external social environment {social capital} (Liao and Welsch 2005; Fuentes Fuentes et al. 2010; Light and Dana 2013; Autio et al. 2014; Neira et al. 2016; Nieto and González-Álvarez 2016; Afandi et al. 2017) for understanding the behaviour of entrepreneurs with regard to entry-exit. Thus, Welter (2011) identified cognition and social capital perspective as the central issue in the realm of researches encircling entrepreneurship.

This study aims to bring out the determinants of agricultural entrepreneurship (early-stage entrepreneurs) in the factor-driven economies. In other words, this study's main motto is to find a person's tendency to create an agri-business in the developing economies, putting the focus on industry-specific research, undertaken with the objective of examining the process of enterprise creation and how it is affected by several factors related to perception, social capital and demographics. Moreover, this study aims to contribute to the literature of entrepreneurship as the cognitive and social psychology framework imbibed in it is majorly used in entrepreneurship research. Hence, with the addition of perception, attitudes and personality traits, a new perspective of research has been added to the sector, bringing it in line with the pioneering research of Arenius and Minniti (2005) whose work inculcated background, perceptual and economic factors while studying nascent entrepreneur.

There are several dimensions in which this study is going to contribute. First and foremost, this study substantiates an industry-specific approach for studying the venture creation process. Secondly, the data used by the researchers are of the best quality, as it is drawn from the most extensive dataset (Global Entrepreneurship Monitor) across the globe with respect to measuring entrepreneurship and its relation with social capital and cognition. The geographical scope of this research is concentrated on factor-driven or developing economies. Additionally, assessing the impact of cognition on decision making of an individual for new enterprise creation in a rural setting, within a highly regulated environment, is also an aim of this study. The reason for becoming an entrepreneur, specifically in agriculture sector, is explained in the findings. Finally, what drives venture creation in agriculture industry is also discussed in the findings, providing a way ahead for policymakers to promote entrepreneurship within the agriculture sector.

The next section discusses the framework providing theoretical background and support for the hypotheses. What follows is the description of variables and methodology used. The implications of the study are discussed in the final section, along with its limitations and scope for future researches

FRAMEWORK OF THE STUDY AND HYPOTHESES

Entrepreneurship is one such process which shows its impact on different social levels and is operative on multiple stages. Also, there is a low success rate and a more significant number of challenges and barriers in the path of enterprise creation (Aldrich and Martinez 2007; Brixy et al. 2012). As the stages unfold, different challenges and barriers are encountered in the process of venture creation, creating a need to study them in a different manner (Robichaud et al. 2007).

Thus, researchers working in the area of entrepreneurship, are studying the multifold stages and levels of the process and adding up various theories and approaches to compile their work (Welter 2011; Zahra et al. 2007). The researches based on cognitive approaches are gaining researchers' attention in studying the factors affecting entrepreneurial activity. Recently, researchers' more attention has been on approaches related to cognitive and social capital to study entrepreneurship research in order to understand factors affecting entrepreneurial activity. As the present study tries to forecast early-stage entrepreneurial activity, it also follows the trend and inculcates cognitive factors.

In accordance with the discussed theoretical parameters of cognition, this study examines the impact of cognitive factors on early-stage agricultural entrepreneurship in developing countries. Entrepreneurial opportunity, perception of risk and self-efficacy are included in the cognitive factor.

Cognitive Factors

There was a kind of agreement in earlier researches with respect to the cognitive process being of utmost importance while studying entrepreneurial activity (Shaver and Scott 1992; Baron 1998; Baron 2004; Krueger et al. 2000; Mitchell et al. 2002; Linen et al. 2011). Thus, literature related to how entrepreneurial behaviour is shaped by the cognitive process is reviewed in this study. Digging up the literature, the factors which are found related to perception are individual perception, social perception and economic perception.

The most researched approach in the past, explaining the uniqueness of an entrepreneur was 'trait approach', and it was quite popular among scholars and policymakers (Gartner 1989). An array of researches can be found where trait approach has been used, and the following traits were studied the most; the need for independence, achievement motivation, risk-taking propensity and locus of internal control (McClelland 1967; Collins and Moore 1964; Brockhaus 1980; Borland 1975; Jennings and Zeithaml 1983). Progressing further, the researchers also clubbed trait factors along with demographic factors like gender, education, religion, experience, age, socioeconomic status etc., to better understand the entrepreneurial behaviour (Cooper and Dunkelberg 1987; Cooper et al. 1994; Reynolds et al. 1994; Storey 1994; Dahlqvist et al. 2000; Wagener et al. 2010). Though this cocktail of traits and demographics was proving beneficial to the policymakers (Santos 2004), criticism started pouring in, hitting its perspectives of theory, concepts and methodology (Gartner 1989; Robinson et al. 1991; Krueger et al. 2000). This mixed approach of traits and demographics was discarded, and its failure was related to its incapability in studying the influence of context and environment on entrepreneurial behaviour. According to Shaver and Scott (1992), the psychology of cognition explains that when individuals interact with situation-specific stimuli, they are bound to behave in a specific manner. Also, this was the popular view from the era following the 1960s (Shaver and Scott 1992; Krueger and Carsrud 1993; Guzman and Santos 2001; Mitchell et al.2002; Mitchell et al.2004; Baron 2004; Liñán and Chen 2009). This approach explains that an individuals' behaviour is moulded by certain psychology-related processes namely motivations, attitudes, learning (Krueger 2003), helping them in gaining, storing, regaining, conveying and converting needful information. Regarding entrepreneurship, structures of knowledge used by individuals for making decisions with respect to growth and enterprise creation are defined as entrepreneurial cognitions (Mitchell et al. 2002).

For assessing the impact of cognition on entrepreneurship, motivation and certain other cognitive factors have been studied. Motivation is considered as the prime reason for new venture creation (Shane et al. 2003), and some earlier studies consider motivation as the most important element of cognition (Collins and Moore 1964; McClelland 1967). An important theory of motivation named 'social learning theory' (SLT) was presented by Bandura (1977, 1982) and it explains that stimuli \within the environment, the feedback process and observational learning lead to the development of an individuals' behaviour. In 1991, this idea got emphasis from Ajzen who presented his new theory and stated that if reasoned action (Fishbein and Ajzen, 1975) and 'perceived behavioural control' theories are added up, predicting human behaviour will be much more feasible. This new theory was given the name 'theory of planned behaviour', and since then, it is extensively used in observing the role of perceptual factors on entrepreneurial behaviour (Shapero and Sokol 1982; Krueger 1993; Kolvereid 1996; Krueger 2003; Liñán and Chen 2009). Therefore, the central cognitive factors pumping entrepreneurial activity are perceptions, referred to as explanation or interpretation of the environment by the individual. This study incorporates three factors of perception, namely opportunity perception, fear of failure and self-efficacy as an aide to explain why entrepreneurs create new venture.

Opportunity Perception

Since the year 2000, when Shane and Venkatraman published their seminal work on entrepreneurship, opportunity perception has been considered as a major predictor of the entrepreneurial intention. Opportunity perception is found to be a motivator for new enterprise development (Shane et al. 2003; McMullen and Shepherd 2006), but according to Baron (1998), it may also result in misconceptions. Ajzen (1991), in his theory of planned behaviour, specified that an individuals' behaviour affects his belief and attitude. When opportunities appear, individuals observe them, and if they find any possibility of opportunity, a positive attitude is established towards behaviour (Ajzen 1991). Therefore, the argument that individuals with positive attitude are more likely to develop a new enterprise hold foot and is supported by recent researches. (Honjo 2015; Arafat and Saleem 2017). Thus, the following hypothesis is articulated for opportunity perception:

Hypothesis 1: Opportunity Perception results in an increase in venture creation propensity in the Agriculture sector of developing countries.

Fear of Failure

Laussel and Le Breton (1995) put forth two approaches to fear of failure; psychological approach and economic approach. The economic approach highlights the adverse effect of risk perception on entrepreneurial decision making (Kihlstrom and Laffont 1979; Arenius and Minniti 2005). A number of studies observed this adversity in economic approach and proposed, to overcome this adversity, it is suggested to reduce the perception of risk (Morales-Gualdrón and Roig 2005; Arenius and Minniti 2005; Langowitz and Minniti 2007; Minniti and Nardone 2007; Wagner 2007; Koellinger et al.2013; Noguera et al. 2013; Urbano and Alvarez 2014; Arafat and Saleem 2017). Moving on to the psychological approach, few researchers dubbed it as a socio-cultural trait (Vaillant and Lafuente 2007; Gómez-Araujo et al. 2015) having an adverse effect on a persons' opinion by setting social benchmarks like failure is a shameful experience (Tezuka 1997; Hessels et al. 2011). Tsai et al. (2016) argued that fear of failure is also affected by perceived capability. Thus, the possibility to face an unpredictable state like new enterprise creation is shrunken due to fear of failure. Following hypotheses is formulated for the risk of failure:

Hypothesis 2: Fear of failure negatively impacts the propensity to create a new venture in the Agriculture sector of developing countries.

Self-Efficacy

Self-efficacy is the aplomb a person has while indulging in an activity (Bandura 1977), and it acts as help in shaping an individuals' belief regarding attainment of certain goals (Cromie2000). An uprise in the confidence of an individual is associated with the creation of a new business in several studies (Krueger and Brazeal 1994; Segal et al. 2002). A positive relation between self-efficacy and intention to create a new enterprise has been observed (Chen et al., 1998), complementing the claim of Boyd and Vozikis (1994) that self-efficacy leads the path to entrepreneurship. Likewise, the available literature on entrepreneurial intention is filled up with studies focused on self-efficacy and its testing with regard to entrepreneurship (Shapero and Sokol 1982; Krueger,1993; Liñán and Chen 2009; Krueger et al. 2000; Arenius and Minniti 2005; Langowitz and Minniti 2007; Roy et al. 2017). Additionally, its evidence can also be found in various meta-analysis (for example, Schlaegel and Koenig 2014) with emphasis on its healthy approach in forecasting entrepreneurial activities. Thus, the formulated hypothesis for self-efficacy is:

Hypothesis 3: Self-efficacy has a positive impact on venture creation propensity in the Agriculture sector of developing countries.

Demographic Factors (Control Variables)

Falling in line with earlier studies, this study also incorporates the demographic factors as important control variables. The literature provides evidence regarding the importance of demographic variables in influencing the behaviour of entrepreneurs and enterprise sustainability (Davidsson and Honig 2003; Arenius and Minniti 2005). Thus, variables of demography like gender, education, age, education, household income and entrepreneurial experience, are also included in this study to explain the entrepreneurial intention. The ability to take risk declines with age and risk-taking is considered as an integral part of the entrepreneurial process; thus age is proposed to have a negative relation with venture creation propensity (Brixy and Hessels 2010). Several global reports present a picture with wide gap between male and female entrepreneurs (Markussen and Røed 2017) and a deeper gap in the factor-driven economies (Vossenberg 2013). Thus, it is suggested that women constitute the disadvantaged group where the creation of a new business is concerned, though, the researchers believe that women-led businesses have an equal chance of survival and growth (Kalleberg and Leicht 1991). Individuals possessing higher educational skills are considered better to process and utilise opportunities due to their confidence in sensing and evaluating entrepreneurial opportunities. Clubbing the earlier entrepreneurial knowledge with high education results in better management of the next enterprise. Also, learning lessons from mistakes result in more confidence and opportunity exploitation for next venture creation (Davidsson and Honig 2003; McKelvie and Wiklund 2010).

Figure 2. Research model
Source: Authors

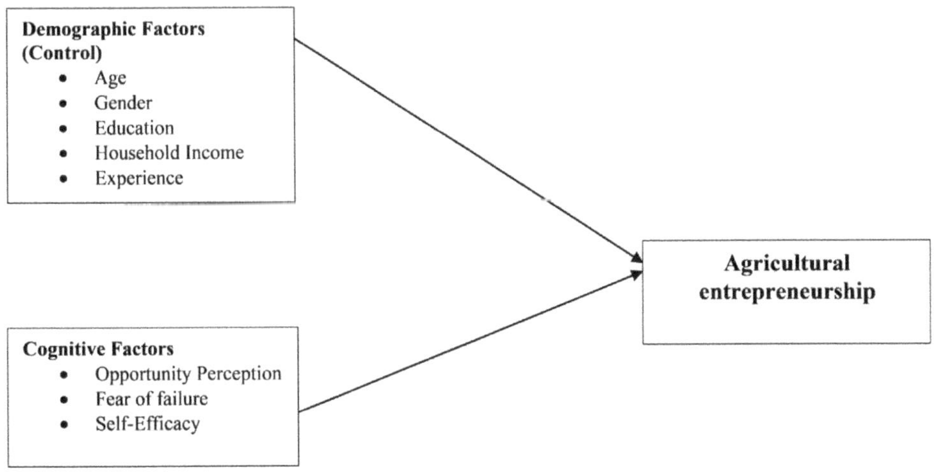

RESEARCH METHODOLOGY

Data

For the purpose of analysis, the data used in this study is from the most distinguished source of entrepreneurship studies, the Global Entrepreneurship Monitor (GEM). The GEM is the largest research organisation with respect to entrepreneurship and publishes entrepreneurial panel data on global scale. The GEM developed a thorough questionnaire for data collection which covers variables such as intellectual capital, perception, social capital and demographic variables. These variables help in investigating a persons' propensity to create a new venture in agriculture sector. The Adult Population Survey including 2,44,471 observations with respect to world-wide businesses under all classifications and categorisation is downloaded from the GEM website (http://gemconsortium.org). Those respondent are selected who are in early stages of venture- creation in developing countries agriculture sector. From the abovementioned data, a set of 848 samples of adults (18–64 years old) was selected. Reynolds et al. (2005) have published the details regarding the data collection process of GEM giving a detailed insight. Coming to the variables, a total of 488 variables are covered in the survey, and for this study, only those variables are considered which have relevance with the study. The following table provides a description of dependent and independent variables.

Table 2. Description of variables

Variable	Label	Description
Dependent	Agricultural Entrepreneurship	Binary variable which takes the value of 1 for agriculture entrepreneurs, those who are actively involved in starting a business and belongs to category "Agriculture, forestry, and fishing" and 0 otherwise. (The definition of the agriculture sector follows the International Standard Industrial Classification (ISIC) –Revision 4.)
Independent	Cognitive Factors	
	Opportunity Perception	Binary variable which takes the value of 1 if the individual see good opportunities to start a firm in the area where they live and 0 otherwise.
	Fear of Failure	Binary variable which takes the value of 1 if he/she indicates that fear of failure would prevent them from setting up a business and 0 otherwise.
	Self-Efficacy	Binary variable which takes the value of 1 if the individual considers that he/she has the knowledge, skills and experience to start up a business and 0 in the other case.
Control	Demographic Factors	
	Age	Age of Individual
	Gender	Binary variable takes value 1 for males and 2 for females.
	Education	The respondents were asked to state the highest level of education they had attained. Their responses were harmonized in all the participating countries into a five-category variable. The five categories are "0=no education", "111 = some secondary", "1212 =secondary degree", "1316 =post-secondary", and "1720=university bachelor's degree or higher". In the logistic regression analysis, the first group is used as the reference category.
	Household Income	The response categories were "lowest 33, middle, and upper 33 percentile, the first income group was considered for reference category.
	Entrepreneurial Experience	The response categories were "1=full: full or part time", "2 =part time only", "3=retired, disabled", "4= homemaker", "5=student", "6 = not working, other". In the logistic regression analysis, the first category was selected as reference

Logistic Regression

In order to measure the effect of a series of independent variables on a single binary (non-metric) dependent variable, this study implies the use of logistic regression technique. It is an econometric technique which helps in forecast and justification of a two-group (binary) categorical dependent variable and thus, is appropriate for this study. Hair et al. (1998) contended that there exists similarity among the logistic regression variate and multiple regression variate.

$$Y_1 = X_1 + X_2 + X_3 + \text{............} + X_n$$

(binary non-metric) (non-metric and metric)

Logistic regression helps in establishing the best-fitting model depicting the relationship between the dependent and independent variables. The coefficients generated through the logistic regression, forecast a logit transformation of the probability of the presence of relationship characteristics. Further, this technique ignores the distribution of data (Greene, 2002). So, for this empirical study the technique of logistic regression has been used due to these arguments:

1. The dependent variable (agricultural entrepreneurship) is binary.
2. Most of the independent variables are also binary or categorical.

RESULTS AND DISCUSSION

Results

The descriptive statistics of the data of factor-driven economies is presented in Table 2. Out of the total, 41% are focusing mainly on new start-ups. In sensing the available opportunities in their area, around 75% of the total respondents have a positive attitude, while only 17% have a fear of failure in an agri-business start-up. A vast majority (82%) feels confident with respect to their ability in starting a new business, and 69% are in touch with existing entrepreneurs. The average age of the respondents is 36 years, while 64% have entrepreneurial experience in some capacity. The correlation results (Table 3) also support some of the hypotheses. Among the cognitive factors, opportunity perception and self-efficacy are significantly and positively related to early-stage entrepreneurial activity. On the other hand, fear of failure has no significant correlation with the propensity to become an entrepreneur. Knowing an entrepreneur is also not significantly correlated with early-stage entrepreneurship in developing economies. Furthermore, among the demographic variables, age is significantly but negatively while gender and entrepreneurial experience are significantly and positively correlated with early-stage entrepreneurship.

The regression results (Table 4) depicts only fear of failure as insignificant among all the cognitive factors. Therefore, the presumed relationship between early-stage entrepreneurship and fear of failure has not been provided support by this result. Thus, the second hypothesis can be rejected as the result contradicted with the presumption. Going by general understanding of the subject, the result may sound ambiguous, but few other studies examining this relationship, Ramos- Rodríguez et al. 2012 (hotel and restaurant industry), Pindado and Sánchez 2017 (agriculture industry) and Neira et al.,2017; Arafat and Saleem 2017 a,b (generic entrepreneurship), have reported similar results showing the same relationship among both the variables.

Table 3. Descriptive statistics

	N	Min	Max	Mean	SD
1.Agricultural entrepreneurship	848	0	1	.41	.492
2. Age	841	18	64	35.82	11.573
3. Gender	848	1	2	1.32	.468
4. Income level	778	33	68100	19234.39	29380.720
5. Entrepreneurial Experience	848	0	1	.64	.480
6. Education Level	844	0	1720	401.95	578.646
7. Opportunity Perception	763	0	1	.75	.431
8. Fear of Failure	846	0	1	.17	.377
9. Self-Efficacy	841	0	1	.82	.387

On the contrary, a positive and significant relationship is observed among opportunity perception and early-stage entrepreneurship, which support the first hypothesis that perception of entrepreneurial opportunities results in venture creation. The odds ratio, in this case, is 1.741, indicating a 74% likelihood for venture creation in agriculture industry among the individuals who are perceiving entrepreneurial opportunities. This result bears similarity with the results of studies examining similar relationship in the context of agricultural entrepreneurship (Pindado and Sánchez 2017) and other type of entrepreneurship researches (Arenius and Minniti

Table 4. Correlation

	1	2	3	4	5	6	7	8	9	10
1. Agri-entrepreneur	1									
2. Age	-0.22***	1								
3. Gender	.168***	-.077*	1							
4. Education Level	-0.051	-0.072*	-0.095**	1						
5. Income Level	-0.018	0.045	-0.098**	0.074*	1					
6.Entrepreneurial Experience	0.517***	-0.074	0.026	-0.069*	0.036	1				
7. Opportunity Perception	0.08*	-0.041	-0.022	0.008	0.051	0.012	1			
8. Self-Efficacy	0.064*	0	0.009	0.105**	0.047	0.033	0.064*	1		
9. Fear of Failure	-0.005	0.013	0.05	0.025	-0.006	-0.055	-0.035	-0.197***	1	

*Correlation is significant at 0.1 level of significance
**Correlation is significant at 0.05 level of significance
***Correlation is significant at 0.01 level of significance

2005; Linan et al.2011; Ramos-Rodríguez et al. 2012; Ahmad et al.2014; Honjo 2015; Tsai et al. 2016; Arafat and Saleem, 2017 b).

Table 5. Logistic regression

	B	S.E.	Wald	df	Sig.	Exp(B)
Demographic Factors						
1. Age	-.054	.009	36.563	1	.000	.947
2. Gender	.924	.223	17.102	1	.000	2.519
3. Household Income			1.554	2	.460	
Middle 33 percentile	.289	.237	1.489	1	.222	1.335
Upper 33 percentile	.080	.243	.109	1	.742	1.083
4.Entrepreneurial Experience	3.428	.304	127.454	1	.000	30.818
5. Education Level	.000	.000	.010	1	.920	1.000
Cognitive Factors						
1.Opportunity Perception	.555	.231	5.775	1	.016	1.741
2. Fear of Failure	.136	.275	.243	1	.622	1.145
3. Self-efficacy	.509	.256	3.964	1	.046	1.664

The third variable under cognitive factors is self-efficacy, which does have a positive and significant relationship with early-stage entrepreneurial venture creation resulting in confirmation of hypothesis 3. The results show an odds ratio of 1.664 denoting that those individuals who are confident about their skills and abilities have a 66% more likelihood to start a new agricultural venture. This result is also in line with the results of previous researches (Arenius and Minniti 2005; Ramos-Rodríguez et al. 2012; Tsai et al. 2016; Arafat and Saleem 2017b).

Among the demographic variables, findings related to age, gender and entrepreneurial experience show highly significant results pointing towards a strong relationship between early entrepreneurship and the respective variables. Though the relationship between age and early entrepreneurial activity is negative, pointing towards a decline in propensity to become an entrepreneur with age. The odds ratio shows that the people in higher age groups are one time less-likely to become an entrepreneur than their young counterparts.

The odds ratio of gender represents the typical orthodox mindset of the developing nations with males 2.5 times more probable to become an entrepreneur rather than females.

Prior entrepreneurial experience is the dominant demographic variable when it comes to developing nations. Its odds ratio depicts that people with prior entrepreneurial experience are 30 times more probable to become an entrepreneur. Results of household income and education level are insignificant and thus depict no relationship between them and early-stage entrepreneurship.

Discussion

Since the results belong specifically to factor-driven economies, some of the findings are contradictory to previous researches, paving the way for further discussion. Out of cognitive factors, fear of failure, which was confirmed by previous researches as a deterrent to venture creation (Shinnar et al. 2012; Noguera et al. 2013; Koellinger et al. 2013; Martin-Sanchez et al. 2017) is found to be insignificant in the case of developing economies. To support these findings, literature does have prior evidence, where Nabi and Liñán (2013) pointed that if there is a dearth of opportunities, it will be rare to find, even risk-taking individuals, engaging in new venture creation. Another study (Tsai et al. 2016) also highlighted that low fear of failure does not necessarily guarantee new venture creation, supporting this study's results.

Opportunity perception is considered as the primary driver of early-stage entrepreneurship in the available literature, and this study is also conducted on the same assumption. The results supported the assumption as the two variables are positively and significantly related. Though the numbers are less than assumed, still the results are considered complementing the existing literature. As for the suggestions regarding the policymakers, they are advised to develop entrepreneurial alertness helping individuals identify the opportunities which are missed otherwise (Kirzner 1979). Also, the policymakers should identify prospective entrepreneurs and familiarise them with the benefits of new venture creation along with designing some industry-specific plans related to the agriculture sector in developing countries.

The results also show that individuals confident in their adeptness and activities have a higher propensity to become an agricultural entrepreneur in developing economies. This finding is imperative for policy formulation as it suggests them to indulge in activities like training and orientation of such individuals, helping them create a new venture.

This study suggests that the government and specific institutes of entrepreneurship to built a platform where people from different walks of life may come together to exchange cogitations and resources for building an environment amicable to entrepreneurship. Since age is negatively related to venture creation in factor-driven economies, government and institutes related to policy formulation should address the young population of such nations with motivational talks and speeches from successful entrepreneurs in order to acquaint them with information regarding successful venture creation. Also, as it is evident from results that there exist gender biases when it comes to new venture creation in factor-driven economies, the policymakers are advised to educate and promote women to participate in the process of enterprise creation. Women entrepreneurs are successful in all the spaces they are operating and are proving beneficial in nation-building.

Education level is not significantly related to enterprise creation in developing countries. This finding is contradictory to what was observed in the recent literature. Scholars (Arafat et al. 2018; Pindado and Sanchez, 2017; Pindado, Sanchez, Verstegen&Lans 2018), in their work on agricultural entrepreneurship, found that education level is strongly related to thepropensity to become an entrepreneur. This unexpected finding could be explained in two ways. First, this may be operationalisation of construct as GEM's education level not captures the type of education followed (e.g. food technology or rural development) which may influence the personal preferences to have a proactive orientation to identify specific business opportunities within the sector(Pindado et al. 2018). Another exaplantion is that the education system of these developing economies is effective and do not enable people to become entrepreneur.

On the other hand, the most significant demographic variable related to enterprise creation is prior entrepreneurial experience. This finding is consistent with previous research (Arafat et al. 2018; Pindado and Sanchez, 2017). This may be because of the developing phase of these economies. As people in these economies might have experienced failure or less opportunities in sectors other than the agriculture.

In general, this study contributes to the literature by being the first of its kind. This study incorporates high-quality, globally accepted GEM data of factor-driven economies. It helps in analysing the impact of cognitive factors on agricultural entrepreneurship in the developing countries. Further, the industry-specific approach is applied for this research which specifies that agricultural entrepreneurship operates in a slightly different ecosystem than other entrepreneurship studies. It helps in assessing the accurate impact of cognition on enterprise creation in the agriculture industry in developing countries.

LIMITATIONS AND CONCLUSION

Limitations of the Study

This study follows suit with other secondary data based researches and carries its limitations. Here the limitations are concerned with the GEM database from where this research extracts its data. To start with, the database lacks question on early entrepreneurial activity, cognition and social capital. Moving on, the format of data is binary which acts as a road-block for the use of several valuable statistical tools like Exploratory Factor Analysis (EFA) and Structural Equation Modelling (SEM). Furthermore, use of 5-point scale data would have provided base for application of more convenient techniques like EFA, CFA, SEM. However, the GEM data has always been praised for its uniqueness and is cogitated very significantly in assessing early-stage entrepreneurial behaviour. Lastly, considering only factor-driven economies for this research can also be considered as a limitation as it neglects numerous other possible dimensions.

Conclusion

Understanding the role of cognitive factors in promoting early-stage agricultural entrepreneurship in developing countries is the main focus of this study. For achieving its aim, this study formulated a model from the data of factor-driven economies published by GEM, using a sample of 848 individuals. Instead of walking the old path by measuring the performance of the agricultural business (e.g. Ali 2016), this study moves forward by assessing the impact of perception of opportunities, fear of failure and self-efficacy on the propensity to become an agriculture entrepreneur. Using logistic regression, the authors observed that only opportunity perception and self-efficacy are important drivers of new venture creation in the agriculture sector of developing countries. Other than these, prior entrepreneurial experience also results in enterprise creation.

This study put forth several relevant suggestions for the policymakers in order to promote entrepreneurship among farmers in the developing countries. The policymakers should identify prospective entrepreneurs and familiarise them with the benefits of new venture creation along with designing some industry-specific plans related to the agriculture sector in developing countries. Policymakers are also advised to indulge in activities like training and orientation of confident individuals, helping them create a new venture. Along with the policymakers, some specific institutes of entrepreneurship are also advised to built a platform where people from different walks of life may come together to exchange cogitations and resources for building an environment amicable to entrepreneurship. Since age is negatively related to venture creation in factor-driven economies, government and institutes related to policy formulation should address the young population of such nations with motivational talks and speeches from successful entrepreneurs in order to acquaint them with information regarding successful venture creation. Also, as it is evident from results that there exist gender biases when it comes to new venture creation in factor-driven economies, the policymakers are advised to educate and promote women to participate in the process of enterprise creation. Women entrepreneurs are successful in all the spaces they are operating and are proving beneficial in nation-building.

Researchers can further undertake this research by studying countries selected on specific parameters like region, economic development level, country union etc. Also, comparison based on the abovementioned parameters provides a platform for future studies. Similarly, the use of more recent GEM data will provide a more comprehensive research base for further study. Additionally, assessing the position of women entrepreneurs in developing economies is an exciting avenue for future research. This study can also be replicated for studying the family owned farm businesses as empirically studied byDeRosa, McElwee& Smith (2019).

REFERENCES

Acs, Z. J., Desai, S., & Hessels, J. (2008). Entrepreneurship, economic development and institutions. *Small Business Economics*, *31*(3), 219–234. https://doi.org/10.1007/s11187-008-9135-9

Afandi, E., Kermani, M., & Mammadov, F. (2017). Social capital and entrepreneurial process. *The International Entrepreneurship and Management Journal*, *13*(3), 685–716. https://doi. org/10.1007/s11365-016-0421-8

Ajzen, I. (1991). *The theory of planned behaviour*. Organizational Behaviour and Human Decision.

Aldrich, H., & Zimmer, C. (1986). Entrepeneurship through social networks. *The Art and Science of Entrepeneurship*. doi:10.2139/ssrn.1497761

Aldrich, H. E., & Fiol, M. (1994). Fools rush in? The institutional context of industrial creation. *Academy of Management Review*, *19*(4), 645–670. https://doi.org/10.5465/AMR.1994.9412190214

Aldrich, H. E., & Martinez, M. A. (2007). Many are called, but few are chosen: An evolutionary perspective for the study of entrepreneurship. In Entrepreneurship (pp. 293–311). Concepts, Theory and Perspective. https://doi.org/10.1007/978-3-540-48543-8_14.

Ali, J. (2016). Performance of small and medium-sized food and agribusiness enterprises: Evidence from indian firms. *The International Food and Agribusiness Management Review*, *19*(4), 53–64. https://doi. org/10.22434/IFAMR2016.0024

Alsos, G. A., Carter, S., & Ljunggren, E. (2011). *The handbook of research on entrepreneurship in agriculture and rural development*. Edward Elgar Publishing.

Amit, R., Glosten, L., & Muller, E. (1993). Challenges to theory development in entrepreneurship research. *Journal of Management Studies, 30*(5), 815–834.

Arafat, M. Y., & Saleem, I. (2017a). Examining start-up intention of Indians through cognitive approach: A study using GEM data. *Journal of Global Entrepreneurship Research, 7*(1), 13. https://doi.org/10.1186 /s40497-017-0073-3

Arafat, M. Y., & Saleem, I. (2017b). Determinants of entrepreneurial startups in India: A study using GEM data. In S. Misra, S. Shukla, & G. Batthini (Eds.), *Twelfth biennial conference on Entrepreneurship. Paper presented at twelfth biennial conference on entrepreneurship, entrepreneurship development Institute of India, Ahmedabad 22–24 February* (pp. 656–665). Delhi: Bookwell.

Arafat, M. Y., Saleem, I., Dwivedi, A. K., & Khan, A. (2018). Determinants of agricultural entrepreneurship: A GEM data-based study. *The International Entrepreneurship and Management Journal*, •••, 1–26.

Arenius, P., & Kovalainen, A. (2006). Similarities and differences across the factors associated with women's self-employment preference in the Nordic countries. *International Small Business Journal, 24*(1), 31–59.

Arenius, P., & Minniti, M. (2005). Perceptual variables and nascent entrepreneurship. *Small Business Economics, 24*(3), 233–247.

Autio, E., Kenney, M., Mustar, P., Siegel, D., & Wright, M. (2014). Entrepreneurial innovation: The importance of context. *Research Policy, 43*(7), 1097–1108. https://doi.org/10.1016/j.respol.2014.01.015

Baker, T., Gedajlovic, E., & Lubatkin, M. (2005). A framework for comparing entrepreneurship processes across nations. *Journal of International Business Studies, 36*(5), 492–504. https://doi.org/10.1057/palgrave.jibs.8400153

Bandura, A. (1977). Self-efficacy: Toward a unifying theory of behavioural change. *Psychological Review, 84*(2), 191–215.

Barbieri, C., & Mahoney, E. (2009). Why is diversification an attractive farm adjustment strategy? Insights from Texas farmers and ranchers. *Journal of Rural Studies, 25*(1), 58–66. .jrurstud.2008.06.001. doi:10.1016/j

Barnes, A. P., Hansson, H., Manevska-Tasevska, G., Shrestha, S. S., & Thomson, S. G. (2015). The influence of diversification on long-term viability of the agricultural sector. *Land Use Policy, 49*, 404–412.

Baron, R. A. (1998). Cognitive mechanisms in entrepreneurship: Why and when entrepreneurs think differently than other people. *Journal of Business Venturing*, *13*(4), 275–294. https://doi.org/10.1016 /S0883-9026(97)00031-1

Baron, R. A. (2004). The cognitive perspective: A valuable tool for answering entrepreneurship's basic "why" questions. *Journal of Business Venturing*, *19*(2), 221–239.

Baron, R. A. (2007). Behavioural and cognitive factors in entrepreneurship: Entrepreneurs as the active element in new venture creation. *Strategic Entrepreneurship Journal*, *1*(1/2), 167–182.

Borland, C. M. (1975). *Locus of control, need for achievement and entrepreneurship* (Doctoral dissertation). ProQuest Information & Learning.

Boyd, N. G., & Vozikis, G. S. (1994). The influence of self-efficacy on the development of entrepreneurial intentions and actions. *Entrepreneurship Theory and Practice*, *18*, 63–63.

Britz, W., van Ittersum, M., Lansink, A. O., & Heckelei, T. (2012). Tools for integrated assessment in agriculture. State of the art and challenges. *Bio-Based and Applied Economics*, *1*(2), 125–150.

Brixy, U., & Hessels, J. (2010). *Human capital and start-up success of nascent entrepreneurs*. EIM Research Reports H, 201013.

Brixy, U., Sternberg, R., & Stüber, H. (2012). The selectiveness of the entrepreneurial process. *Journal of Small Business Management*, *50*(1), 105–131. https://doi.org/10.1111/j.1540-627X.2011.00346.x

Brockhaus, R. H. Sr. (1980). Risk taking propensity of entrepreneurs. *Academy of Management Journal*, *23*(3), 509–520.

Brunjes, J., & Revilla, J. (2013). 'Recession push' and 'prosperity pull' entrepreneurship in a rural developing context. *Entrepreneurship and Regional Development*, *25*, 251–271.

Burge, W. D. (2017). *The relationship between the human and social capital characteristics of nascent entrepreneurs and expected job growth in the United States* (Doctoral dissertation). The University of Southern Mississippi.

Carter, S. (1998). Portfolio entrepreneurship in the farm sector: Indigenous growth in rural areas? *Entrepreneurship and Regional Development*, *10*(1), 17–32.

Carter, S., & Rosa, P. (1998). Indigenous rural firms: Farm enterprises in the UK. *International Small Business Journal, 16*(4), 15–27.

Chen, C. C., Greene, P. G., & Crick, A. (1998). Does entrepreneurial self-efficacy distinguish entrepreneurs from managers? *Journal of Business Venturing, 13*(4), 295–316.

Cinar, E. M., Du, Y., & Hienkel, T. (2018). Chinese entrepreneurship attributes: A comparative GEM data analysis. *Journal of Entrepreneurship in Emerging Economies, 10*(2), 217–248. https://doi.org/10.1108/JEEE-03-2017-0016

Collins, O. F., & Moore, D. G. (1964). The enterprising man. East Lansing: Michigan State University. Coleman, J. S. (1988). Social Capital in the Creation of human capital. *American Journal of Sociology, 94*, S95–S120. https://doi.org/10.1086/228943

Cooper, A. C., & Dunkelberg, W. C. (1987). Entrepreneurial research: Old questions, new answers and methodological issues. *American Journal of Small Business, 11*(3), 11–24.

Cromie, S. (2000). Assessing entrepreneurial inclinations: Some approaches and empirical evidence. *European Journal of Work and Organizational Psychology, 9*(1), 7–30.

Cuervo, Á., Ribeiro, D., & Roig, S. (2007). Entrepreneurship: Concepts, Theory and Perspective. Introduction. In Entrepreneurship (pp. 1–20). Springer.

Dahlqvist, J., Davidsson, P., & Wiklund, J. (2000). Initial conditions as predictors of new venture performance: A replication and extension of the Cooper et al. study. *Enterprise and Innovation Management Studies, 1*(1), 1–17.

Daugstad, K., Rønningen, K., & Skar, B. (2006). Agriculture as an upholder of cultural heritage? Conceptualizations and value judgements—A Norwegian perspective in international context. *Journal of Rural Studies, 22*(1), 67–81.

Davidsson, P. (2005). The types and contextual fit of entrepreneurial processes. *International Journal of Entrepreneurship Education, 2*, 4–407.

Davidsson, P., & Honig, B. (2003). The role of social and human capital among nascent entrepreneurs. *Journal of Business Venturing, 18*, 301–331. https://doi.org/10.1016/S0883-9026(02)00097-6

De Massis, A., Kotlar, J., Wright, M., & Kellermanns, F. W. (2018). *Sector-based entrepreneurial capabilities and the promise of sector studies in entrepreneurship.* doi:10.1177/1042258717740548

Deakins, D., Bensemann, J., & Battisti, M. (2016). Entrepreneurial skill and regulation: Evidence from primary sector rural entrepreneurs. *International Journal of Entrepreneurial Behaviour & Research*, *22*(2), 234–259.

DeRosa, M., McElwee, G., & Smith, R. (2019). Farm diversification strategies in response to rural policy: A case from rural Italy. *Land Use Policy*, *81*, 291–301. https://doi.org/10.1016/j.landusepol.2018.11.006

Emerson, R. M. (1972). Exchange theory, part II: Exchange relations and networks. *Sociological Theories in Progress*, *2*, 58–87.

FAO, Food and Agricultural Organization of the United Nations. (2016). In *FAO Statistical Yearbook 2012, Part 1—The setting*. Available at www.fao.org/docrep/015/i2490e/i2490e00.htm

Fishbein, M., & Ajzen, I. (1975). *Belief, Attitude, Intention and Behaviour: An Introduction to Theory and Research*. Addison-Wesley.

Fuentes Fuentes, M. M., Ruiz Arroyo, M., Bojica, A. M., & Fernández Pérez, V. (2010). Prior knowledge and social networks in the exploitation of entrepreneurial opportunities. *The International Entrepreneurship and Management Journal*, *6*(4), 481–501. https://doi.org/10.1007/s11365-010-0136-1

Gartner, W. B. (1989). Some Suggestions for Research on Entrepreneurial Traits and Characteristics. *Entrepreneurship Theory and Practice*, *14*(1), 27–38.

Gómez-Araujo, E., & Lafuente, , I. E., Vaillant, Y., & Gómez Núñez, L. M. (2015). The differential impact of selfconfidence, reference models and the fear of failure in young entrepreneurs. *Innovar (Universidad Nacional de Colombia)*, *25*(57), 157–174.

Grande, J. (2011). New venture creation in the farm sector–critical resources and capabilities. *Journal of Rural Studies*, *27*(2), 220–233.

Grande, J., Madsen, E. L., & Borch, O. J. (2011). The relationship between resources, entrepreneurial orientation and performance in farm-based ventures. *Entrepreneurship and Regional Development*, *23*(34), 89–111. https://doi.org/10.1080/08985620903183710

Granovetter, M. S. (1992). Problems of explanation in economic sociology. In N. Nohria & R. G. Eccles (Eds.), *Networks and organizations: Structure, form and action* (pp. 25–56). Harvard Business School Press.

Greene, W. H. (2002). *Econometric analysis* (5th ed.). Prentice Hall.

Grootaert, C., & Van Bastelaer, T. (2001). Understanding and measuring social capital: A synthesis of findings and recommendations from the social capital initiative. *Social Capital Initiative Working Paper*, *24*(24), 1–31. doi:10.1227/00006123-197907010-00058

Guzman, J., & Santos, F. J. (2001). The booster function and the entrepreneurial quality: An application to the province of Seville. *Entrepreneurship and Regional Development*, *13*(3), 211–228.

Hair, J. F., Black, W. C., Babin, B. J., Anderson, R. E., & Tatham, R. L. (1998). Multivariate data analysis. Prentice Hall.

Hansson, H., Ferguson, R., Olofsson, C., & Rantamäki-Lahtinen, L. (2013). Farmers' motives for diversifying their farm business -the influence of family. *Journal of Rural Studies*, *32*, 240–250. https://doi. org/10.1016/j.jrurstud.2013.07.002

Hessels, J., Grilo, I., Thurik, R., & van der Zwan, P. (2011). Entrepreneurial exit and entrepreneurial engagement. *Journal of Evolutionary Economics*, *21*(3), 447–471.

Honjo, Y. (2015). Why are entrepreneurship levels so low in Japan? *Japan and the World Economy*, *36*, 88101. https://doi.org/10.1016/j.japwor.2015.08.002

Jennings, D. E., & Zeithaml, C. P. (1983). Locus of Control: A Review and Directions for Entrepreneurial Research. In Academy of Management Proceedings (No. 1, pp. 417–421). Briarcliff Manor: Academy of Management.

Kalleberg, A. L., & Leicht, K. T. (1991). Gender and organizational performance: Determinants of small business survival and success. *Academy of Management Journal*, *34*(1), 136–161. https://doi.org/10.2307 /256305

Kihlstrom, R. E., & Laffont, J. J. (1979). A general equilibrium entrepreneurial theory of firm formation based on risk aversion. *Journal of Political Economy*, *87*(4), 719–748.

Kirzner, I. M. (1979). *Perception, opportunity, and profit: Studies in the theory of entrepreneurship*. University of Chicago Press.

Klyver, K., & Hindle, K. (2007). The role of social networks at different stages of business formation. *Small Enterprise Research*, *15*(1), 22–38.

Koellinger, P., Minniti, M., & Schade, C. (2013). Gender differences in entrepreneurial propensity. *Oxford Bulletin of Economics and Statistics*, *75*(2), 213–234.

Kolvereid, L. (1996). Prediction of Employment Status Choice Intentions. *Entrepreneurship Theory and Practice*, *21*(1), 47–58.

Krueger, J. R. F. N., Reilly, M. D., & Carsrud, A. L. (2000). Competing models of entrepreneurial intentions. *Journal of Business Venturing*, *15*(5–6), 411–432. doi:10.1016/S0883-9026(98)00033-0

Krueger, N. F. (2003). The cognitive psychology of entrepreneurship. In *Handbook of entrepreneurship research* (Vol. 1, pp. 105–140). doi:10.1007/0-387-24519-7_6

Krueger, N. F., & Brazeal, D. V. (1994). Entrepreneurial potential and potential entrepreneurs. *Entrepreneurship Theory and Practice*, *18*, 91–91.

Krueger, N. F., & Carsrud, A. L. (1993). Entrepreneurial intentions: Applying the theory of planned behaviour. *Entrepreneurship and Regional Development*, *5*(4), 315–330.

Langowitz, N., & Minniti, M. (2007). The entrepreneurial propensity of women. *Entrepreneurship Theory and Practice*, *31*(3), 341–364.

Lans, T., Seuneke, P., & Klerkx, L. (2013). Agricultural entrepreneurship. In E. G. Carayannis (Ed.), Encyclopedia of creativity, invention, innovation and entrepreneurship (pp. 44–49). Springer. https://doi.org/10.1007/978-1-4614-3858-8_496.

Lans, T., Seuneke, P., & Klerkx, L. (2017). Agricultural entrepreneurship. In *Encyclopedia of creativity, invention, innovation and entrepreneurship* (pp. 1–7). Springer.

Lans, T., Van Galen, M. A., Verstegen, J. A. A. M., Biemans, H. J. A., & Mulder, M. (2014). Searching for entrepreneurs among small business owner-managers in agriculture. *NJAS Wageningen Journal of Life Sciences*, *68*, 41–51.

Larson, A., & Starr, J. A. (1993). A network model of organization formation. *Entrepreneurship Theory and Practice*, *17*(2), 5–16.

Laussel, D., & Le Breton, M. (1995). A general equilibrium theory of firm formation based on individual unobservable skills. *European Economic Review*, *39*(7), 1303–1319. (94)00092-E doi:10.1016/0014-2921

Liao, J., & Welsch, H. (2005). Roles of social capital in venture creation: Key dimensions and research implications. *Journal of Small Business Management*, *43*, 345–362. X.2005.00141.x doi:10.1111/j.1540-627

Light, I., & Dana, L. P. (2013). Boundaries of social Capital in Entrepreneurship. *Entrepreneurship Theory and Practice*, *37*(3), 603–624. https://doi.org/10.1111/etap.12016

Lin, N. (2003). *Social capital, a theory of social structure and action*. Cambridge University Press.

Lin, N., Ensel, W. M., & Vaughn, J. C. (1981). Social resources and strength of ties: Structural factors in occupational status attainment. *American Sociological Review*, *46*, 393–405.

Liñán, F., & Chen, Y. W. (2009). Development and cross-cultural application of a specific instrument to measure entrepreneurial intentions. *Entrepreneurship Theory and Practice*, *33*(3), 593–617.

Liñán, F., Santos, F. J., & Fernández, J. (2011). The influence of perceptions on potential entrepreneurs. *The International Entrepreneurship and Management Journal*, *7*(3), 373.

Mancuso, J. R. (1975). The entrepreneur's quiz. In C. M. Baumack & J. R. Mancuso (Eds.), *Entrepreneurship and venture management*. Prentice Hall.

Markussen, S., & Røed, K. (2017). The gender gap in entrepreneurship–the role of peer effects. *Journal of Economic Behavior & Organization*, *134*, 356–373.

Martin-Sanchez, V., Contín-Pilart, I., & Larraza-Kintana, M. (2017). The influence of entrepreneurs' social referents on start-up size. *The International Entrepreneurship and Management Journal*, *14*, 173–194. https://doi.org/10.1007/s11365-017-0454-7

Maula, M., Autio, E., & Arenius, P. (2005). What drives micro-angel investments? *Small Business Economics*, *25*(5), 459–475.

McClelland, D. C. (1967). *Achieving society* (Vol. 92051). Simon and Schuster.

McDougall, P. P. (1989). International versus domestic entrepreneurship: New venture strategic behavior and industry structure. *Journal of Business Venturing*, *4*(6), 387–400. https://doi.org/10.1016/0883-9026(89)90009-8

McElwee, G. (2006). Farmers as entrepreneurs: Developing competitive skills. *Journal of Developmental Entrepreneurship*, *11*(03), 187–206.

McElwee, G. (2008). A taxonomy of entrepreneurial farmers. *International Journal of Entrepreneurship and Small Business*, *6*(3), 465–478.

McKelvie, A., & Wiklund, J. (2010). Advancing firm growth research: A focus on growth mode instead of growth rate. *Entrepreneurship Theory and Practice*, *34*(2), 261–288. https://doi.org/10.1111/j.15406520.2010.00375.x

McMullen, J. S., & Shepherd, D. A. (2006). Entrepreneurial action and the role of uncertainty in the theory of the entrepreneur. *Academy of Management Review*, *31*, 132–152. https://doi.org/10.5465 /amr.2006.19379628

Minniti, M., & Nardone, C. (2007). Being in someone else's shoes: The role of gender in nascent entrepreneurship. *Small Business Economics*, *28*(2), 223–238.

Mitchell, R. K., Busenitz, L., Lant, T., McDougall, P. P., Morse, E. A., & Smith, J. B. (2002). Toward a theory of entrepreneurial cognition: Rethinking the people side of entrepreneurship research. *Entrepreneurship Theory and Practice*, *27*(2), 93–104.

Mitchell, R. K., Busenitz, L., Lant, T., McDougall, P. P., Morse, E. A., & Smith, J. B. (2004). The distinctive and inclusive domain of entrepreneurial cognition research. *Entrepreneurship Theory and Practice*, *28*(6), 505–518.

Morales-Gualdrón, S. T., & Roig, S. (2005). The new venture decision: An analysis based on the GEM project database. *The International Entrepreneurship and Management Journal*, *1*(4), 479–499.

Nabi, G., & Liñán, F. (2013). Considering business start-up in recession time: The role of risk perception and economic context in shaping the entrepreneurial intent. *International Journal of Entrepreneurial Behaviour & Research*, *19*(6), 633–655.

Nahapiet, J., & Ghoshal, S. (1998). Social capital, intellectual capital, and the organizational advantage. *Academy of Management Review*, *23*(2), 242–266.

Navis, C., & Ozbek, O. V. (2016). The right people in the wrong places: The paradox of entrepreneurial entry and successful opportunity realization. *Academy of Management Review*, *41*(1), 109–129. https://doi.org/10.5465/amr.2013.0175

Neira, I., Calvo, N., Fernández, L., & Portela, M. (2017). Entrepreneur: Do social capital and culture matter? *The International Entrepreneurship and Management Journal*, *13*(2), 665–683.

Nieto, M., & González-Álvarez, N. (2016). Social capital effects on the discovery and exploitation of entrepreneurial opportunities. *The International Entrepreneurship and Management Journal*, *12*(2), 507530. https://doi.org/10.1007/s11365-014-0353-0

Noguera, M., Alvarez, C., & Urbano, D. (2013). Socio-cultural factors and female entrepreneurship. *The International Entrepreneurship and Management Journal*, *9*(2), 183–197.

Ostgaard, T. A., & Birley, S. (1994). Personal networks and firm competitive strategy—Astrategicor coincidental match? *Journal of Business Venturing*, *9*(4), 281–305. doi:10.1016/0883-9026(94)90009-4

Pindado, E., & Sánchez, M. (2017). Researching the entrepreneurial behaviour of new and existing ventures in European agriculture. *Small Business Economics*, *49*(2), 421–444. https://doi.org/10.1007/s11187-0179837-y

Pindado, E., Sánchez, M., Verstegen, J. A., & Lans, T. (2018). Searching for the entrepreneurs among new entrants in European Agriculture: The role of human and social capital. *Land Use Policy*, *77*, 19–30.

Portes, A. (1998). Social capital: Its origins and applications in modern sociology. *Annual Review of Sociology*, *24*(1), 1–24.

Rajalahti, R., Woelcke, J., & Pehu, E. (2005). *Developing research systems to support the changing agricultural sector* (Vol. 14). Agriculture & Rural Development Department, World Bank.

Ramos-Rodríguez, A. R., Medina-Garrido, J. A., & Ruiz-Navarro, J. (2012). Determinants of hotels and restaurants entrepreneurship: A study using GEM data. *International Journal of Hospitality Management*, *31*(2), 579–587. https://doi.org/10.1016/j.ijhm.2011.08.003

Reynolds, P., Bosma, N., Autio, E., Hunt, S., De Bono, N., Servais, I., & .... (2005). Global entrepreneurship monitor: Data collection design and implementation 1998–2003. *Small Business Economics*, *24*(3), 205231.

Reynolds, P., Storey, D. J., & Westhead, P. (1994). Cross-national comparisons of the variation in new firm formation rates. *Regional Studies*, *28*(4), 443–456.

Reynolds, P. D., Bygrave, W. D., Autio, E., & Arenius, P. (2004). *GEM 2003 global report*. Babson College.

Robichaud, Y., Zinger, J. T., & LeBrasseur, R. (2007). Gender differences within early stage and established small enterprises: An exploratory study. *The International Entrepreneurship and Management Journal*, *3*(3), 323–343. https://doi.org/10.1007/s11365-007-0039-y

Robinson, P. B., Stimpson, D. V., Huefner, J., & Hunt, H. K. (1991). An attitude approach to the prediction of entrepreneurship. *Entrepreneurship Theory and Practice*, *15*(4), 13–31.

Roy, R., Akhtar, F., & Das, N. (2017). Entrepreneurial intention among science & technology students in India: Extending the theory of planned behaviour. *The International Entrepreneurship and Management Journal*, 1–29.

Santiago-Roman, J. (2013). *Entrepreneurism in Puerto Rico: Job creation and entrepreneurial development*. Northcentral University.

Santos, F. J. (2004). Convergencia, desarrollo y empresarialidadenelproceso de globalizacióneconómica. *Revista de Economía Mundial*, *10*(11), 171–202.

Schlaegel, C., & Koenig, M. (2014). Determinants of entrepreneurial intent: A meta-analytic test and integration of competing models. *Entrepreneurship Theory and Practice*, *38*(2), 291–332.

Segal, G., Borgia, D., & Schoenfeld, J. (2002). Using social cognitive career theory to predict self-employment goals. *New England Journal of Entrepreneurship*, *5*(2), 47–56.

Senger, I., Borges, J. A. R., & Machado, J. A. D. (2017). Using the theory of planned behaviour to understand the intention of small farmers in diversifying their agricultural production. *Journal of Rural Studies*, *49*, 32–40.

Seuneke, P., Lans, T., & Wiskerke, J. S. (2013). Moving beyond entrepreneurial skills: Key factors driving entrepreneurial learning in multifunctional agriculture. *Journal of Rural Studies*, *32*, 208–219.

Shane, S., Locke, E. A., & Collins, C. J. (2003). Entrepreneurial motivation. *Human Resource Management Review*, *13*, 257–279. https://doi.org/10.1016/S1053-4822(03)00017-2

Shane, S., & Venkataraman, S. (2000). The promise of entrepreneurship as a field of research. *Academy of Management Review*, *25*(1), 217–226.

Shapero, A., & Sokol, L. (1982). The Social Dimensions of Entrepreneurship, in The Encyclopaedia of Entrepreneurship. In *Encyclopedia of Entrepreneurship* (pp. 72–90). doi:10.1093/oxfordhb/9780199546992.003.0019

Shaver, K. G., & Scott, L. R. (1992). Person, process, choice: The psychology of new venture creation. *Entrepreneurship Theory and Practice*, *16*(2), 23–46.

Shinnar, R. S., Giacomin, O., & Janssen, F. (2012). Entrepreneurial perceptions and intentions: The role of gender and culture. *Entrepreneurship Theory and Practice*, *36*(3), 465–493.

Sieger, P., Fueglistaller, U., & Zellweger, T. (2016). *Student entrepreneurship 2016: Insights from 50 countries*. Academic Press.

Storey, D. (1994). *Understanding The Small Business Sector*. Routledge.

Svendsen, G. L. H., & Svendsen, G. T. (2004). *The creation and destruction of social capital: Entrepreneurship, co-operative movements, and institutions*. Edward Elgar Publishing.

Tezuka, H. (1997). Success as the source of failure? Competition and cooperation in the Japanese economy. *Sloan Management Review*, *38*(2), 83.

Trnka, M., Olesen, J. E., Kersebaum, K. C., Skjelvåg, A. O., Eitzinger, J., Seguin, B., & … . (2011). Agroclimatic conditions in Europe under climate change. *Global Change Biology*, *17*(7), 2298–2318.

Tsai, K.-H., Chang, H.-C., & Peng, C.-Y. (2016). Refining the linkage between perceived capability and entrepreneurial intention: Roles of perceived opportunity, fear of failure, and gender. *The International Entrepreneurship and Management Journal*, *12*(4), 1127–1145.

Urbano, D., & Alvarez, C. (2014). Institutional dimensions and entrepreneurial activity: An international study. *Small Business Economics*, *42*(4), 703–716.

Vaillant, Y., & Lafuente, E. (2007). Do different institutional frameworks condition the influence of local fear of failure and entrepreneurial examples over entrepreneurial activity? *Entrepreneurship and Regional Development*, *19*(4), 313–337.

Veciana, J. M. (2007). Entrepreneurship as a scientific research programme. In A. Cuervo, D. Ribeiro, & S. Roig (Eds.), *Entrepreneurship. Concepts, Theory and Perspective* (pp. 23–71). Springer.

Vesala, H. T., & Vesala, K. M. (2010). Entrepreneurs and producers: Identities of Finnish farmers in 2001 and 2006. *Journal of Rural Studies*, *26*(1), 21–30. https://doi.org/10.1016/j.jrurstud.2009.06.001

Vik, J., & McElwee, G. (2011). Diversification and the entrepreneurial motivations of farmers in Norway. *Journal of Small Business Management*, *49*(3), 390–410. X.2011.00327.x. doi:10.1111/j.1540-627

Vossenberg, S. (2013). Women entrepreneurship promotion in developing countries: What explains the gender gap in entrepreneurship and how to close it. *Maastricht School of Management Working Paper Series, 8*, 1–27.

Wagener, S., Gorgievski, M., & Rijsdijk, S. (2010). Businessman or host? Individual differences between entrepreneurs and small business owners in the hospitality industry. *Service Industries Journal*, *30*(9), 1513–1527.

Wagner, J. (2007). What a difference a Y makes-female and male nascent entrepreneur in Germany. *Small Business Economics*, *28*(1), 1–21.

Welter, F. (2011). Contextualizing entrepreneurship—Conceptual challenges and ways forward. *Entrepreneurship Theory and Practice*, *35*(1), 165–184. https://doi.org/10.1111/j.15406520.2010.00427.x

Yaseen, A., Saleem, M. A., Zahra, S., & Israr, M. (2018). Precursory effects on entrepreneurial behaviour in the agri-food industry. *Journal of Entrepreneurship in Emerging Economies*, *10*(1), 2–22.

Zahra, S. A. (1999). 'Governance, Ownership, and Corporate Entrepreneurship: The Moderating Impact of Industry Technological Opportunities', Academy of Management Journal, 39 (6), December, 1713-35. *International Library of Critical Writings in Economics*, *106*, 325–347.

Zahra, S. A., & Wright, M. (2011). Entrepreneurship's next act. *The Academy of Management Perspectives*, *25*(4), 67–83. https://doi.org/10.5465/amp.2010.0149

Zahra, S. A., Wright, M., & Abdelgawad, S. G. (2014). Contextualization and the advancement of entrepreneurship research. *International Small Business Journal*, *32*(5), 479–500. https://doi.org/10.1177 /0266242613519807

Zamberi Ahmad, S., Roland Xavier, S., & Rahim Abu Bakar, A. (2014). Examining entrepreneurial intention through cognitive approach using Malaysia GEM data. *Journal of Organizational Change Management*, *27*(3), 449–464.

Chapter 2
Transformations of Urban Agro Ecology Landscapes in Territory Transition

José G. Vargas-Hernández
https://orcid.org/0000-0003-0938-4197
Postgraduate and Research Department, Tecnológico Mario Molina, Zapopan, Mexico

ABSTRACT

This chapter has the objective to analyze the transformation process of the urban agro ecology landscape in territory transition. It begins questioning the implications that the agro ecological practices and territorial transformation and transition have on food systems sovereignty and security as well as other effects on land uses, climate change, environmental services, etc. The method used is based on an analytical review of the literature to elaborate a critical perspective of benefits and challenges. It is concluded that agro ecology is the key element in the construction of food system sovereignty and security which requires the transition towards the urban agro ecology based on the transformation of social and political power structures moving away from corporate control towards community governance aimed to achieve improvement ecosystem services and environmental sustainability of the city.

INTRODUCTION

Agro-ecological transition requires a process-goal seeking approach to operate the changes projected and targeted on environmental sustainability and ecosystem services. The economic assessments in terms of positive or negative externalities,

DOI: 10.4018/978-1-6684-2349-3.ch002

environmental costs and benefits of the ecosystem services in agro-ecological transitions are minimal. Steering the agro-ecological transition involves all the stakeholders concerned on a collaborative learning process through which the capacities and capabilities of communities and individuals propose the planning of change and implementation of actions.

The agro-ecological transition is complex interdependent between ecological, economic, social, and technological components and multilaterally power-driven interplay on stakeholders. Technical knowledge is relevant for the agro-ecological transition, as the resources are dedicated to monitoring the changes on the ecosystem services bundles and valuations and the agro-ecological transition, proxies, social perceptions of changes etc. Stakeholders must identify the technical knowledge, acceptability, regulatory and administrative issues, and the agro-ecological transition's external economic, institutional and cultural dynamics.

As a contemporary-scientific perspective on agro-ecological transition, permaculture is an agro-ecological related discipline grouped within life sciences and has a minority of scholarly literature and work. The scholarly work on permaculture design systems has a high level of abstraction engaged in topics beyond biophysical and agricultural but weak ties to agroecology.

As the critical element in the construction of food system sovereignty, agroecology requires the transformation of social and political power structures moving away from corporate control towards community governance.

This analysis of urban agroecology landscape transformations in territory transition begins with a section on the agro-ecological landscapes, to continue with the agro-ecological practices and the implications on the urban territorial transformation and transition agroecology. Finally, there advanced some relevant conclusions.

AGRO-ECOLOGICAL LANDSCAPES

The agro-ecological rationale of small-scale farming represents community-based local agriculture to continue feeding people in most rural and urban landscapes. Urban agroecology is related to the growing, processing, distributing, and marketing of organic food through intensive plant cultivation and animal husbandry in cities. Urban agroecology is a broad term to describe grassroots and institution-led projects of food cultivation and animal husbandry on urban land, which creates alternatives of commons to capitalist organizations reshaping urban landscapes.

Urban sustainable agroecology refers to food cultivation and animal husbandry on urban and peri-urban land, reshaping city landscapes and forms of recreation of Commons as an alternative of urban life organization. Urban agro-ecological actors and stakeholders can open their lands and build socio-economic communities for mutual support to build more natural open green spaces in the rural and urban interfaces.

Urban agroecology is a feature of the urban economy that is not integrated into the urban land use system and remains outside, contributing to income, employment and healthy food supply. Urban agroecology is growing or producing healthy food in urban land used for this purpose in heavily populated settlements. Feeding the world's growing population while caring for the public health and environment, urban agroecology can be the answer to solve the problems if land use, financial incentives, infrastructure and support systems are properly planned and implemented. Urban agroecology and animal production occurring within inner cities and peripheries is a productive process linked to feeding the urban population.

Urban agroecology is integrated and embedded into the urban socio-economic and ecological system, managing the use of natural resources, competing with other uses of land and water, connecting with the urban food systems based on the needs of healthy nutrition of resident consumers.

Agroecology produces food by restoring degraded landscapes of smallholder producers to maximize biodiversity and ecosystem services provided by natural resources. Ecosystems are shaped by the benefits of agro ecological landscapes involving different interests of various stakeholders and actors. Some stakeholders' agro-ecological initiatives, practices and policies to improve biodiversity and natural resource management in natural landscape elements embedded in food systems are the territorialized agro-environmental measures

Hart, McMichael, Milder, & Scherr, (2015).Agroecology demands changes in specific policies and practices to transform the social unjust structures, behaviours and cultures.

From a significant landscape perspective, biodiversity conservation moves agroecology toward sustainable priority more than food crisis. The impacts of climatic changes on food production creating a food crisis require a solution through a sustainable agroecology approach. Climate change impacts agriculture, increasing production uncertainty, land use degradation, and floods and droughts while affecting the management of sustainable natural resources and their contribution to food insecurity. Establishing economically sustainable urban agroecology farming and gardening to make a living by creating productive urban landscapes ensures access to the nutrients and increases the city's food security and self-reliance.

The social component in agro-ecological contexts goes beyond food production, such as the demand for quality of consumers guided by ethics (Boogaard, Bock, Oosting, Wiskerke, &Zijpp, 2010) and the landscapes valued as aesthetic and educational resourcesLindemann-Matthies, Briegel, Schüpbach, &Junge, (2010). Agroecology has multiple dimensions in policy development supporting climate change, organic food production and practices, ecosystems services resilience, etc.

AGRO-ECOLOGICAL PRACTICES

Agroecology applies ecological concepts, principles and practices to the design and management framework for sustainable agricultural ecosystems. Numerous farms around the world are following organic and agro ecological practices. Agro-ecological farms base their production on innovative methods, not using chemical inputs such as fertilizers. Agroecology refers to a scientific discipline, a social movement, and agricultural practices (Wezel, &Soldat, 2009).

Urban agroecology is a promising concept and practice to foster the food production while enhancing the resilience and sustainability of urban ecosystems. Small farms adopting agro-ecological farming techniques and practices increase the ecosystem resilience to irregular weather patterns and are the better keepers of the environmental sustainability (Cohn, Cook, Fernandez, Reider, & Steward, 2006).

Urban agroecology is engaged in food production and other related activities practised by various stakeholders with different socio-economic backgrounds and motivations (Mougeot, 2006; FAO, 2007; Duží et al., 2014)

New concepts, trends and practices in urban agroecology for food production permeate multiple disciplines reflecting new alternative sustainable and environmentally models of production, distribution, supply and consumption. These new models establish new forms of relationships among agro-ecological food producers and consumers in urban geographical proximity sharing responsibilities for inputs and outputs in supply chains (Mundler and Laughrea, 2016), food systems (Holloway, Kneafsey, Venn, Cox,Dowler, Tuomainen, 2007; Kirwan, Ilbery, Naye, & Carey, 2013; Hiroki, Garnevska, &McLaren, 2016), community-supported agriculture – CSA (Hvitsand, 2016) and alternative food networks – AFNs (Renting, Marsden and Banks, 2003; Maye, 2013).

Efforts to combat malnutrition and hunger, improve livelihoods for small-scale farmers, upgrade urban ecosystems and transition to low-carbon and natural resources practices require higher levels of urban agro ecological food production.

The development of urban agroecology food policies focuses on replacing conventional practices for food production and food waste with community-supported participation for food sovereignty through significant structural change. Agroecology practices contribute to challenge and tackle the causes of existing power structures. The biggest agro-ecological challenge is the access to land and water is not always affordable to be used inefficient irrigation practices. Besides policies and investments, the agroecology practices require fiscal incentives, ecosystem services, market opportunities, supply management mechanisms, other institutional support, etc., are required to enable the farmers to acquire long term resources.

Agroecology is considered a science, a practice, and a social movement. *Agroecology* is a science that focuses on the mechanisms, technologies, processes, practices and socioeconomic-ecological and environmental dynamics of diversified agricultural systems. Agroecology is a science-based transformative movement to radically counter ideologies, strategies, policies, and practices to maximize agricultural yields over other economic, socioecological, environmental, and bio-cultural objectives.

Agroecology is related in a dialectical process to alternative agro-food movements, engaged in co-production of knowledge and practice, shaping each other, identifying opportunities and developing policy. Agroecology develops an extensive knowledge about food security, adaptation to local practices and climate mitigation. Agroecology is a social movement, practice and scientific knowledge combined with the local knowledge and experience for a more just and sustainable world.

There is enthusiasm for agroecology science, practices and movements among scientists and academics, farmers and agricultural producers, farmers' organizations, alliances between farmers and non-farmers, food justice organizations, food system actors, and consumers, in general, to achieve the socio-economic and environmental transformations through changing policies to food system governance. Agro-ecological practices produce food integrating ecosystem services as fundamental elements (Wezel, Casagrande, Celette, Vian, Ferrer, &Peigné, 2014). Agroecology is an alternative to conventional agriculture and refers to science, practices and a social movement (Wezel, Bellon, Doré, Francis, Vallod, & David 2009)with the objective to develop healthy and sustainable production systems from socio-economic and ecological dimensions (Altieri 1989).

Agroecology has resurged as place-based knowledge excluded from dominant industrial land-use practices as a response to the degradation left by the agricultural production model. The agro-ecological knowledge and practices are created, acquired and altered continuously by farmers. Agroecology is knowledge-intensive and innovative using participatory methods and exchange networks focusing on farmer knowledge to share practices, expertise, techniques, ideas, ecosystem resilience, weather patterns, and management schemes to minimize dependence on agro-chemical inputs.

Knowledge and practical expertise of agro ecological farming offer insights to scientists. The integrated agro-ecological systems include diverse forms of scientific, technological and indigenous knowledge learning, experiences and practices connected to the food sovereignty movement through transparent and democratic governance. Developing agroecology in any context is based on sustainable alternative agricultural practices sustained by traditional Indigenous knowledge systems for food provisioning practices. Food provisioning activities involves fisheries, foraging and other forms besides agriculture practices aligned with the agro-ecological principles.

Radical urban agro-ecological politics is a critical space of dialogue, education, and practice aimed to transform people's lives through collective work experiences by creating potentially material changes in self-valorizing dynamics of more natural ecosystem spaces in the inner city and peri-urban areas. Urban and peri-urban agroecology studies focus on intensive agro-ecological practices in providing food to urban inhabitants. Agroecology has moved beyond self-sufficiency subsistence-based traditional systems using dynamic practices on natural resources, technology-influenced and other sources of income, sometimes driven by cash crop business. Agroecology practices have increased in urban and peri-urban contexts with a potential promotion of food justice and the advancement of environmental sustainability.

Natural resources supporting agroecology practices based on knowledge and experimentation developed by small scale farmers tend to be more effective to enhance biodiversity, sustainability of soils and water and reducing climate change. Small scale owners and producers of urban agro ecological land help break production, commercial and business monopolies and transform the food system.

Certain urban agroecology practices are healthier and have positive effects than others considering other factors such as regulations, management practices, population density, ecological sensitivity, potential risks, etc. Ecological methods to reduce risks related to agro-chemicals are the use of animal wastes management, proper use of wastewater and irrigation practices. Urban agro-ecological growers adopt the same agro-ecological concepts that already have accepted other food justice and security movements, despite the risks associated with agro-ecological practices. Agro-ecological soil and land management practices can overcome identified constraints, enhance farm yields, and contribute significantly to food security. For example, the shade cover of the coffee agroecosystems is directly related to the mitigation of variability in soil moisture and microclimate for the coffee crop.

The agro-ecological food movement needs new approaches to agricultural production aimed to solve the political confrontations to the economic system while scaling out the perpetuation of land speculation, corporate control, market practices that preclude poor consumers from accessing sustainable food goods, labour injustices, etc. Urban food movements embrace agroecology as a base principle in international and local contexts and at a grassroots non-profit by incorporating agro-ecological practices to address urban hunger.

Agroecology practices protect and maintain biodiversity and ecosystem services through processes such as erosion control and pollination, soil organic matter and water quality and quantity, for example. Stakeholders propose to drive soil erosion control by increasing hedge density and the no-till option. Stakeholders to control erosion may turn to no-till agro-ecological practices or to increase the density of hedges. The agroecosystem intensive in permaculture requires water management through a network of surface impoundments, contour ditches, berms and basins (Lancaster and Marshall 2008; Holmgren 2004).

Any agricultural and agro-ecological system consists of a belief system, knowledge and normative framework and practices to apply and land use goals (Norgaard 1984; BerkesColding&Folke 2000). Agro-ecological practices small-producers can be scaled up to achieve optimal levels of quality compliance and meet higher production standards. Some of the agro ecological practices range from improving the biodiversity and ecosystem services, minimizing environmental impacts, increasing the efficiency of input use in processes and products (Duru and Thérond 2015).

Agroecology is a holistic approach that uses principles that can be applied globally to appropriate local practices based on specific economic, social, cultural and environmental contexts. Indigenous peoples practise a holistic perspective of agroecology. A holistic approach to agroecology can be treated in dimension to explain the potential for living in harmony with other people and nature, leading to human rights and the right to food. The agroecology practice is a whole-systems approach focused on agriculture and developing local food systems based on traditional knowledge and practices, alternative agriculture experiences.

The closest proxy of agroecology is certified organic agricultural production despite failing to explain some dimensions of agro-ecological practices. Agroecology is characterized by integrating crops and livestock of pasture-raised animals relying on organic inputs such as cattle and other ruminants using grazing practices as a form of agro ecological practices. The urban collective agro-ecological gardens and orchards can be used as cultural and educative contexts with activities that provide elements to define theoretical principles, methods and objectives connecting with practices. Agroecology education is an integral concept with different dimensions to foster sustainable and educative practices supported by theories and methods of natural and social knowledge framed by a complex understanding of complex socioecological systems.

Agroecology practices have scaled-up extensive benefits, such as crop rotation, intercropping, organic fertilizers and composting, biological controls, etc., and promoting new techniques, making them more accessible and affordable. Agro-ecological practices improve the agrosystems sustainability based on ecosystem services and biodiversity conservation. Agro-ecological practices influence the flows of the ecosystem services, which impact productivity (Dale and Polasky 2007, Duru and Thérond 2015). The influence of agro-ecological practices at existing farm scale on the ecosystem services are critically needed (Porter, Costanza, Sandhu,Sigsgaard, &Wratten, 2009; Sandhu, Wratten, & Cullen, 2010)to develop a holistic approach to the socio-ecological components of agro-ecological systems

Urban agro ecological practices foster ecosystem services at the urban landscape. Urban space landscape supporting urban agroecology practices favour patterns of urban economic sustainable development. Small-intensive urban agro-ecological crops and farming may be practised in land and housing states, public spaces, schoolyard greenhouses, rooftop gardens and beehives, restaurant-supported salad gardens, allotments, balcony and windowsill vegetable growing guerrilla gardening, and other initiatives (Hou, Johnson, JM,& Lawson, 2009; Mougeot, 2005; Nordahl, 2009; Redwood, 2008).

The concept of agroecology designates future farming systems tooted in science and practice strongly connected to the principle of the right to adequate and healthy food achieved when every human being has physical and economic access or means for procurement. Urban small-scale agro-ecological farm practices, in their own or leased land, more than a business should meet the ecological principles to realize the long term benefits. Agroecology contributes to urban economic development and sustainable development because it is most effectively practised in small urban plots of land and is also more labour-intensive. Agro-ecological production methods and practices are facilitated by the labour of local and migrant farmworkers.

Ecological principles integrated into agro ecological practices increase healthy food production and ecosystem services (Seufert, Ramankutty, & Foley, 2012, Ponisio, M'Gonigle, Mace, Palomino, Valpine, &Kremen, 2014; Garbach, K., J. Milder, DeClerck, Montenegro, Driscoll, & Gemmill-Herren, 2017).Agro-ecological practices are based on sustainability principles that result from the transition from unsustainable agricultural practices depending on the ecological system and the local conditions (Altieri and Nicolls 2005; Reijntjes, Haverkort, & Waters-Bayer1992). Agro-ecological practices have to be implemented to reduce the leaching of nutrients and pesticides to the groundwater or surface water and control contamination. The evolving agro-ecological principles have to be revised in practice which eventually leads to an updating and building a practical guide as the basis for a dialogue on what agroecology means and its practices within the different stakeholder's dynamic agroecology movement.

Agroecology addresses economic, environmental, and social dimensions with new conceptual frameworks and operational tools to be used in sustainable land use planning and policymaking decisions which can be used in agro-ecological contexts (Zhang, Ricketts, Kremen, Carney, & Swinton, 2007, Power 2010, Duru and Thérond 2015, Landis 2017, etc.) and practices (Sandhu et al. 2010, Barral et al. 2015, Rapidel, Ripoche, Allinne, Metay, Lamanda, Blazy, Valdés-Gómez, & Gary, 2015). Agro-ecological policy in some regulatory frameworks is shortsighted to integrate science, practices and movements without recognizing its transformative potential. Policies and regulations are implemented to force the stakeholders to change agro-ecological practices towards an agro-ecological transition and conservation of natural resources and agrobiodiversity.

Urban agroecology is related to urban land use planning integrated into sustainable urban development to regulate and stimulate economic growth, social equity and inclusion, ecological and environmental sustainability. Testing soil and water quality where contamination and pollution occur in urban agro ecological plots and aquaculture is a practice to continue the viability subject to the city planning abilities to develop strategies for effective monitoring and treatment to separate toxic wastes from sewage.

Agro-ecological extension and non-governmental organizations may introduce ecological space-intensive and risk-reducing practices, water and energy-saving technologies, development and assessment technology processes, etc. Agro-ecological research has developed many technologies, inputs and practices such as organic improvement of seeds, optimal use of water and planting density that are helping the small farmers to increase production and reduce the adverse effects of industrial agriculture. Some debates surrounding agro-ecological research based on policy and practice concern the provision of ecosystem services or yield maximization, conventional or organic production, intensification or intensification.

Agro-ecological scientific research in urban contexts may focus on the productivity of agro-ecological practices and environmental services. Research projects focusing on transformative agro-ecological practices and sustainable agriculture are a low proportion of all agricultural research funding, providing insight into the research priorities. Urban agroecology research and extension institutions, together with and to small-scale urban farmers, have to develop and disseminate technologies and agro-ecological methods of food production that do not harm the environment, such as water-saving, irrigation systems, cultivars and production practices.

The government's policy framework for small farming should encourage agro-ecological practices more than industrial agriculture systems through an incentive structure. Local stakeholders adopt agro-ecological practices to be involved, engaged, and embedded in agro-ecological socio-technical networks and regimes (Geels 2004). The Multi-stakeholders planning approach is applied to policy design and programming to integrate urban land uses planning to develop multifunctional sustainable urban agroecology. Urban agro ecological producers are informally organized, and their power and voices are weak with poor participation in urban policymaking and planning in development plans and programs.

Environmental services require different agro-ecological management scenarios and trajectories to design change strategies and trace the outcomes of landscape evolution. Agro-ecological management in urban and peri-urban areas should be linked to local authorities with projects to protect the landscape. A concentric model of land use planning is the zones of use intended to determine the distances and the required management to maximize agro-ecological labour productivity (Mollison and Holmgren 1978; Mollison 1988; Mars 2005; Holmgren 2004; Bell 2005; Hemenway 2009; Bane 2012).

Vacant land in open spaces is a critical asset for urban agroecology. Its availability, suitability and accessibility for urban producers through the demarcation of urban areas of land use, protecting urban green areas, creating buffer zones between areas of land uses conflicts, etc. Urban vacant land in urban areas can be inventoried using GIS or community mapping to analyze suitability for agroecology and stimulate the landowners to give longer terms for lease out their land to local producers. This stimulus can be tax reductions, creating allotment gardens, etc.

Human geography considers landscape planning of urban agroecology supported by cultural studies to sustainable and healthy urban agroecology. Urban agroecology remains a marginal field of human geography beyond the sustainable, food and health issues.

Keyline planning is an innovative design applied to agricultural landscapes and adopted into the developing permaculture framework (Mulligan and Hill 2001, p. 202; Mollison and Holmgren 1978; Mollison 1979). The Keyline landscape planning system, the biochar, the aerobic compost tea and the herb spiral are alternative agro-ecological techniques adopted by permaculture practices (Mollison and Holmgren 1978; Yeomans 1954, Soleil 2012, Avis 2012, Mollison 1988).

Despite the relevance of urban farming and gardening for the provisioning of food to urban populations, local landscape diversity and land use features of agroecosystems are very vulnerable to natural predatory practices and pests that require biological control services.

AGRO-ECOLOGICAL TERRITORIAL TRANSFORMATION AND TRANSITION

Agroecology is a radical transformation of agriculture guided by ecological change and promoted with economic, social, political and cultural changes. The geographical-spatial urban and peri-urban territoriality is an essential component of urban agroecology, taking into consideration the development pressures if these spaces experiencing conflicts of land uses that leads to new synergies (Wästfelt and Zhang, 2016, Fanfani, 2006 in Fanfani, 2013). An agroecology territory is defined as the place engaged in transition process toward sustainable agro-ecological food systems.

Agroecology has a transformative vision co-created by social movements and demonstrated in platforms and mechanisms. A new ecosystem based paradigm is emerging for territorial food system complementing the global food supply chains aimed to create a sustainable food system, improve its local management and providing food security (FAO 2011). The foodshed as a strategy is a broadest definition of local food, the geographical and territorial area in which the urban conglomerate is located and loosely delineating the boundaries from where can be sourced the food (Thompson *et al.* 2008, 4). Agroecology links together theoretical and practical approaches to transform local food systems into environmental sustainability, human health and wellbeing.

Food system transformation takes place in an economic and cultural context able to support transitions to more sustainable practices of development. Traditional and local agro ecological knowledge and practice of farmers converge with technological applications contested by social movements supported by principles of participation, equity and justice, create a process of transformation towards sustainable food systems. Transformation of food systems based on agroecology mitigate climate change. Agro-ecological social movements have in common a commitment to economic and social transformation by different discursive and practice processes to develop alternative food systems framed as political agroecology.

A critical geography of urban agroecology has started to explain that the contemporary urban forms are connected to the spheres of land tenure regimes such as land privatization, commons and rural to urban migrations (Bradley, 2009; Fairlie, 2009), the transformations of local agro-ecological systems and the provision and commodification of urban food and the development planning systems of urban living spaces and environments (Van der Schans and Wiskerke, 2012). Changes of thinking and action in the global food system has a high impact on agroecology, providing farm-scale driven change processes building on the transformation nature of human development, ethical system and culture.

A more holistic territorial food system may support healthy lively-hoods, wellbeing and environmental sustainability. Agroecology as a driver in a local scale territory interconnects relationships of stakeholders (Sebillotte2000) towards a transition to sustainable agro-ecological food systems. A local regime transformation in agroecology can be supported by participatory action research approaches constructed between society, policy and science to facilitate the transition through an inclusive e integrative ecosystem services assessments.

Agroecology is seen as a resistance to an economic model of food system in the forms of the green revolution and bioeconomics to transform and repair materials the food system that has been devastated by industrial food production. The mainstreaming agroecology is being co-opted by the industrial food system to justify the environmental discourse linked to social transformation through the use of terms such as sustainable, climate-smart agriculture, ecological intensification.

Transnational corporate agribusiness and institutions are consolidating unprecedented power in food production, distribution and marketing, which represent a threat to small food producers and agribusiness, food social movements and consumers. Agroecology may transform inequitable power and dynamics in the ago food system. Climate-smart agroecology address the challenges of adaptation, mitigation and food security through policy tools to facilitate transformative transitions supported by science-based practices.

Incentives to support and protect agro ecological practices requires a strategy for building policy to transform the existing industrial agricultural model of production and develop the institutional capacities involving private, public, academic and social sectors. Gliessman (2015, a, b) propose a framework for food system change based on five levels. The first three levels are aimed to transform the industrial and conventional agroecosystems and the following two levels to deepen a broader food system.

The task of agro-ecological transition requires the integration of different sectors in a research program based on rigorous analysis of permaculture theory and practice aimed at the design of agro system. Agro-ecological transition requires a context to make meaningful practices using new knowledge and techniques (Sanford 2011).

A territorial agro-ecological transition requires the implementation of agro-ecological practices aimed to integrate ecosystem services to the landscape scales favored by exchanges of knowledge, extension services delivered, sharing resources, inputs, machinery, technology, etc., and incentives provision. Agro-ecological practices derived from a wide range of methodological and subject approaches. The action research methodology identifies agro-ecological stakeholders' actions in an agro-ecological territory scale, as the relevant space for action (Dalgaard, Hutchings, and Porter 2003).

Agroecology is a science embedded with practice to struggle for food sovereignty. Agroecology is at a crossroads in the global struggle for food sovereignty strengthened by food providers and social movements that require a transformation of practice, policy and research. Agroecology is politically focused to transform control structures of power in society, such as the global markets for self-governance by communities. Urban food strategies have a common element in the governance context of the food system transformation as the stake that cities have to produce food in connection with the urban environment (Unger and Wooten 2006, 11). Development of sustainable agroecology requires structural changes, technological innovation, institutional transformations and regulations, farmer-to-farmer and to consumer networks, etc.

The collapse of the Soviet Union in 1991 brought an immediate decline in Cuban agricultural production forcing small farmers to transform practices. Cuba had to design and implement a model of agricultural development not using heavy machinery and chemical fertilizers, technicians and farmers shared experiences, expertise and innovative practices supported by networks and organizations to achieve an increase in food production. Cuban agriculture highly dependent on Soviet Union inputs faced a challenge to provide food security when suddenly confronted with reduction by responding to alternative agro-ecological technologies to sustain productivity (Rosset, 1997a, 1997b)

Organizations and individual farmers involved on the food sovereignty approach have adopted agro ecological practices. The adoption of agroecology by institutions is considered and treated as a technical matter more than the transformative economic and political change of the system, easily to be co-opted such as has happened with organic agriculture and sustainable development.

Urban ecology requires practical and political engagement for social transformation. Urban agroecology linked to food sovereignty projects are material, political and cultural engagements, dialogs and actions in the context of political and practical social transformation. In the struggles for food sovereignty, engaged actors together have to re-make meaning in a common political project of urban ecology across the different levels, scales and places for social transformation. Agroecology provides a radical space for social and ecological transformation. Local-level initiatives engaged in urban agroecology are connected to transformative political thinking and action. When this connection is weak, the transformative potential of urban agroecology is weak and marginal.

Territories, collective rights and access to the commons are fundamental pillars of agroecology. Urban sprawl, widespread urbanization and urban decentralization are some factors that have contributed to break-up the territory and fragmented the city in networks, eroding the natural and agro-ecological spaces and enlarging the urban footprint.

Addressing agro-environmental issues through innovative approaches at different territorial scales provides potential integration of agro-ecological principles into policy. Designing agro ecological transitions should be based on shared concepts from various knowledge sources and learning supported by participatory implementation, co-constructed feedbacks and assessments processes aimed to build a territorial perspective.

An agro ecological transition to complex and sustainable agroecosystems is required to achieve the transformation of conventional farming systems (Gliessman 2009) based on practices that connects the social and ecological environment. Transformation of agroecosystems imply collective engagements in ecosystem services assessments from a more holistic perspective (Mills 2012).

The territorial agro-ecological transition results of the stakeholders and individual decisions and actions integrating landscape environmental protection. The expectations of the stakeholders regarding the agroecology management of the territory to steer the agro-ecological transition implies the trade-offs among the ecosystem services. Stakeholder agro-ecological actions have to be well organized in agroecology territories using participative and transdisciplinary action-oriented methodology (Méndez, Bacon, and Cohen 2013).

To assure transition, improving the territorial agro-ecological biodiversity and natural resource management are supported on the initiatives from the cooperation between stakeholders and local authorities in charge, non-governmental organizations, nature conservation organizations, etc. The stakeholders' community combine decisions and actions to adapt agricultural practices, biodiversity and natural resources conservation and embedded food systems in an agroecology territory.

Community-supported agroecology is involved in transformative actions of producers and other stakeholders towards sharing common values and building bridges in forging a new global food economy and strong local food sovereignty movement. Development of sustainable agro-ecological food systems at local territorial scales is an alternative to feed the growing world population, considering quantitative production, environmental, economic and social issues.

The geographical-spatial urban and peri-urban territoriality is an essential component of urban agroecology, taking into consideration the development pressures if these spaces experiencing conflicts of land uses that leads to new synergies (Wästfelt and Zhang, 2016, Fanfani, 2006 in Fanfani, 2013). Changing from natural agro-ecological landscape into urban agro-ecological landscape in a spatial-temporal dominated with artificial spatial structures is related to an evolutional transformation that has created landscape fragmentations and ecological declination. Spatial analysis of land use is applied for urban agroecology as a component of multifunctional urban green infrastructures. Urban land-use transformations are related with the processes of economic diversification and proportionally increasing around alternative agricultural and agro ecological land.

Natural resources and improvement of biodiversity conservation and management at landscape scale is a relevant issue in the transition process to agroecology territories. Applying agro-ecological practices enhance resources conservation and biodiversity based on the specialization and diversification of agro-ecological farming systems. Conservation of natural resources and biodiversity in crops, livestock, soil and land are the basis for ecosystem services in established agro ecological territories.

Crop-diversification techniques and practices, along with irrigation technology systems aim to strengthen agro ecology-based productive systems. Agroecology promotes diversification of crops production and use of biodiversity, farming practices and methods, preservation and efficient use of natural resources, minimum employment of artificial and chemical inputs, etc.

Agroecology is innovative knowledge-intensive science and practices converging from agronomy and ecology, organized and participative. Agroecology is knowledge-intensive that requires investment in knowledge creation, dissemination and sharing through networks as a knowhow practice between farmers.

An agroecology territory exists in a transition toward sustainable agro-ecological practices and where biodiversity and resource conservation is linked to territory with an embedded food system, and supported by stakeholders. The stakeholders in an agro-ecological territory are the responsible to adapt the agro-ecological practices, conservation of natural resources and biodiversity for the transition towards an agro-ecological food system. The land-sharing model is based on agroecology and biodiversity conservation on the same territory (Vandermeer and Perfecto2007; Perfecto and Vandermeer, 2010).

Agro-ecology territories establish natural landscape elements to serve as ecological corridors in inner and around cities, to fulfill ecological functions and services while they have also a relevant function for producers of food and to protect against soil erosion, reduction of nutrients and biological agro-ecological conservation control. Urban landscape performance can be improved by integrating design principles of urban agroecology and urban forestry.

All the stakeholders involved in urban agroecology may collaborate to create an inventory of all land uses which can be dedicated to urban agroecology and map urban agroecology initiatives to identify the areas in a database. Technical and social dimensions of agroecology require political struggles to be transformed. In any transition toward agroecology territories, the scale of embedded food systems are socio-technical networks to facilitate the links and interactions between stakeholders, natural resources and artifacts, encompassing the landscapes scales, such as the terroir products and geographical labels, etc. The resilient system is diverse at multiple scales where the elements have different but complementary roles in the agro ecological landscape contributing to the diverse ecosystem services and ecological functions provided to society at large.

Urban agroforestry cultivates perennial woody fruit- and nut-producing species provides food while offering environmental, cultural and recreational services in urban landscapes (Nordahl 2009). The net benefit of agroforestry practices is higher to non-fertilized production and yields higher returns in areas where infrastructure is poor (Ajayi et al. 2007). Urban agroforestry ecology remains separate in their science and practice without integrated ecological multifunctional concepts to improve urban landscape sustainability for urban food forestry. Urban agroforestry, the cultivation of perennial woody plants and crop or animal farming serving multiple purposes such as food security and alleviate poverty (Garrity 2004) has been rarely practiced in cities (Smith 1929; Lelle and Gold 1994; Sachez 1995) to provide ecosystem services, soil regeneration and biodiversity habitat (Belsky 1993; Nair 1993; Kumar 2006; Blanco and Lal 2010).

Integrating scientific principles from agroecology and urban agroforestry into urban forestry may improve nutrition and health. Urban food forestry is defined as the strategic use of woody perennial food-producing species in urban landscapes to improve the sustainability and resilience of urban communities (Clark & Nicholas, 2013). Urban food agroforestry integrates ecosystem services into landscapes through the strategic use of multifunctional species that embody services of urban forestry and urban agroecology, including biodiversity habitat, climate regulation, oxygen production, air and water quality, erosion control (Konijnendijk 2003; Nowak 2006; Nowak and Dwyer 2007), community involvement and social capital, food security, public health and microenterprise opportunities (Brown and Jameton 2000; FAO/ WHO 2008; Dubbeling et al. 2009; Lovell 2010; De Zeeuw et al. 2011). Food agroforestry planting in multi-story designs consider the plant heights (Jacke and Toensmeier 2005; Crawford 2010).

Agroecology principles are related to the permaculture principles that have corollaries articulated at high abstract level (Mollison 1988, Mollison and Slay 1997, Holmgren 2004). Extrapolation from ecological knowledge and principles, current land use may be replaced with permaculture systems in any context (Mollison and Slay 1997, p. 1). Access to public and community land use for local agroecology to produce sustainable food in urban and peri-urban areas and promote integration with urban development planning and policymaking. Policy on urban agroecology based on different scenarios are a specific mix of the perspectives and giving emphasis according to the locations, actors and populations in the different territories.

Permaculture is defined as "Consciously designed landscapes which mimic the patterns and relationships found in nature, while yielding an abundance of food, fiber and energy for provision of local needs" (Holmgren, 2004, p. xix). Permaculture is the harmonious integration of landscape and people providing their food, energy, shelter, other material and non-material needs in a sustainable way (Mollison, 1988). Permaculture contributes to agro-ecological transition design and practice enhanced by exchanging knowledge with related disciplines such as agroecology, ecological engineering, agroforestry, etc.

Permaculture is a conscious design and maintenance of agro-ecological productive ecosystems maintaining the diversity, stability, and resilience of natural ecosystems. The agro-ecological social movements connected to current practices of the science of agroecology and its role in research, have on the agenda the policy landscape. Agroecology and permaculture design and practice overlap in the agro-ecological transition and calls for collaborative research to assess the impacts in sites and field trials through comparative analysis of agroecosystems.

Ecosystem services valuation may lead to sustainable agro ecological landscapes by improving the quality of the environment and reducing inequalities (Jacobs et al. 2016). Permaculture is an international movement that refers to a design system, and a set of associated practices. Permaculture's movement is a conceptual framework for integrating knowledge and practice to design the agro-ecological system transition supported by collaboration between the different stakeholders involved with land users and researchers.

Permaculture is a movement that disseminates and practices a design system framework adapted for the application (Guitart, Pickering, Byrne 2012; Wezel and Soldat 2009). Permaculture contributions incorporates the principles to the agroecologist frameworks for the agro system design of the transition project towards an agroecosystem configuration. Agroecosystem design oriented by permaculture practices is based on site-specificity and spatial configurations of microclimate and interactive components from field-scale polycultures and multiple functions to agroecosystem-scale land-use diversity. Permaculture practices are adopted traditional agro-ecological systems, tropical home gardens and food forest (Mollison and Holmgren 1978).

Permaculture is an active movement in food sovereignty that has the potential to provide solution to the complexity of food systems by involving residents in suburban landscapes with access to soil with nutrients, water and sunlight and making efficient use of urban infrastructure and energy to grow vegetables, fruit and livestock.

Food sovereignty in local communities provides political linkages between regional and local authorities and farmers aimed to redesign the food system, secure land tenure, provide water access, detoxify from use of chemicals in agro-ecological food production and meet the demands for healthy food.

Land use diversity in the permaculture scientific literature includes integrated terrestrial and aquatic systems, perennial and annual plant and animal production and when benefits generate benefits synergies (Mollison and Holmgren 1978; Mollison 1988; Bane 2012, Frei and Becker 2005; Jamu and Piedrahita 2002; Berg 2002; Gomiero, Giampietro, Bukkens, Paoletti, 1999; Kadir, Clonts, Jolly, 1990; Talpaz and Tsur 1982; Devendra and Thomas 2002; Rukera, Mutanga, Micha, 2012; Dey, Paraguas, Kambewa, Pemsl, 2010; Pant, Demaine, Edwards, 2005; Dalsgaard and Oficial 1997).

Changes in the design of food system process may guide the policies and strategies for the transformation of sustainable environment based on agroecology. Urban food strategies and policies aimed to interrelate the food chain and the food cycle develop synergies and transformative innovations between food producers and consumers between urban and rural areas, focusing on a systemic approach of the spatial, socio-economic and ecological perspectives (Lang and Barling, 2012, Sonnino 2009, Viljoen and Wiskerke 2012, Marsden and Sonnino 2012).

CONCLUSION

Urban agroecology projects must be articulated in the food sovereignty and sustainable food system movements for social transformation reached through innovative practices but organized through shifts of power relationships for economic, social, political, cultural and institutional change. Agroecology must transform rather than conforms to the industrial food system supported by the agro-business and institutions. Generation and transfer of agro ecological knowledge and practices must be facilitated by government institutions, non-profit and grassroots organizations.

The transition toward agroecology territory practices must take into consideration the territorial conservation of natural resources and biodiversity and the development of a food system. An integrated land use planning and management collaboration for urban agro ecological practices between local and city governments must apply a holistic and ecosystem approach on natural resources and landscape features. Participative management is a strategy to accomplish the transformation of the local food system. The stakeholders involved in the agro ecological transition must have a shared knowledge and understanding of a multilevel system framework to make decisions on the biophysical-economic-socio-environmental implications embedded in the social valuation.

Transition to agroecology sustainable systems should be designed for biodiversity, resilience and autonomy of local stakeholders and actors with disparate interests, expectations and values but integrated under the ecosystem services valuation aimed to develop shared knowledge for a concerted action.

Local-based agro-ecological practices must take into account the stakeholder's capacity considering the global regulations and forces to undertake local changes at specific agro ecology territories (Pimbert,Thompson, Vorley, Fox, Kanji, &Tacoli, 2001). Plot agroecology practice must not be contentious nature to be practiced within institutional frameworks governing and regulating farming activities and access to available and vacant land and water, framed by conflict resolution and interaction among the stakeholders.

Urban agro-ecological transformation requires public awareness of how can benefit cities, even though agro-food production practices have been present. Integrated ecosystem services valuations monitor the agro-ecological transition, although the policy and governance can be targeted in issues such as the equity in a context of power imbalances

REFERENCES

Ajayi, M. T. (1987). Effects of spacing/number of plants per stand and fertilizer placement on performance of two maize varieties intercropped with cowpea in the forest zone of Nigeria. *Ghana Journal of Agricultural Science*, *20*, 39–46.

Alsagoff, S. A. K., Clonts, H. A., & Jolly, C. M. (1990). An integrated poultry, multi-species aqua culture for Malaysian rice farmers: A mixed integer programming approach. *Agricultural Systems*, *32*(3), 207–231.

Altieri, M. A. (1989). Agroecology: A new research and development paradigm for world agriculture. *Agriculture, Ecosystems & Environment*, *27*(1-4), 37–46. doi:10.1016/0167-8809(89)90070-4

Altieri, M. A., &Nicholls, C. I. (2005). *Agroecology and the search for a truly sustainable agriculture*. United Nations Environmental Programme, Environmental Training Network for Latin America and the Caribbean.

Avis, R. (2012) Compost teas and extracts: brewin' and bubblin' basics. *Res. Inst. - Permac. Forums Courses Inf. News*. https://permaculturenews.org/2012/07/11/compost-teas-and-extracts-brewin-and-bubblin-basics/

Bane, P. (2012). *The permaculture handbook: garden farming for town and country*. New SocietyPublishers.

Barral, Rey Benayas, Meli, & Maceira. (2015). Quantifying the impacts of ecological restoration on biodiversity and ecosystem services in agroecosystems: a global meta-analysis. *Agriculture, Ecosystems and Environment, 202*, 223–231. doi:10.1016/j.agee.2015.01.009

Bell, G. (2005). *The permaculture garden*. Chelsea Green.

Belsky, J. (1993). Household food security, farmt rees, and agroforestry: A comparative study in Indonesia and the Philippines. *Human Organization, 52*(2), 130–141. doi:10.17730/humo.52.2.308kw181875xpt0r

Berg, H. (2002). Rice monoculture and integrated rice-fish farming in the Mekong Delta, Vietnam—Economic and ecological considerations. *Ecological Economics, 41*(1), 95–107. doi:10.1016/S0921-8009(02)00027-7

Blanco-Canqui, H., & Lal, R. (2010). Soil and water conservation. In *Principles of soilconservation and management* (pp. 1–19). Springer.

Boogaard, B. K., Bock, B. B., Oosting, S. J., Wiskerke, J. S., & van der Zijpp, A. J. (2011). Social acceptance of dairy farming: The ambivalence between the two faces of modernity. *Journal of Agricultural & Environmental Ethics, 24*(3), 259–282. doi:10.100710806-010-9256-4

Bradley, M. (2009). Down with the fences: Battles for the commons in South London. *The Land, 7*, 34–38.

Brown, K. H., & Jameton, A. L. (2000). Public health implications of urban agriculture. *Journal of Public Health Policy, 21*(1), 20–39. doi:10.2307/3343472 PMID:10754796

Clark, K. H., & Nicholas, K. A. (2013). Introducing urban food forestry: A multifunctional approach to increase food security and provide ecosystem services. *Landscape Ecology, 28*(9), 1649–1669. doi:10.100710980-013-9903-z

Clark, K. H., & Nicholas, K. A. (2013). Introducing urban food forestry: A multifunctional approach to increase food security and provide ecosystem services. *Landscape Ecology, 28*(9), 1649–1669. doi:10.100710980-013-9903-z

Cohn, A. (2006). *Agroecology and the Struggle for Food Sovereignty in the Americas: A Collaborative Project of the International Institute for Environment and Development (IIED), the IUCN Commissionon Environmental, Economic and Social Policy (CEESP) and the Yale School of Forestry & Environmental Studies (Yale F&ES)*. IIED.

Crawford, M. (2010). *Creating a forest garden*. Green Books.

Dalgaard, T. H., & Porter, J. R. (2003). Agroecología: escalamiento e interdisciplinariedad. Agricultura. *Ecosistemas y Medio Ambiente, 100*(1), 39-51.

Dalsgaard, J. P. T., & Oficial, R. T. (1997). A quantitative approach for assessing the productive performance and ecological contributions of smallholder farms. *Agricultural Systems, 55*(4), 503–533. doi:10.1016/S0308-521X(97)00022-X

De Zeeuw, H., Van Veenhuizen, R., & Dubbeling, M. (2011). The role of urban agriculture in building resilient cities in developing countries. *Thai Journal of Agricultural Science, 149*(S1), 153–163. doi:10.1017/S0021859610001279

Devendra, C. (2002). Crop–animal systems in Asia: Implications for research. *Agricultural Systems, 71*(1-2), 169–177. doi:10.1016/S0308-521X(01)00042-7

Dey, M. M., Paraguas, F. J., Kambewa, P., & Pemsl, D. E. (2010). The impact of integrated aquaculture–agriculture on small-scale farms in Southern Malawi. *Agricultural Economics, 41*(1), 67–79. doi:10.1111/j.1574-0862.2009.00426.x

Doll, W. J., & Torkzadeh, G. (1987). The relationship of MIS steering committees to size of firm and formalization of MIS planning. *Communications of the ACM, 30*(11), 972–978. doi:10.1145/32206.32213

Duru, M., & Therond, O. (2015). Designing agroecological transitions; A review. *Agronomy for Sustainable Development, 35*(4), 1237–1257.

Duží, B., Tóth, A., Bihuňová, M., & Stojanov, R. (2014). Challenges of urban agriculture. Highlights on Czech and Slovak experience. In Current Challenges of Central Europe: Society and Environment (pp. 82–107). Charles University.

Fairlie, S. (2009). A short history of enclosure in Britain. *TheLand, 7*, 16–31.

Fanfani, D. (2006). Ilgoverno del territorio e del paesaggioruralenellospazio "terzo" periurbano. Il parco agricolo come strumento di politiche e di progetto. *Ri-Vista. Researchforlandscapearchitecture*, *6*(2), 54–69.

Fanfani, D. (2013). Local development and "agri-urban" domain: Agricultural park as promotion of an "active ruralship". *Planum: The Journal of Urbanism*, *27*(2), 38–47.

FAO. (2007). *Profitability and Sustainability of Urban and Peri-urban Agriculture*. FAO.

FAO. (2011). *Food, agro ecology and cities: Challenges of food and nutrition security, agro ecology and ecosystem management in an urbanizing world*. http://www.fao.org/fileadmin/templates/FCIT/PDF/FoodAgriCities_Oct2011.pdf

FAO/WB. (2008). *Urban agriculture for sustainable poverty alleviation and food security*. Food and Agriculture Organization of the United Nations. http://www.fao.org/fileadmin/templates/FCIT/PDF/UPA_-WBpaper-Final_October_2008.pdf

Frei, M., & Becker, K. (2005, May). Integrated rice-fish culture: Coupled production saves resources. *Natural Resources Forum*, *29*(2), 135–143. doi:10.1111/j.1477-8947.2005.00122.x

Garbach, K., Milder, J. C., DeClerck, F. A., Montenegro de Wit, M., Driscoll, L., & Gemmill-Herren, B. (2017). Examining multi-functionality for crop yield and ecosystem services in fives ystems of agroecological intensification. *International Journal of Agricultural Sustainability*, *15*(1), 11–28. doi:10.1080/14735903.2016.1174810

Garrity, D. P. (2004). Agroforestry and the achievement of the Millennium Development Goals. *Agroforestry Systems*, *61*(1), 5–17. doi:10.1023/B:AGFO.0000028986.37502.7c

Gliessman, S. (2015). *Agroecology: A growing field*. Academic Press.

Gliessman, S. R. (2009). The framework for conversion. In S. R. Gliessman & M. Rosemeyer (Eds.), *The conversion to sustainable agriculture: principles, processes, and practices* (pp. 3–16). CRC Press., doi:10.1201/9781420003598-c1

Gliessman, S. R. (2015). *Agroecology: the ecology of sustainable food systems* (3rd ed.). Academic Press.

Gold, M. A. (1994). Agroforestry systems for temperate climates: lessons from Roman Italy. *Forest & Conservation History*, *38*(3), 118–126.

Gomiero, T., Giampietro, M., Bukkens, S. G., & Paoletti, M. G. (1999). Environmental and socio economic constraints to the development of freshwater fish aquaculture in China. *Critical Reviews in Plant Sciences*, *18*(3), 359–371. doi:10.1080/07352689991309261

Guitart, D., Pickering, C., & Byrne, J. (2012). Past results and future directions in urban community gardens research. *Urban Forestry & Urban Greening*, *11*(4), 364–373. doi:10.1016/j.ufug.2012.06.007

Hemenway, T. (2009). *Gaia's garden: a guide to home-scale permaculture*. Chelsea Green Publishing.

Hiroki, S., Garnevska, E., & McLaren, S. (2016). Consumer perceptions about local food in New Zealand, and the role of lifecycle-based environmental sustainability. *Journal of Agricultural & Environmental Ethics*, *29*(3), 479–505. doi:10.100710806-016-9616-9

Holloway, L., Kneafsey, M., Venn, L., Cox, R., Dowler, E., & Tuomainen, H. (2007). Possible food economies: A methodological framework for exploring food production–consumption relationships. *Sociologia Ruralis*, *47*(1), 1–19. doi:10.1111/j.1467-9523.2007.00427.x

Holmgren, D. (2004). *Permaculture: principles and pathways beyond sustainability*. Holmgren Design, Hepburn.

Hvitsand, C. (2016). Community supported agriculture (CSA) as a transformational act—Distinct values and multiple motivations among farmers and consumers. *Agroecology and Sustainable Food Systems*, *40*(4), 333–351. doi:10.1080/21683565.2015.1136720

Jacke, D., & Toensmeier, E. (2005). Edible forest gardens, volume II: ecological design and practice for temperate-climate permaculture. Chelsea Green Publishing.

Jamu, D. M., & Piedrahita, R. H. (2002). An organic matter and nitrogen dynamics model for the ecological analysis of integrated aquaculture/agriculture systems: II. Model evaluation and application. *Environmental Modelling & Software*, *17*(6), 583–592. doi:10.1016/S1364-8152(02)00017-8

Kirwan, J., Ilbery, B., Maye, D., & Carey, J. (2013). Grassroots social innovations and food localisation: An investigation of the Local Food programme in England. *Global Environmental Change*, *23*(5), 830–837. doi:10.1016/j.gloenvcha.2012.12.004

Kneafsey, M., Venn, L., Schmutz, U., Balázs, B., Trenchard, L., Eyden-Wood, T., ... Blackett, M. (2013). Short food supply chains and local food systems in the EU. A state of play of their socio-economic characteristics. *JRC Scientific and Policy Reports, 123*, 129.

Konijnendijk, C. C. (2003). A decade of urban forestry in Europe. *Forest Policy and Economics, 5*(2), 173–186. doi:10.1016/S1389-9341(03)00023-6

Kumar, B. M. (2007). Agroforestry: The new old paradigm for Asian food security. *Journal of Tropical Agriculture, 44*, 1–14.

Lancaster, B., & Marshall, J. (2008). *Water-harvesting earthworks*. Rainsource.

Landis, D. A. (2017). Designingagriculturallandscapesforbiodiversity-basedecosystemservices. *Basic and Applied Ecology, 18*, 1–12. doi:10.1016/j.baae.2016.07.005

Lang, T., & Barling, D. (2012). Food security and food sustainability: Reformulating the debate. *TheGeographicalJournal, 178*(4), 313–326.

Lindemann-Matthies, P., Briegel, R., Schüpbach, B., & Junge, X. (2010). Aesthetic preference for a Swiss alpine landscape: The impact of different agricultural land-use with different biodiversity. *Landscape and Urban Planning, 98*(2), 99–109. doi:10.1016/j.landurbplan.2010.07.015

Lovell, S. T. (2010). Multifunctional urban agriculture for sustainable land use planning in the United States. *Sustainability, 2*(8), 2499–2522. doi:10.3390u2082499

Mars, R. (2005). *The basics of permaculture design*. Chelsea Green Publishing.

Marsden, T., & Sonnino, R. (2012). Human health and wellbeing and the sustainability of urban–regional food systems. *Current Opinion in Environmental Sustainability, 4*(4), 427–430. doi:10.1016/j.cosust.2012.09.004

Mauro, F., & Hardison, P. D. (2000). Traditional knowledge of indigenous and local communities: International debate and policy initiatives. *Ecological Applications, 10*(5), 1263–1269. doi:10.1890/1051-0761(2000)010[1263:TKOIAL]2.0.CO;2

Maye, D. (2013). Moving alternative food networks beyond the niche. *International Journal of Sociology of Agriculture and Food, 20*(3), 383–389.

Méndez, V. E., Bacon, C. M., & Cohen, R. (2013). Agroecology as a transdisciplinary, participatory, and action-oriented approach. *Agroecology and Sustainable Food Systems, 37*(1), 3–18.

Milburn, L. A. (2010). Greening Cities, Growing Communities: Learning From Seattle's Urban Community Gardens by Jeffrey Hou, Julie M. Johnson, and Laura J. Lawson. 2009. Seattle: University of Washington Press and the Landscape Architecture Foundation. 232 pages, 140 illustrations, 130 in color. $40.00, paperback. ISBN-13: 978-0-295989-28-0. *Landscape Journal, 29*(2), 238–239. doi:10.3368/lj.29.2.238

Mills, J. (2012). Exploring the social benefits of agri-environment schemes in England. *Journal of Rural Studies, 28*(4), 612–621. doi:10.1016/j.jrurstud.2012.08.001

Mollison, B. (1988). Permaculture: A Designers' Manual. Tagari Publications.

Mollison, B., &Slay, R. M. (1988). *Permaculture: a designers' manual* (Vol. 1). Tyalgum: Tagari Publications.

Mollison, B., Slay, R. M., Girard, J. L., & Girard, J. L. (1991). *Introduction to permaculture*. Tagari Publications.

Mollison, B. C., & Holmgren, D. (1978). *Permaculture 1: A perennial agriculture system for human settlements*. University of Tasmania.

Mougeot, L. J. (Ed.). (2005). *Agropolis: The social, political, and environmental dimensions of urban agriculture*. IDRC.

Mougeot, L. J. (2006). *Growing better cities: Urban agriculture for sustainable development*. IDRC.

Mulligan, M., & Hill, S. (2001). *Ecological pioneers: a social history of Australian ecologica lthought and action*. Cambridge UniversityPress.

Mundler, P., & Laughrea, S. (2016). The contributions of short food supply chains to territorial development: A study of three Quebec territories. *Journal of Rural Studies, 45*, 218–229. doi:10.1016/j.jrurstud.2016.04.001

Nair, P. K. R. (1993). *An introduction to agroforestry*. Kluwer Academic Publishers. doi:10.1007/978-94-011-1608-4

Nordahl, D. (2009). *Public Produce: The New Urban Agriculture*. Island Press.

Norgaard, R. B. (1984). Traditional agricultural knowledge: Past performance, future prospects, and institutional implications. *American Journal of Agricultural Economics, 66*(5), 874–878. doi:10.2307/1241018

Nowak, D. J. (2006). Institutionalizing urban forestry as a "biotechnology" to improve environmental quality. *Urban Forestry & Urban Greening, 5*(2), 93–100. doi:10.1016/j.ufug.2006.04.002

Nowak, D. J., & Dwyer, J. F. (2007). Benefits and costs of urban forest ecosystems. In J. E. Kuser (Ed.), *Urban and community forestry in the Northeast* (pp. 25–46). Springer. doi:10.1007/978-1-4020-4289-8_2

Pant, J., Demaine, H., & Edwards, P. (2005). Bio-resource flow in integrated agriculture–aquaculture systems in a tropical monsoonal climate: A case study in Northeast Thailand. *Agricultural Systems*, *83*(2), 203–219. doi:10.1016/j.agsy.2004.04.001

Perfecto, I., & Vandermeer, J. (2010). The agro ecological matrix as alternative to the land-sparing/agriculture intensification model. *Proceedings of the National Academy of Sciences of the United States of America*, *107*(13), 5786–5791. doi:10.1073/pnas.0905455107 PMID:20339080

Pimbert, M. P., Thompson, J., Vorley, W. T., Fox, T., Kanji, N., & Tacoli, C. (2001). Global restructuring, agri-food systems and livelihoods. Gatekeeper Series 100. London: International Institute for Environment and Development.

Ponisio, L., M'Gonigle, L. K., Mace, K. C., Palomino, J., de Valpine, P., & Kremen, C. (2014). Diversification practices reduce organic to conventional yield gap. *Proceedings. Biological Sciences*, *282*(1799), 20141396. doi:10.1098/rspb.2014.1396 PMID:25621333

Porter, J., Costanza, R., Sandhu, H., Sigsgaard, L., & Wratten, S. (2009). The value of producing food, energy, and ecosystem services within an agroecosystem. *AMBIO: A Journal of the Human Environment*, *38*(4), 186–193. doi:10.1579/0044-7447-38.4.186

Power, A. G. (2010). Ecosystem services and agriculture: Tradeoffs and synergies. *Philosophical Transactions of the Royal Society of London. Series B, Biological Sciences*, *365*(1554), 2959–2971. doi:10.1098/rstb.2010.0143 PMID:20713396

Rapidel, B., Ripoche, A., Allinne, C., Metay, A., Deheuvels, O., Lamanda, N., Blazy, J.-M., Valdés-Gómez, H., & Gary, C. (2015). Analysis of ecosystem services trade-offs to design agroecosystems with perennial crops. *Agronomy for Sustainable Development*, *34*(4), 1373–1390. doi:10.100713593-015-0317-y

Redwood, M. (2008). *Agriculture in Urban Planning: Generating Livelihoods and Food Security*. Earthscan.

Reijntjes, C. B., Haverkort, B., & Waters-Bayer, A. (1992). *Farming for the future. Anintroduction to low-external input and sustainable agriculture*. Macmillian Press.

Renting, H., Marseden, T. K., & Banks, J. (2003). Understanding alternative food networks: Exploring the role of short food supply chains in rural development. *Environment & Planning A*, *35*(3), 393–411. doi:10.1068/a3510

Rosset, P. M. (1997a). Cuba: Ethics, biological control, and crisis. *Agriculture and Human Values*, *14*(3), 291–302. doi:10.1023/A:1007433501248

Rosset, P. M. (1997b). Alternative agriculture and crisis in Cuba. *Technology in Society*, *16*(2), 19–15.

Rukera, S.T., Mutanga, O., & Micha, J-C. (2012). Optimization of an integrated rabbit-fish-rice system for sustainable production in Rwanda. *Rwanda J, 24*.

Sachez, P. A. (1995). Science in Agroforestry. *Agroforestry Systems*, *30*(1-2), 5–55. doi:10.1007/BF00708912

Sandhu, H. S., Wratten, S. D., & Cullen, R. (2010). Organic agriculture and ecosystem services. *Environmental Science & Policy*, *13*(1), 1–7. doi:10.1016/j.envsci.2009.11.002

Sanford, A. W. (2011). Ethics, narrative, and agriculture: Transforming agricultural practice through ecological imagination. *Journal of Agricultural & Environmental Ethics*, *24*(3), 283–303. doi:10.100710806-010-9246-6

Sebillotte, M. (2000). Territoires: De l'espace physique au construit social. Les enjeuxpourdemainet les apports de la recherche. *OCL Oléagineux Corps Gras Lipides*, *7*(6), 474–479. doi:10.1051/ocl.2000.0474

Seufert, V., Ramankutty, N., & Foley, J. A. (2012). Comparing the yields of organic and conventional agriculture. *Nature*, *485*(7397), 229–232. doi:10.1038/nature11069 PMID:22535250

Smith, J. R. (1929). *Tree crops: a permanent agriculture*. New York Brace & Co.

Soleil, S. (2012) The Biochar Miracle. In *Permac. Res. Inst. - Permac.* Forums Courses Inf. News. http://permaculturenews.org/2012/02/03/the-biochar-miracle/

Sonnino, R. (2009). Feeding the city: Towards a new research and planning agenda. *International Planning Studies*, *14*(4), 425–435. doi:10.1080/13563471003642795

Talpaz, H., & Tsur, Y. (1982). Optimising aquaculture management of a single-species fish population. *Agricultural Systems*, *9*(2), 127–142. doi:10.1016/0308-521X(82)90027-0

Thompson, E., Jr., Harper, A. M., & Kraus, S. (2008). *Think Globally—Eat Locally: San Francisco Foodshed Assessment, American Farmland Trust.* Accessed June 23, 2009 at: https://www.farmland.org/programs/states/ca/Feature%20Stories/San-Francisco-Foodshed-Report.asp

Unger, S., & Wooten, H. (2006). *A food systems assessment for Oakland, CA: Toward a sustainable food plan.* Mayor's Office of Sustainability.

Van der Schans, J. W., & Wiskerke, J. S. C. (2012). Urbanagri-culture in developed economies. In A. Viljoen & J. S. C. Wiskerke (Eds.), *Sustainable Food Planning: Evol-ving Theory and Practice* (pp. 245–258). Wageningen Academic Publishers.

Vandermeer, J., & Perfecto, I. (2007). The agricultural matrix and a future paradigm forconservation. *Conservation Biology, 21*(1), 274–277. doi:10.1111/j.1523-1739.2006.00582.x PMID:17298536

Vickers, N. J. (2017). Animal communication: Wheni'mcallingyou, willyouanswertoo? *Current Biology, 27*(14), R713–R715. doi:10.1016/j.cub.2017.05.064 PMID:28743020

Viljoen, A., & Wiskerke, J. S. C. (Eds.). (2012). *Sustainable Food Planning: Evolving Theory and Practice* (pp. 349–364). Wageningen Academic Publishers. doi:10.3920/978-90-8686-187-3

Wästfelt, A., & Zhang, Q. (2016). Reclaiming localization and revitalising agriculture: A case study of peri-urban agricultural change in Gotheburg, Sweden. *Journal of Rural Studies, 47*, 172–185. doi:10.1016/j.jrurstud.2016.07.013

Wezel A, Bellon S, Doré T, Francis C, Vallod D, David C (2009) Agroecology as a science, a movement and a practice. A review. *Agron Sustain Dev, 29*, 503–515. doi:10.1051/agro/2009004

Wezel, A., Casagrande, M., Celette, F., Vian, J. V., Ferrer, A., & Peigné, A. (2014). Agroecological practices for sustainable agriculture. A review. *Agronomy for Sustainable Development, 34*(1), 1–20. doi:10.100713593-013-0180-7

Wezel, A., & Soldat, V. (2009). A quantitative and qualitative historical analysis of the scientific discipline agroecology. *International Journal of Agricultural Sustainability, 7*(1), 3–18. doi:10.3763/ijas.2009.0400

Yeomans, P. A. (1954). *The Keyline plan.* PA Yeomans.

Zhang, Ricketts, Kremen, Carney, & Swinton. (2007). Ecosystem services and dis-services to agriculture. *Ecological Economics, 64*, 253–260. doi:10.1016/j

KEY TERMS AND DEFINITIONS

Agro Ecological Practice: It is the practical application of ecological concepts and principles to the study, design and management of ecological interactions in agricultural systems (for example, the relationships between biotic and abiotic elements).

Agro Ecology: Is the discipline that is responsible for administering the ecological principles of the production of food, fuels, fibers and pharmaceuticals. This encompasses a wide range of approaches and they consider it a science and a way of seeing life, whether organic, conventional, intensive or extensive.

Landscape: Extension of land seen from a specific place and considered as a spectacle.

Territory: Extension of land that belongs to a state, province, or other type of political division.

Transition: Step or change from one state, way of being, etc., to another. Intermediate state between an older one and another that is reached in a change.

Urban Agro Ecology: Is called agroecological practices that are developed in or near cities. ... Urban agroecology usually develops on the roofs of buildings, on the walls of houses and on the balconies and terraces of buildings.

Chapter 3
Scholastic Review of Food, Memory, and Identity:
Classical Migrant Experience

Femi Emmanuel Oni
https://orcid.org/0000-0001-9994-413X
University of KwaZulu-Natal, South Africa

Lloyd Baiyegunhi
University of KwaZulu-Natal, South Africa

ABSTRACT

Food is more than nutrition; it has veritable socio-cultural meanings, and it encapsulates all manner of associations. This chapter reviews several experiences of migrants that are relevant, using different approaches, creating a link between food, identity, and memory of migrants as well as looking at the sequential movement of food and its interactions by reviewing extant literatures in the global and African contexts. Migration and migrants are evident across the borders of countries around the universe. It was revealed that migrants are encumbered with different experiences-accepting and repelling in the course of migration, as it is glaring that there is a conglomerate between food, memory, and identity.

INTRODUCTION

Many migrants covering almost all continents have moved along with their food irrespective of the distance of their migration, trying to retain their identity about food. This food migration has also created a psychological effect on the migrants

DOI: 10.4018/978-1-6684-2349-3.ch003

thereby reminding them of the past and creating a memory of home, making them home - sick.

Food is more than nutrition; it has veritable socio-cultural meanings and it encapsulates all manner ofa ssociations. A strong emotional state is associated with the foods we consume, and food is one of the essential ways of life yet what we consume often still seems trite. The tangible nature of food vanishes when it is consumed compared to other material commodities, but it leaves a tendency of a memorable social instinct to all else in life and the mind which include the past and the imminent. However, food has the capacity of mending time even though it vanishes when consumed (Mintz 2003).

Powerful association exists between food and memory. It was stated in Holtzman (2006) that their ironclad association is frequently offered to us initially insufficiently, so that a piece of this association at some occasion is of a hovering hint, although in unique ways. Food as an aspect can refer to a more realistic sense as substance that is required by us as organisms when consumed to acquire energy. However, it stands to be an inherently multifaceted and multi-coated topic with several dimensions, which include social, psychological, physiological, symbolic phases, and with culturally cobbled up interpretations that are sometimes contradictory. On the other hand, Holtzman (2006) further enunciated that memory tends to be much more coded and complex when explored. It can be unerringly referred to as a collection of special processes that are not only limited to special dynamism but also have the intention to comprehend for very special reasons, as it is been generated from colossal public planning to the schmaltz evoked by the biscuit that is soaked in tea. Based on this assertion, certain foods and plants can play the role of reconnecting with a missing past.

According to Koc and Welsh (2001), food is more than a basic source of nutrients; it is also an integral and intrinsic aspect of our culture, crucial to our identity. Identities, nevertheless, changes with time and place, but on a social basis produced and reproduced around a given communal strata, bringing about certain aberration in the existence and fictional physical restrictions in the experiences of individuals. It was further stated by Koc and Welsh (2001)that certain personal and collective influences which shape our food choices are pertinent and cannot be overemphasized due to their constant occurrence and relevance to life.

Academic history, as shown in Fischler (1988), stated that a cardinal element of our culture, food is also equidistant to our identity. The personality of people in their daily hustling and bustling can be shaped to assume various identities, outlining who they are and how they can live their lives can also be reflected in food as cultural identity. This cultural identity is bespoken in routine practices, which include but not limited to religious activities, a custom of passage, conversation, vacation activities, clothing, art, literature and, music. Building on this, Fox (2003) provided an argument thatour food choices can play a symbolic role in defining us when it comes to religion, ethnicity, and social class as certain food have connotations. From this statement it cannot be concluded that food and identity are inert, as the prevailing issue of globalization have evident that there are increase in human synergy and the super-imposed of foods and its preparatory methods which can be as a result of compromise in the course of migration.

As highlighted in Koc and Welsh (2001), the process of migration has both positive and negative effect on migrants. It introduces possibilities for metamorphosis in ways of life and changes in individual thought and identity. Vallianatos and Raine (2008) further elucidated that how and when these changes occur has to do with an individual's personal precedence in combination with the general, historical, political, and spatial context that surround migration.

The influx of immigrants can affect the economy of a country in diverse ways that in turn can influence the choices and consumption pattern of migrants in such a way that the migrant will have to adapt to the living conditions of the environment found themselves. According toVenkatesha, Venkatarajan, &Premlata, (2016), economic growth in a country causes a reasonable shift in food preferences and consumption pattern and affects nutritional cachet of the people. Also, in order to maintain a better health and elimination of malnutrition, a diversified and balanced food basket is required as stated by (Sangeetha, Sharma, Burman, & Lenin. 2013) which should be ensured in the process of agricultural production. During the past 50 years, agricultural production environment, practices and culture have changed and crop breeding programmes have made farmers less dependent on seasons (Kearney 2010). Based on this, associations exist between cultural, individual activities and food, which can affect consumption, marketing and production of food

This paper reviews several experiences of migrants'relevant using different approaches, creating a link between food, identity and memory of migrants by reviewing extant literatures on food identity in the global and African contexts.

THE NATURE OF FOOD AND FOOD STUDY AS PORTRAYED BY MIGRANTS

Food study has become a vested interwoven and multidisciplinary field of study that looks into the convoluted association that exist among environments, culture and principally food from various angles in sciences, social sciences and even in the field of humanities, to mention just a few. Various areas, which cover engineering, art, sociology, education, economics, health, social justice, literature, anthropology, and history, have provided several insights in food studies. More than a few institutes in various parts of the world are now offering programs in this field.

Almerico (2014), reported that food studies look at people's associations pertaining to food and reveals several and salient information about them. He further stated that the choices we make concerning food could trot out individual and collective principles, passions, personalities, upbringing knowledge, and conjectures. Similarly, Hauck-Lawson (2004), a dietitian brings together the idea of food voice. She propounded that a person's identity or emotion can be expressed in what one consumes or chooses not to consume in a manner that words alone cannot. The choices of food that one makes can suggest family lineage, migrations, ingestions, frictions, transformation, and individual as well as collective identity.

In contrast to the above literatures, Brown (2014), looked at food studies relating meaningful experiences of an individual from the psychological perspective to have a symbolic meaning. A scenario of these symbolic meanings that include food preferences is seen in our mutual elucidations. It was stated that "*psychological needs crisscross with social factors when foods are used more for the meaning they represent more than the nutrients they offer or provide*" (Almerico 2014: 4).

According to Kniazeva and Venkatesh (2007), the first detailed food analysis in consumer behaviour relating to the lived experiences of people and its relevance to other foodstuffs was tackled by (Levy 1981; Levy and Rook 1999), the explored interviews with married women shows the application of a structural approach following Lévi-Strauss (1963) propositions. The foundation of the assumption was because events conducted have symbolic meanings. Based on this, food was seen as a process of searching for meaning in analysing the story of allegory that was gathered from the respondents. He asserted that consumers adapt to roles and symbols that conform to their identities based on the roles and symbols that are prescribed by the accompanying society that reflect the nature of symbols. This was used to distinguish between identities by exploring the terminologies that are involved in cooking and eating. This forms the basis for identifying symbolic distinctions and an important tool used by other scholar in the area of food study. This was applied among various study on specify foods, their mode of preparations and the ideas they carry.

In Heisley and Levy (1991), there are three different levels that are embedded in consumer behaviour which are used in the symbolic framing of food which have been helpful in the quest to open up the meanings hidden in foods. They allude a certain attachment to a particular kind of food, its manufacturing procedure and preparation steps, and modes of eating whether taken slowly or fast, which connotes particular symbols. This view has helped consumers negotiate their identities to satisfy their complex demands in this contemporary world - a procedure that clinches food as one of the most communicative and universal tools. As a result of this, postmodernism of consumer cultural category can easily be understood and represented in food.

In the first level, well-articulated and collectively shared meanings and images are strongly related to specific foods that make this level straightforward. Examples highlighted include hamburgers symbolising teenagers' food and fondue as a fun meal (Levy 1981; Heisley and Levy 1991). These are reflected and expressed when these foods are consumed. Similarly, Wallendorf and Arnould (1991) in a study of thanksgiving rituals, gave some concurring examples that reflect the first level when they found that tradition and a bonded family can be reflected in Jell-0, messages that depict common America origins are sent by root vegetables, and the message of superiority of nature above commerce is conveyed by wide use of butter over margarine.

There was a shift in the second level, where devotion was moved toward ways by which food was manufactured. The technological material that was employed was the bases of the meanings that was inferred from the food and not the food itself. Accordingly, this in turn forms the units that are derived from the analysis. Words used include domestic, man-made, regular, organic, packed, treated, and stationary foods. Instances cited show signs of "industrialization", "modernization and progress", and "natural foods" signify the return to nature, a withdrawal from over modification, and genuineness. (Kniazeva 2002, 2006; Levy 1981). Also, local foods, likewise, operate as a symbolic barricade against interruption of the market into the local area (Moisio, Risto, Arnould, & Price, 2004).

The third level is the most complex and comprehensive as it looks into the multiple dimensions of foods, attributes ascribe to food and the combine meaning generated from the previous two levels. The symbols recognised the linkage that exist between the properties, patterns and values of food and justify behaviour of consumer. Instances from this level show that family unity is symbolically associated with eating at home and conformism, while festivity and separation are symbolically associated with eating out(Bove, Sobal, &Rauschenbach, 2003; Kemmer, Anderson, & Marshall, 1998). In consumer terminology, self-indulgence, femininity, taking control and categorization of food is formed in this stage.

Almerico (2014) contributed that food symbolism gets deep into our social psyche. Certain expressions tending toward food are used to give certain interpretation to some expression like happiness, grief, importance, physique, etc. Statements like, *"he's the big cheese"*, *"she's rolling in the dough"*, *"easy as pie"*, *"a bone to pick"*, and *"he's a good egg"* were used for these expressions. In addition, she further conveyed, one's selection of food or restaurant can be influenced by one's social status, for example "a diet of rice and beans insinuates poverty, whereas steak insinuates wealth" (Almerico 2014, 6).

On the African continent, Shepler (2011), supported the stand point of the symbolic nature of food by exploring the case of a West African country, Sierra Leone. Using the post-war experience (1991 and 2002), Shepler analysed the central material and symbolic role of food in the African context. It was stated in post conflict that the narration of the experience during wartime could be expressed by the central phenomenon of food. She explained daily war experiences using food as a medium and as forefront in the suffering over the remarkable. She also delved into the cultural meaning of food and food idioms to explain and describe sociality, reciprocity, political clientelism, the greed of politicians considered responsible for the war, likewise wartime hardship stories narrated attributed to hunger. In addition, she described that there is shift in moral regime that brought about new ingredients and new method of preparation innovation during the displacement and the new cosmopolitanism experienced in the everyday strategies of food finding under rebel control and in refugee camps.

In a comprehensive work carried out by (Saleh, Amanatidis, &Samman, 2002) on food and migration, using a medical approach, their investigation shows that diet alterations result from migration using the situation of Ghanaian migrants that reside in Sydney, Australia in relation to Type II diabetes and obesity. Based on the research, information gathered shows each of the subject dietary data obtained with the aid of questionnaire, carries anthropometric and duplicate fasting blood glucose level measurements. According to the WHO criteria using the information gathered, men and women on the percentage of 20 and 11 were diabetic, ascribing 22% and 20% of sustaining impaired fasting glucose to each respectively. There was an increase in their body mass upon migration to Australia because of variation in their dietary consumption patterns. The comparison of their food intake as to when they reside in Ghana to after migration to Australia confirms that they consume 14.2 and 12.6 percent less in serving of fruits and 3.8 and 5.3 percent less in serving of fish for men and women respectively per week. It was also noticed that tropical root crops common to them had almost exclusively been replaced with potato starch, which has increased the risk of Type II diabetes, and obesity among the migrant population inferring consumption changes can be evoked because of migration, which can in turn impair our health condition.

In like manner Renzaho and Burns (2006) ascertained that African migrants have a tendency of acculturation in their diet which might take a chronological progression of substitution, supplementation and modification of recipes. Locating their indigenous traditional foods become almost impossible because of this they were forced to adapt to new foods due to the situation experienced. The foods that form part of their new lifestyle upon their arrival constitutes of junk food and cereals obtained from fast food store with a few of new fruits and vegetables. Noticeable changes fostered because of migration lead to the consumption of readymade food from eateries like McDonald's. In addition, there are certain reasons for eating out from these eateries, which might be because of crave for favourite food, routine family outing, special occasions like birthdays and anniversaries and no time to cook. In the process of trying to conform to these dietary changes, a host of them may be predisposed to rapid weight gain and chronic diseases that are not consistent with good health. It was concluded that swift modernization and the foreign culture network in a compound way with traditional eating and socialization does more of shaping Sub Sahara Africa migrants.

All these assumptions cannot be generalized because culture and identity vary around the world and perception also differs. In addition, consumption habits that affect health cannot be concluded based on individual differences alone. Change has occurred over time that gives limitations to some of these assertions taking into consideration the case of rice and beans as a diet that connotes poverty (Almerico 2014). The prices of these food items vary based on the production level and seasons which affect the prices and therefore renders the symbolism void that moment and creates a gap in his assumption

Food as An Expression of Identity

Consumption has different ways of helping us express our identity. Food is one of the fascinating areas of consumption that has received great attention in this context. Identity in food can be expressed by certain people who restrict themselves to certain types of meals for obvious reasons. For example, vegetarians, consumers of organic produce, and equally, those who do not take up any of these identity variations. However, food can be used to express group identity, for instance, as a household eating collectively, distribution of food and receiving of communion.

Kittler, Sucher, &Nelms. 2012, from a general point of view introduced the word food habits (synonymous to food culture) to define how food is used ranging from how it is preferred, purchased, and peddled to who makes, ministers, and munches it. It was specified that food habit is a process that is significant and exclusive to human beings as animals and other creatures are void of potential to perform these processes. They argued on why so much is expended by people with regard to time, energy, money, and creativity in food consumption and could draw an inference that forms a noticeable reflection and exemplifies the impression of identity and food is, "*you are what you eat*" (Almerico 2014, 3)

Coincidentally, Valentine (1999) also suggested the significance food in expressing family identities, in collaboration with gender identities within a household relating to food. It was conveyed that a woman is always saddled with the duty of making preparation of food for the family and also make selection of the food their partner and children liked side-lining her own preference. Consequently, her own food identity is maligned. An accurate instance of this situation was drawn by Slama and Wolfe (1999). They viewed a theatre piece of a woman who treated the needs of her children and husband over her own and later realised they had no value for the priority she gave to them. This made her identity shattered because her own identity was meaningless to her. The approach has created a suicidal of identity on the part of the woman which I think is not an ideal way of forming family identity.

In the African context, food also plays a constructive role in identity formation among migrants leaving a trace in the identity construction of Africans in the diaspora. Gasparetti (2009), looked at food from the perspective of its cultural meanings and its multi-power character in shaping the identity of Senegalese migrants in Italy. He probed their experiences and concluded that when constructing and re-ascertaining their identity, they considered categorization and vivification that are imposed on them by the host community, making no difference between them and the host community, therefore overcoming the internal differences that exist. The home country, Senegal becomes the most important and common reference point in the process of creating and defining the collective identity of Senegalese in the context of migration. In this process, their food becomes an object of concern in linking them to their home country thereby nourishing the Senegalese in diaspora.

He further strengthened his study by stating that eating a kind of food "*tie boujenn*",one of their popular dishes collectively strengthens their wider sense of belonging and reinforces the internal loyalty of the group. A host of their food can be purchased in the African shops manned by fellow Africans, where their women source all the necessary ingredients needed to make their dish "*tie boujenn*" and the other ones they consume. "*Tie boujenn*", a special delicacy for most of people in Senegal, can be termed as not only welcome for consumption but also creates a sense of belonging as Lévi-Strauss (1966) would say greater importance is ascribed to the cultural meaning that is constructed in Senegal in the migration context.

In addition, women in the region of Senegal emblematize the fanciful status of the mother and their homeland feeding of their children, due to the ideal depiction of woman built by the entire Senegalese community. They are responsible for providing food for all Senegalese brothers as a welfare package, safeguarding the adhesion and recognition procedures of their assembly in all jurisdiction of their stay, both at home and in the street, where "*tie boujenn*" is been sold.

Similarly, in order to create a flow for inclusion and exclusion, Walker (2012) noted that the relationship that exists between commensality and foods is quite an important marker of identity and kinship among Comorians in disbandment on the Comorian island of Ngazidja. Commensality generates cohesion in Ngazidja, where food differences are minimal; in the disbandment, these food differences form an important aspect and the identities are said to be invoked in the foods themselves. It was explored how these entities formed and maintained by Comorian identities in France and in Zanzibar as well as in Ngazidja, and how the opportunity of both sharing and denial of sharing has generated and transcended borders between them and their neighbours and between the Comorians in disbandment and at home. A firm decision was reached by analysing the special Ngazidja food "ntsambu", creating a rupture within the broader Comorian identity which circuitously was often offered to be too Comorian to be eaten by Comorians in disbandment.

This is why Hahn (2008) argued that the popularity currently gained by the concept of "cultures of consumption" has an indicting understanding of consumption, as it is not only the fulfilment of needs to satisfy hunger, but also as a means of expressing social identity. He then stated that description given to consumption until now has only been done based on the achievements and specific problems that is been faced in the developed societies, that in the "less affluent world" that is the local areas, the fundamental assumption of consumer culture is of little or no relevance at all. This assumption was further faulted that consumption was merely geared towards the fulfilment of basic needs only and fallen short in some areas to give a concrete explanation of economic constraints and unequal conditions of life in those countries. He was able to give a rejoinder to this criticism by showing how consumption in the African contexts may be proved useful in order to analyse abridging concepts about consumption. However, history and contemporary interaction of local, inter-region and global forces is very much needed in order to understand consumption in the Africa context. Global commodities are highly sophisticated and well utilized by many African consumers. In his attempt to use cultural assumption as a conceptual framework, he gave some detail analysis on the specificity of quantitatively reduced belongings in many African families. In this context he coined that the wide range of cultural practises has a reflection on the different modes of transformation and consumption of globally circulating consumer commodities. Based on this, the expression of and the driving force for social change is being indicated in the agency of the consumers.

The Connection Between Food and Memory

Holtzman (2006) states the centrality of sensuousness of food in the understanding much of its power as a tool that drives memory. However, Sutton (2001), created an important stand point for considering the connection that between food and memory by compilation of efforts exerted to tackle the subjects of memory from various ideals and perspectives. In his curiosity to muzzle out an accurate, appropriate and theoretical framework to look into less explored areas he coined a foundation which he termed "Proustian anthropology." Based on his observation, the information gathered shows that participants remembered past events through food citing an example of an apricot fruit that was eaten in the exploration of an abandoned synagogue during one of their occupation. He was able to point out an important dimension that focuses on the many of the varied phenomena that are usually labelled memory. He was able to give different examples, how a pear in

August was used to illustrate the seasonal food cycle, shaping prospective memory by looking ahead of past events; how past occurrences are being brought back to the present through food; how time in some instances is being held by consistently repeating daily habits; how reference memories have been used to understand the current anthropological interest. Meanwhile narratives of past generosity are been used to frame social relationships; and how there is a connection between past and present meals as a result of understanding one meal from the orientation derived from previous meals. There is a great insight in the broad-ranging treatment of memory into the phenomena that were termed memory, although also to some extent of omission in the above-mentioned incertitude concerning the disparities among the varying phenomena, we term memory.

Vast literatures have considered exploring the study of food and memory and have come out with reasonable conclusions. From the angle of individual memory, Batsell *et al.* (2002), brought to light the experience of childhood in the United States, how it has fostered the clear-out of individual plate from the convincing "flashbulb memories," remembering in clear detail the experiences of early childhood: a stage when little or none may be remembered. Conversely from a collective approach, Powles (2002) argued on the social memory of Angolan migrants in the settlement of Meheba, Zambia using collective memory of displaced immigrants, it was shown that experiences were honed most emotionally as a result of the physical experience of the non-appearance of fish which became a bad memory.

According to Almerico (2014, 5), the statement "*The memory as well as the spaghetti was delicious*" gives an interpretation that the memory of food connects to the heritage and ways of life of the family, and also gives a further backing that food carries more weight than the basic source nutrients. Emotionally, sensuously, and ethnically there is a trace of connections, found in food. Lupton (1994), similarly used the memory-work as an innovative qualitative research approach to look at implication of structuring eating habits as an emotional embodied memory that surrounds a particular food to unravel the meanings that surrounds food practices in industrialised societies. The memory was observed, showing the contribution of food as an important aspect to social relationships and cultural observations, explaining the individual adherence of certain eating habits and the circumvention of others that can be pointed towards a further relevance of memory-work to explain the meanings and symbolic nature of food in the developed societies.

According to Arvela (2013), food is also a powerful cultural marker. A clear distinction was made with regards to food at home and food abroad bringing into play the imaginary bridges that can keep someone connected with the memory of faces, practices, customs, tastes and smells that are common which was left behind when abroad and food as what is taken for just mere reasons when taken at home. These have shown the crucial role of food culture in a strange environment in diaspora. Substantial examples include the study of Fiss (2001), he recognized that Portuguese immigrants that reside in Brazil eat "*bacalhau*" a local meal to remain connected emotionally to their motherland, its customs, traditions and the relevant othersin the they left behind. Also, Richardson (2003) after critically examining Australians' predilection for vegemite and how much expatriates missed this homemade food whilst away, he concluded that "most Australians living abroad have come to notice that their needs are available locally and they have been saved the stress of overloading their suitcase when travelling".

Cook (2008) conveniently looked at food and memory. He postulated that certain food can evoke memories of the past when it is prepared, cooked or experienced. The choice of this food when eaten or shared can serve as an act of nostalgia that makes consumers to remember their families, friends and places of history. Based on this, individual have the tendency of connecting, irrespective of physical and temporary separation, when food associated with home is consumed. Correspondingly, another literature of interest that supports Cook (2008) idea is Parasecoli (2014) who examined food with the connection of the past and found that memories of the past, a more or less traumatic displacement, and interactions inside and outside their familiar circles all contribute to shaping migrants' new culinary competences in their physical, emotional, and cognitive dimensions.

Gasparetti (2009), using his prerogative said, Senegalese migrants always "feel at home" when abroad as a result of the evocative power of food representing the place and space for memories. Also, the major key player in the development of syncretic identity for migrants from Senegal among other immigrant groups most often Africa is food whereby serving as a contributing factor to the establishment of new affairs outside the group. Extending Gasparetti's spatial reference, Christensen (2001), viewed the place of cooking as a fountain for memory; using his mother's experience to state that, "to open the covering of a garlic and dice its contents into grains has allowed a kind of rebirth making her a daughter again, to re-enter the female world of her puberty."

Food Preference and Choice Among Migrants

Migrants are saddled with the challenge of living out of their comfort zone when it comes to issues of food and other essential commodities which has led to making choices from the alternatives as a result of their preference. Food preference can be seen as the way in which migrants make selection based on the availability of different comestibles on the ground of social, biological or economic perceptions which may include their taste, value, purity, ease or difficulty of preparation, and the availability of fuel and other preparation tools. Currently, various literature from anthropology looked at food choices as been conditioned by the capitalist or nationalist goals of food gossipmongers (Belasco and Scranton 2014; Nestle 2013; Schlosser 2012) or a situation that is perceived as the result of new discovery in supply chains and economic conditions brought about by colonialism and globalization as well as the international aid process (Lindenbaum 1986). However, the decision of an individual within a social context actually affects what is consumed. According to Solms and Hall (1981), the act of consuming food may represent the ultimate basic locus of identity, conformity, and resistance, similarly is the mouth to body as it serve as the opening. Likewise, food choice in preparation and consumption can also be exercised by those who are said to be otherwise powerless, this have been a subject of consideration in the study to observe the subjugated marginal groups in which slaves are included (Armstrong and Kelly 2000; McKee 1995).

In complex societies as it was an important component of plant domestication that started 10,000 years ago at the end of the Holocene, food preference and its effect on human-nature interactions appears to be predated. As viewed by Harris and Hillman (1989) also in Smith (2001) it is a series of gradual and subtle transitions, whereby human rulership over plant reproduction entails decisions about which plants to harbour and propagate.

According to Vallianatos and Raine (2008), loss of tradition and new food ways has almost always been incorporated into the lifestyle of immigrants in the course of migration. The tendency to loss or incorporates foods depends on the situation surrounding each immigrant. Accessibility to some foods can be restricted, some direct or indirect factors which may include but not limited to physical, political, economic and sociocultural realities of the immigrant's fresh atmosphere. It all depends on whether the foods are simply unavailable, or they have negative associations in the immigrant's new terrace. It is glaring that the shifts in social identity and self-conceptualisation are corresponding to the changes in food ways. Following the constructs of Vallianatos and Raine (2008), the idea of Phinney (1990) can be re-ignited as it was highlighted that, multiple models exist to aid in the analysis and understanding of identity development processes due to the fact that identity is a dynamic construct that develops with time through the acquisition of new behaviours, attitudes and experiences. This was further buttressed by Tse (2001) that models recognise the behavioural aspects of identity, and plays an importance role in identity development processes. Based on this assertion, the most basic sense in the consumption and preparation of food is a physical act therefore; models that are proposed could be a useful tool in the analysis of the role of food and food ways in identity development. Indeed, this was what madeMintz and Du Bois (2002) to argue that the understanding of broad societal processes when studying food is critical and made a declaration that it can be useful in ongoing debates about anthropological theories and methods.

Critically looking at the work of Parasecoli (2014), one comes to realise that the adaptation process of migrants to the new land is shared, influenced, and constructed through interactions at least within the intimate circles of family, friends, neighbours, co-workers, and the immediate social sphere unless they find themselves alone and refrain from any contact which is out rightly impossible. These connections which are as a result of familiarization are so much of value as they have to adapt to the different and new activities that exist in this new environment which may include abstruse periodic cycles, overseas agendas, and new holidays where their food plays no part. As immigrants collectively expand and reshape their culinary competence to make sense of new situations, the communal repository of memories and experiences related to the place of origin may also influence the way they relate to each other, migrant food preference is sharpened by his environment and has nothing to do with his identity.

Parasecoli (2014) concluded that migrant communities remain effective and significant to individuals and groups precisely because they are founded on bodies and embodied experiences that at the personal, communal, collective, and institutional levels constantly negotiate not only with such ideals such as nation, identity, authenticity, and tradition, but with the pull of more abstract networks that highlight connections while disregarding the emotional impact of embeddedness in specific places. The shift and instable nature of communities, and also their existential relevance can be revealed by the full understanding of food-related does, norms, and ideas among migrants. However, the reflection on food, whose main function is to be destroyed by ingestion, could be of helping hands to facilitate modern theories of globalization in a general and specific, emotional, and deficient dimension of the body, with its advantages and its disadvantages.

CONSUMPTION, MARKETING AND PRODUCTION OF LOCAL FOOD

Food can be either local or international. The uniformity in the definition of local food is still yet to be reached despite the fact it is becoming an increasing topical issue in the current economic and agro-environmental climate (DEFRA 2003; Jones, Comfort, & Hillier. 2004). However, there is a generally accepted definition by farmers' markets used in identifying producers who are permitted to sell there. According to DEFRA (2003), it was summarized that 'food planted, managed, bought and sold within a given geographic usually of a restricted radius is termed local food. Due to the subjectivity of local food to business and consumer, there is hence some realistic litheness that needed to be allowed in the definition. Distance is a subject of consideration when it has to do with consumer perspective of local food. For instance, IGD (2005) found that "local" is a symbol of small arena around where the consumer stays or buy the food. Additionally, the CPRE (2008) proposed a classification that food should be within the jurisdiction of 30 miles between origins and/or be processing and the store.

In contrast to the consumer perspective, an attempt to define local agricultural food system from another perspective brought about the definition of Fonte (2006) as the territorially coordinated production-consumption networks. Terms like "local production systems", "local productive systems", "agri-food local productive systems" (ALPS), "local agro-food clusters" and industrial clusters are used interchangeably with local agric food system. This concept is overlapping as consumption, marketing and production are involved and small business that involves producers and marketers and consumer perceptions are considered appropriate.

Following the concept local food, Dawson (2002) attributed values to consumers in the increasing scale of importance: production, nearness of production to household, and product value which have been personified and prepared by local food initiatives such as "farmers' markets, farm shops, food fairs and organic box schemes" (CPRE 2003; FARMA 2008;

Archer, Sánchez,Vignali, &Chaillot, 2003) In order to align the relationship that exist between the key player in the concept, (Duffy, Fearne, & Healing. 2005)

Tregear *et al.* 2007) conveyed that the dependency of the local food producer on the supporting network of regional and county food groups will aid promotion and marketing of food, hereby aiming at sustainable local business and generating revenue in the local economy, hence more focus is given to benefiting the local producers and the local environment as a result of this diversification from mainstream agri-food.

Van der Veen (2003) then theorized the value of daily food in order to understand social cohesion. As a result of this, it enables us to see the reason agricultural strengthening; the result of elites was not solely provocative production for the purposes of social exaggeration through feasting,

Local Food Consumption, Marketing and Production as A Vehicle for Development Opportunities

Consumer confidence regarding food originality and traceability has been increased as result of concerns that surround food fright. Also, moving from the conventional food supply chain to high quality market based the origin, validity and organoleptic properties of food has generated a growing hope (Morley, Sparkes, & Thomas, 2000; DEFRA 2002). As highlighted in Britain (2002), locally produced foods have gained the opportunity of added values and profits as a result of building public enthusiasm for it by Policy initiatives as one of the responses (Weatherell, Tregear, & Allinson. 2003). In (Knight, Holdsworth, & Mather, 2007), it was argued that produce sourcing can be advantageous when it is being able to be traced back to production by individual grower.

Marketing of locally produced foods and drinks is becoming a growing category in both supermarkets and independent retailers in the UK. As reported by IGD (2010) interest in the marketing of local food doubled from 15% in 2006 to 30% in 2010 as the sellers of these locally produced commodities claimed it. Similarly, Mintel (2010) also noticed 29% interest among the retail customers care who sells locally produced food product. In the process of the recognition of the defining problems, Mintel (2008), recorded an estimated value of R87.36 billion (current exchange rate) in the UK retail sales. Expenditures from food service outlets serving residents and tourists are been excluded from this analysis, what a jumbo and important market for a small food business. Over the year, the research carried out by (IGD 2010; Mintel 2010) showed that support for local farmers and other food producers as well as retailers have increased the buying of locally produced foods interestingly. Adding to this,Hingley, Boone, & Haley, (2010)

Hingley *et al.* (2010) made a remark that great opportunities arise for the SMEs as a result of the emergence of farm shops and other direct selling initiatives, making them profitable from this demand.

Challenges in Local Food Consumption, Marketing and Production

In a study conducted by Watkins (2008), local food production, marketing and consumption possess to have challenges as palpable complication for local food exist as a result of perceived price premium, predominantly when there is recession in luxury food commodities which can be lay-off by the cost-conscious consumers. In the UK, (Weatherell *et al.* 2003; IGD 2005) have suggested that, although consumers have interest in locally produced foods yet they find it more demanding in terms of cost than the other possibilities. Based on this, Mintel (2008) established the fact that the arch obstacle to augmentation of the marketing of locally produced foods is price, as 20% of the buyers' claim it is on the high side when it comes to price. Nevertheless, Mintel (2008) also found that those consumers sourcing for locally produced foods were trying to save cost, whereby bringing about a subject of confusion between the high cost of this local food and the local food itself. Hence, Hingley *et al.* (2010) raised the attention to address the issues that relate to consumers connection regarding their perception concerning price, quality and value for money.

Another challenge noted by Archer *et al.* (2003) was consumer lack of convenience and preference for supermarket in sourcing for these locally produced foods which was attributed to lack of enlightenment and price factors. In addition, customers profile serve as a barrier to local foods. Archer *et al.* (2003) further found that the profile of these customers also affected the farmers' markets buyers. They cited an example of a retired grown up female of over fifty-five years with other majority travelling as far as ten miles at the cost of an average of R55-R180.

Similarly, IGD (2005) also identified the major barriers to the growth of local food market as awareness, accessibility and availability, together with price. Based on these findings, there is a need to place emphasis on the availability, accessibility and price of locally produced foods to boost the future confidence and growth of the food production sector.

CONCLUSION AND SIGNIFICANCE

Having looked at various literatures on food, memory and identity under various headings using different approaches, we have observed that the relationships that exist between food, memory and identity are important in the field of research as it has psychological, economic and social effects on both humans and the environment. Despite enormous research carried out in this area study, it was noticed that the African context is not a major area of concern and no study has been conducted relating food, memory and identity as it has affected agricultural production and developmental processes in a given geographical settings which should be a motivating factor behind a study on memory, identity and food production within the scope of Africa and other continents.

REFERENCES

Adrian, M., Sparkes, A., & Thomas, B. (2000). Strategic and regional initiatives in the Agri-food industry in Wales. *British Food Journal*, *102*(4), 274–289. doi:10.1108/00070700010327689

Amy, B. (2014). *Understanding food: principles and preparation*. Nelson Education.

Andre, R., & Burns, C. (2006). Post-migration food habits of sub-Saharan African migrants in Victoria: A cross-sectional study. *Nutrition & Dietetics: the Journal of the Dietitians Association of Australia*, *63*(2), 91–102. doi:10.1111/j.1747-0080.2006.00055.x

Andrew, D. (2002). *March: Local and regional food*. IGD.

Angela, T., Arfini, F., Belletti, G., & Marescotti, A. (2007). Regional foods and rural development: The role of product qualification. *Journal of Rural Studies*, *23*(1), 12–22. doi:10.1016/j.jrurstud.2006.09.010

Annie, H.-L. (2004). *Introduction.* Taylor & Francis.

Ashraf, S., Amanatidis, S., & Samman, S. (2002). The effect of migration on dietary intake, type 2 diabetes and obesity: The Ghanaian Health and Nutrition Analysis in Sydney, Australia (GHANAISA). *Ecology of Food and Nutrition, 41*(3), 255–270. doi:10.1080/0367-020291909778

Caron, B. (2003). Food choices among newly married couples: Convergence, conflict, individualism, and projects. *Appetite, 40*(1), 25–41. doi:10.1016/S0195-6663(02)00147-2 PMID:12631502

Charlotte, W., Tregear, A., & Allinson, J. (2003). In search of the concerned consumer: UK public perceptions of food, farming and buying local. *Journal of Rural Studies, 19*(2), 233–244. doi:10.1016/S0743-0167(02)00083-9

Claude, F. (1988). Food, self and identity. *Information (Basel), 27*(2), 275.

Claude, L.-S. (1963). *Totemism Pa Txt.* Beacon Press.

Claude, L.-S. (1966). Anthropology: Its achievements and future. *Nature, 209*(5018), 10–13. doi:10.1038/209010a0 PMID:5925326

Cook, I. (2008). Geographies of food: Mixing. *Progress in Human Geography, 32*(6), 821–833.

CPRE. (2003). *Down your way? A CPRE briefing on supermarkets and local food.* London: CPRE. [Available from: https://www.cpre.org.uk/library/results/local-foods

CPRE. (2008). *Help local food survive the supermarket onslaught.* CPRE. Available from: http://support.cpre.org.uk/cpre/supermarket_solutions.asp

David. (2001). *Remembrance of repasts: an anthropology of food and memory.* Academic Press.

David & Hillman. (1989). An evolutionary continuum of people-plant interaction. *Foraging and Farming: The Evolution of Plant Exploitation, 11.*

Debbie, K., Anderson, A., & Marshall, D. (1998). Living together and eating together: Changes in food choice and eating habits during the transition from single to married/cohabiting. *The Sociological Review, 46*(1), 48–72. doi:10.1111/1467-954X.00089

Deborah, H., & Levy, S. (1991). Auto driving: A photo elicitation technique. *The Journal of Consumer Research, 18*(3), 257. doi:10.1086/209258

Deborah, L. (1994). Food, memory and meaning: The symbolic and social nature of food events. *The Sociological Review*, *42*(4), 664–685. doi:10.1111/j.1467-954X.1994.tb00105.x

DEFRA. (2002). The strategy for sustainable farming and food: facing the future. London: DEFRA.

DEFRA. (2003). Local food – A snapshot of the sector. Report of the working group on local food. London: DEFRA.

Douglas, A., & Kelly, K. (2000). Settlement patterns and the origins of African Jamaican society: Seville plantation, St. Ann's Bay, Jamaica. *Ethnohistory (Columbus, Ohio)*, *47*(2), 369–397. doi:10.1215/00141801-47-2-369

Eric, S. (2012). *Fast food nation: The dark side of the all-American meal*. Houghton Mifflin Harcourt.

Fabio, P. (2014). Food, identity, and cultural reproduction in immigrant communities. *Social Research*, *81*(2), 415.

FARMA. (2008). *21 Reasons to support local foods*. FARMA. Available from: http://www.farmauk.org/21reasons.htm

Fedora, G. (2009). The cultural meaning of food and its polyvalent role in the construction of identity among Senegalese migrants in Italy. In *Food and Migration Workshop*. School of Oriental and African Studies, University of London. Retrieved from https://www.soas.ac.uk/migrationdiaspora/seminarsevents/food_migration_abstracts/file49

Geoffrey, A., Sánchez, J. G., Vignali, G., & Chaillot, A. (2003). Latent consumers' attitude to farmers' markets in North West England. *British Food Journal*, *105*(8), 487–497. doi:10.1108/00070700310497264

Gina, A. (2014). Food and identity: Food studies, cultural, and personal identity. *Journal of International Business and Cultural Studies*, *8*, 1.

Great, B. (2002). *Farming & Food: A Sustainable Future. Report of the Policy Commission on the Future of Farming and Food*. Cabinet Office.

Hahn. (2008). Consumption, Identities and Agency in Africa: Introduction. *Consumption in Africa: Anthropological Approaches*, 9.

Helen, V., & Raine, K. (2008). Consuming food and constructing identities among Arabic and South Asian immigrant women. *Food, Culture, & Society*, *11*(3), 355–373. doi:10.2752/175174408X347900

Iain, W. (2012). Ntsambu, the foul smell of home: Food, commensality and identity in the Comoros and in the diaspora. *Food & Foodways*, *20*(3-4), 187–210. doi:10.1080/07409710.2012.715962

IGD. (2005). *The local and regional food opportunity*. IGD. Available from: http://www.defra.gov.uk/foodfarm/food/industry/regional/pdf/localregfoodopps.pdf

IGD. (2010). *Local food gains ground with shoppers*. Available from IGD: https://www.igd.com/index.asp?id=1&fid=1&sid=8&tid=16&cid=1427

Jean, P. (1990). Ethnic identity in adolescents and adults: Review of research. *Psychological Bulletin*, *108*(3), 499–514. doi:10.1037/0033-2909.108.3.499 PMID:2270238

John, K. (2010). Food consumption trends and drivers. *Philosophical Transactions of the Royal Society of London. Series B, Biological Sciences*, *365*(1554), 2793–2807. doi:10.1098/rstb.2010.0149 PMID:20713385

John, K., Holdsworth, D., & Mather, D. (2007). Determinants of trust in imported food products: Perceptions of European gatekeepers. *British Food Journal*, *109*(10), 792–804. doi:10.1108/00070700710821331

Jon, H. (2006). Food and memory. *Annual Review of Anthropology*, *35*(1), 361–378. doi:10.1146/annurev.anthro.35.081705.123220

Juerg, S., & Hall, R. (1981). *Criteria of food acceptance: how man chooses what he eats*. Papers presented at an interdisciplinary symposium, Einsiedeln.

Julia, P. (2002). Like baby minnows we came with the current": social memory among Angolan refugees in Meheba settlement, Zambia. *Annual Meeting of the Association of Social Anthropologists of Great Britain and the Common wealth*.

Lucy, T. (2001). Resisting and reversing language shift: Heritage-language resilience among U.S. native biliterates. *Harvard Educational Review*, *71*(4), 676–706. doi:10.17763/haer.71.4.ku752mj536413336

Maria, F. (2006). Slow Food's presidia: what do small producers do with big retailers? In *Between the Local and the Global* (pp. 203–240). Emerald Group Publishing Limited.

Maria, K., & Venkatesh, A. (2007). Food for thought: A study of food consumption in postmodern US culture. *Journal of Consumer Behaviour*, *6*(6), 419–435. doi:10.1002/cb.232

Marion, N. (2013). *Food politics: How the food industry influences nutrition and health* (Vol. 3). Univ of California Press.

Mark, S., & Wolfe, R. (1999). Consumption as self-presentation: A socioanalytic interpretation of Mrs. Cage. *Journal of Marketing*, *63*(4), 135–138. doi:10.1177/002224299906300411

Martin, H., Boone, J., & Haley, S. (2010). Local food marketing as a development opportunity for small UK agri-food businesses. *International Journal on Food System Dynamics*, *1*(3), 194.

McKee, L. (1995). The Earth is their witness. *The Sciences*, *35*(2), 36–41. doi:10.1002/j.2326-1951.1995.tb03183.x

Melanie, W., & Arnould, E. (1991). "We gather together": Consumption rituals of Thanksgiving Day. *The Journal of Consumer Research*, *18*(1), 13. doi:10.1086/209237

Mintel. (2008). *Locally sourced foods*. Mintel. Available from: https://academic.mintel.com/

Mintel. (2010). *Consumer Food Labelling*. London: Mintel.

Mintz, S. W., & Du Bois, C. M. (2002). The anthropology of food and eating. *Annual Review of Anthropology*, *31*(1), 99–119. doi:10.1146/annurev.anthro.32.032702.131011

Moisio, R., Arnould, E., & Price, L. (2004). Between mothers and markets: Constructing family identity through homemade food. *Journal of Consumer Culture*, *4*(3), 361–384. doi:10.1177/1469540504046523

Mustafa, K., & Welsh, J. (2001). *Food, foodways and immigrant experience*. Centre for Studies in Food Security.

Pamela, K. (2012). *Food and Culture*. Cengage Learning.

Paula, A. (2013). Ethnic food: the other in ourselves. In D. Sanderson & M. Crouch (Eds.), *Food: Expressions and Impressions* (pp. 45–56). Inter-Disciplinary Press.

Peter, J., Comfort, D., & Hillier, D. (2004). A case study of local food and its routes to market in the UK. *British Food Journal*, *106*(4), 328–335. doi:10.1108/00070700410529582

Pia, C. (2001). Mac and Gravy. In *Pilaf, Pozole and Pad Thai*. University of Massachusetts Press.

Pradeep, Sangeetha, & Singhb. (2016). Relationship between Food Production and Consumption Diversity in India–Empirical Evidences from Cross Section Analysis. *Agricultural Economics Research Review, 29*, 139.

Rachel, D., Fearne, A., & Healing, V. (2005). Reconnection in the UK food chain: Bridging the communication gap between food producers and consumers. *British Food Journal, 107*(1), 17–33. doi:10.1108/00070700510573177

Regina, F. (2001). A Imigração Portuguesa e as Associações Como Forma de Manutenção da Identidade Lusitana-Sul do Brazil. *Scrita Nova, 94*, 27. http://www.ub.edu/geocrit/sn-94-27/htm

Richardson, K. (2003). Vegemite, Soldiers and Rosy Cheeks. *Gastronomica, 3*(4), 60–62. doi:10.1525/gfc.2003.3.4.60

Robert, B., Brown, A., Ansfield, M., & Paschall, G. (2002). "You will eat all of that!": A retrospective analysis of forced consumption episodes. *Appetite, 38*(3), 211–219. doi:10.1006/appe.2001.0482 PMID:12071687

Robin, F. (2003). Food and eating: an anthropological perspective. Social Issues Research Centre.

Shirley, L. (1986). *Rice and wheat: the meaning of food in Bangladesh*. RS Khare.

Sidney, L. (1981). Intepreting consumer mythology: A structural approach to consumer behavior. *Journal of Marketing*, 49.

Sidney, L., & Rook, D. (1999). *Brands, Consumers, Symbols and Research: Sidney J Levy on Marketing*. Sage Publications.

Sidney, M. (2003). Remembrance of Repasts: An Anthropology of Food and Memory. *American Ethnologist, 30*(3), 474–475. doi:10.1525/ae.2003.30.3.474

Susan, S. (2011). The real and symbolic importance of food in war: Hunger pains and big men's bellies in Sierra Leone. *Africa Today, 58*(2), 42.

Valentine, G. (1999). Eating in: Home, consumption and identity. *The Sociological Review, 47*(3), 491–524. doi:10.1111/1467-954X.00182

Van der Veen, M. (2003). When is food a luxury? *World Archaeology, 34*(3), 405–427. doi:10.1080/0043824021000026422

Venkatarajan, S., Sharma, J., Burman, R., & Lenin, V. (2013). Food Security Vs Nutritional Security—Need for Multi-sectoral Convergence. *International Journal of Agriculture and Food Science Technology, 4*(6), 621.

Warren, B., & Scranton, P. (2014). *Food nations: selling taste in consumer societies.* Routledge.

Watkins, J. (2008). Regional groups instrumental in encouraging local markets. *Farm Business,* 16.

Chapter 4
Is Agripreneurship a Mitigating Measure for Agricultural Issues in India ?

Dezy Kumari
Institute of Economic Growth, India

Mohd Taqi
Aligarh Muslim University, Aligarh, India

Mohd Ajmal
Aligarh Muslim University, Aligarh, India

Aslam Khan
Aligarh Muslim University, Aligarh, India

ABSTRACT

The agriculture sector is considered the mainstay of Indian economy because it is a significant source of raw materials for industries and demands for many industrial products particularly fertilizers, pesticides, agriculture implements, and a variety of consumer goods. Agriculture takes up new scope and shape and is no longer the customary cultivation of crops and nurturing of animals or an enterprise for the rural people. Agripreneurship may be defined as the amalgamation of agriculture and entrepreneurship. It is the choice to assimilate in the quest to make agriculture an enterprise of adjuration in contemporary business engagements. If agriculture must alter from its largely sustenance status to becoming a competitive enterprise in the gauging of entrepreneurs, incorporation of business concepts in maneuvering issues of agriculture is the way to go. The present study is a review work engrossed in using agribusiness to boost job creation and raise productivity and the income bases of farmers.

DOI: 10.4018/978-1-6684-2349-3.ch004

"Agriculture is our wisest pursuit, because it will in the end contribute most to real wealth, good morals, and happiness." — Thomas Jefferson

INTRODUCTION

Agricultural sector is considered the mainstay of the Indian economy. It is a significant source of raw materials for industries to fulfill demands for many industrial products, particularly fertilizers, pesticides, agricultural implements and a variety of consumer goods. Agriculture takes up a greater scope and shape than being solely an enterprise for the rural people, constituting the customary cultivation of crops and nurturing of animals. Agripreneurship may be defined as an amalgamation of agriculture and entrepreneurship. It is the choice to assimilate in the quest to make agriculture an enterprise of adjuration in contemporary business engagements. If agriculture has to undergo a transition from largely being a means of sustenance to a competitive enterprise in the eyes of entrepreneurs, incorporation of business concepts in maneuvering issues of agriculture is the way to go. The present study is a review that contemplates the idea of using agribusiness to boost job creation, raise productivity and the income bases of farmers. It is taken into consideration because of the fact that a large population is engaged in agriculture, and the growth of agribusiness and agripreneurship will go a long way in powering the growth of the economy, diminishing the prevalent unemployment therein and ensuring food security. Agripreneurship can accelerate social and economic development, effect a reduction in the poverty index, ensure food security along with a good nutrition. Also, it will help in diversifying the income and economy bases, generating employment and entrepreneurial opportunities.

The Indian economy is principally an agrarian economy. Agriculture is pivotal to the nation's food security, apart from equipping about 52 percent of the working population of the country with livelihood. India consists of 17.5% of the world population on 2.4% of its land. At the time of Independence, agriculture accounted for more than half of the national income, with more than seventy percent of the total population dependent on it for livelihood (Pandey, 2013). The industrial sector is crucially dependent on the agricultural sector to provide it with vital resources, including fertilisers, pesticides, agricultural implements and a variety of consumer goods (Bairwa, Kushwaha, Meena, Lakra, & Kumar 2014a).

The dynamic socio-economic, political, environmental and cultural dimensions over the world resulting in farmer and nation options for abidance and for imperishable ensuring success in changing their respective economic environments has become increasingly crucial. It is also worth noting that the emergence of the free market economies globally has resulted in the development of a new spirit of enterprise 'Agripreneurship' and the increased individual need for responsibility for running their own businesses (Alex &Lwakuba 2011). The term Agripreneurship is a synonym of entrepreneurship in agriculture and interrelates to agribusiness establishment in agriculture and allied sector. Agripreneurship metamorphoses agricultural activity into an entrepreneurial activity.

Dollinger (2003) defines entrepreneurship in agriculture as the formation of innovative economic organization for the purpose of broadening or heightening under conditions of risk and uncertainty in agriculture. Gray (2002) defines an entrepreneur as an individual who carries on a business with the aim of enlarging the business, leadership and managerial skills necessary for attaining those goals (Bairwa&Lakra 2014). In the aspect of growing unemployment and poverty in rural areas and steady growth of agriculture, there is a requisite of entrepreneurship in agriculture for more productivity and profitability. The Agripreneurship program is necessary to develop entrepreneurs and management workforce to serve the agricultural Industry across the world (Bairwa, Kalia, Meena, Lakra, &Kushwaha, 2014b). Agripreneurship is greatly influenced mainly by the economic condition, education and culture (Singh, 2013).

In India, agriculture is suffering challenges that are yet to be addressed, such as insufficient infrastructure, strain in accessing credit and absence of training for small holder farmers on opportunities that the enterprise provides, among other constraints in modern farming techniques. Mitigating these issues will abet in improving Indian food security, develop agribusiness, grow the GDP and raise foreign exchange earnings (PwC, 2016). Other challenges are soaring food prices that are technically beyond the reach of the common man, changes in climatic systems that have changed patterns of agricultural practices and apliant crops, rapid urbanisation that has redo production and consumption patterns; the list is endless.

These developments have accompanied changes in the food markets, innovated new opportunities and challenges for the farmers, especially the smallholders. With the identification of the significance of agriculture in the economic development constancy and growth, market and business-oriented agriculture seem to be pre-eminent factors that will introduce dynamism in the enterprise. Agripreneurship is a key in this regard (Nwasiwe 2017). The dwindling prospects and coincidences of Indian agriculture place vital need on the development of a system that can support the agricultural sector that is already well constrained by scanty technologies and institutional weaknesses, among others. The agriculture sector is remain competitive in the global economy, new ideas must be developed and operate for value creation in a sustainable manner devised. (Uneze, 2013)

Presently, agriculture is taking up new shape and scope and no more a customary cultivation of crop and nurturing of animals or an enterprise for the rural population. Other facets like value addition, high-tech agriculture, global marketing, organic farming etc have redefined agriculture (Palanivelu and Madhupriya, 2013). These have guided to ameliorate performance of the sector and intensified human resource development initiative. Initially, many people viewed agriculture as an enterprise that is saddled with hard work and little profit, this is a misapprehension. Agriculture is a vigorous enterprise, offering numerous profitable opportunities for engagement along the value chain.

India's agriculture policies have had multiple decrees, including a production imperative (national food security), a consumer imperative (keeping food prices low for a large low-income population), and a farmer welfare imperative. Economic transformation in developing nations is actuated by increases in agricultural incomes and stimulating industrial growth. The latest example is China (Gulati and Fan 2007). Raising farm productivity is scarce for long-term raising farmer incomes in India, as land fragmentation means that many Indian farmers are farming plots of such small sizes (especially those below 1 hectare) that even doubling their incomes would leave them with scanty earnings.

The commercialization of agriculture will involve the development of input supply chains, application of technologies, market linkages and financial services that engage the farmers. These are crucial for job creation, economic diversity, and sustainable economic growth (Ado, 2017). Therefore, it is of great consequence to develop the current agriculture system in the context of product development, value addition, and, as a business that is knowledge-driven.

The addition of business knowledge to agriculture bears agribusiness and the fusion of entrepreneurial principles into agribusiness have developed agripreneurship. An agripreneur like an entrepreneur must be skilled to detect and innovate business opportunities that he or she can exploit. Agripreneurship is an employment approach that will ensure self-reliance and self-sufficiency. Its development through training is an imperative factor in the promotion of Micro, Small and Medium Enterprises particularly, for the first generation agripreneurs (Nagalakshmi and Sudhakar, 2013). There are numerous opportunities from the farm to the table begging to be exploited. Agripreneurs must be determined and continual, intuitive and organized with good management skills. He or she must be an innovator who can induct changes by developing and serving new markets. Agripreneurship is the profit-oriented union of agriculture and entrepreneurship; it turns a farm into an agribusiness (Birwa, Kushwaha, Lark, & Mena, 2014).

OBJECTIVES OF THE STUDY

- To compile and study the information relating to agripreneurship in Indian context.

- To address the awareness level regarding the agripreneurial activities that the potential agripreneur may undertake.
- To understand the extent of employment opportunities in the area of agripreneurship.
- To identify the issues and challenges hindering the agripreneurs to emerge in India.
- To discuss the phenomena and scope of agribusiness scenario in an economic context.

The present study is entirely a review one that made use of secondary data that were culled from different reliable sources. It focused mainly on agripreneurship at the place of agriculture under the scheme of India's economic development with a population of more than 130 billion. India's rural economy is significantly dependent on agriculture, with several agri-business opportunities yet to be exploited through agripreneurship initiatives.

DISCUSSION

Agriculture and Entrepreneurship

Entrepreneurship in agriculture is regarded as the development of an innovative economic organization for growth purposes or benefit under conditions of agricultural risk and uncertainty (Dollinger, 2003). Contribution of Agripreneurship to economic growth is well known through some studies. Agripreneurship plays various roles in the growth and development of the national economy through the creation of entrepreneurship that increases the level of income and job opportunities in both rural and urban areas (Bairwa et al., 2012). It should be noted that the advent of free-market economies globally has led to the creation of a new 'Agripreneurship' entrepreneurial spirit and a growing individual desire for accountability to operate their own businesses (Alex &Lwakuba 2011).

In order to grow the entrepreneurship and management workforce, the Agripreneurship programme is important to cater for the agricultural industry around the world (Bairwa et al., 2014b). Agripreneurship is primarily affected by the economical, educational and cultural situation (Singh, 2013). Entrepreneurs are those individuals that show common characteristics such as one-mindedness, motivation, determination, imagination, problem-solving, realistic and goal-oriented. An entrepreneur is an individual who recognizes and takes the risk to seek an opportunity or unmet need. He needs to build those skills, maintain competitiveness and search for new markets (Singh, 2013). Personal qualities of an agri-entrepreneur significantly affect the agribusiness (Brockhaus and Horwitz, 1986; Nandram and Samson, 2000). Hanf and Muller (1997) say that open-minded farm entrepreneurs would identify more problems in a competitive world of rapid technological development than they can rationally solve. Man et al .(2002) categorized entrepreneurial skills in six key areas, including skills in recognizing opportunities, building relationships, organizing, strategic skills, conceptual thinking, and problem solving. The Study citing Kahn (2012) suggested that entrepreneurs are also creative. However, in reference to Wongtschwski et al (2013), the study further argued that farmers, while typically innovative and imaginative, are often constrained by insufficient experiences, access to resources, markets, and skills that could generate practical opportunities for entrepreneurs to succeed.

Table 1.Key concepts of agripreneurship

Terms	Description
Agripreneur	Entrepreneur whose main business is agriculture or agriculture-related Agriculture + Entrepreneur = Agripreneur Agripreneur should be proactive, curious, determined, persistence, visionary, hard working, honest, integrity with strong management and organizational skills. Agripreneur also known as entrepreneurs.
Agripreneurship	Agripreneurship is the profitable marriage of agriculture and entrepreneurship. It is synonym with entrepreneurship in agriculture and refers to agribusiness establishment in agriculture and allied sector.
Agriclinics	provide expert advice and services to farmers on technology, cropping practices, protection from pests and diseases, market trends, prices of various crops in the markets and also clinical services for animal health which would enhance productivity of crops/ animals and increased income to farmers.
Agribusiness Centers	Provide farm equipments on hire, sale of inputs and other services. These centres will provide a package of input facilities; consultancy and other services with the aim of strengthen transfer of technology and extension services and also provide self employment opportunities to technically trained persons.

Source: Baiwara&Lakra 2014

Carr and Rollin (2016) noted in UNCTAD (2015) that the growth of agripreneurial culture in developing countries poses challenges such as inadequate knowledge, limited skills, insecurity, inadequate resources and infrastructure, among others, making it difficult to create effective businesses. It has also stressed the need for effective policies to encourage small-scale agriculture and non-agricultural sectors. Nevertheless, to overcome these challenges and make a profit, an agripreneur like an entrepreneur must possess those innate qualities. He or she must be constructive, ambitious, imaginative and be innovative and creative, with good management and organizational skills. (2011) (Needy). This can be gained through training that

Table 2. Scope of agripreneurship

Terms	Description
Beekeeping	There is an immense scope of honey production and bee-keeping in the country due to its wide area of flora and fauna. This is such a growing venture where even as illiterate and resource poor men/women can start their own with no land required.
Animal Husbandry	India ranks 1st in milk and milk product production. Livestock management and cattle rearing has been the part of day to day life. Rearing of improved breed and their proper's management can give a good return to the farmers.
Fruit and vegetable preservation	The production of various fruits and vegetables preserved items viz., Potato chips, Potato fingers, Potato pappad, Mango and Litchi Squashes, Jam, Jelly Marmalade, Mixed vegetable, Tomato pickles, Tomato sauce, Ketchup etc.
Horticulture based enterprises	India is major producer of vegetables viz., Potato, Onion, Eggplant, and Cauliflower. Other fruits grow are Mango, Guava, Citrus, Banana, Papaya, Ber, Pineapple and Makhana. Flavor of its Spices, red Chilli and Coriander spreads all over the country.

Source: Verma et al (2018)

Table 3. Necessary skills for an agripreneur

Category	Underlying Skills
Cooperation and Networking Skills	Skill relating to cooperating with other farmers and companies
Opportunity Skills	Recognizing the business opportunities Market and customer orientation Awareness of threats Innovation skills Risk management skills
Management Skills	Financial management Skills Human resource management skills Customer management skills General planning skills
Strategic Skills	Skills to receive and make use of feedback Reflection skills Monitoring and evaluation skills Conceptual skills Goal setting skills Strategic decision making skills Strategic planning skills Goal setting skills
Professional Skills	Plant and animal production skills Technical skills

Source: Tripathi and Agarwal (2015)

offers the information, expertise and knowledge necessary to help make informed decisions. They can be bought through formal institutions or through informal means such as public lectures, fieldwork and presentations, workshops, seminars etc (Tripathi and Agawam, 2015). The figure below shows some necessary skills the agripreneur must be equipped with to be competitive and successful as presented by Tripathi and Agarwal (2015).

Through the Green Revolution phase that started in the 1960s, followed by the Yellow, White and Blue Revolutions, our country has seen unprecedented gains in agricultural and food production, recording nearly 300 million tonnes of food grains, 315 million tonnes of fruit and vegetables, 176 million tonnes of milk (the world's highest) and 12 million tonnes of fish, During the past six decades or so, output increases have risen steadily. Furthermore, amid growing climate change uncertainty, the coefficient of output variance has declined considerably, indicating improved resistance to the uncertainties.

Table 4. Production of food grains, pulses, oilseeds, milk, fruits and vegetables, meat (mt) and eggs (billion).

Year	Food-grain	Pulses	Oilseeds	Milk	Eggs	Fruits-veg	Meat
1951-52	50.82	8.41	5.16	17	1.832	166.94*	1.9*
2016-17	275.68	22.95	32.10	165.4	18.139	299.85	7.4
2017-18	284.83	25.23	31.31	176.3	27.950	305.40	11.4

Source: Agriculture Statistics, MoA& FW, GoI

Figure1. Growth trends in Indian agriculture
Source: Agriculture Statistics, MoA& FW, GoI

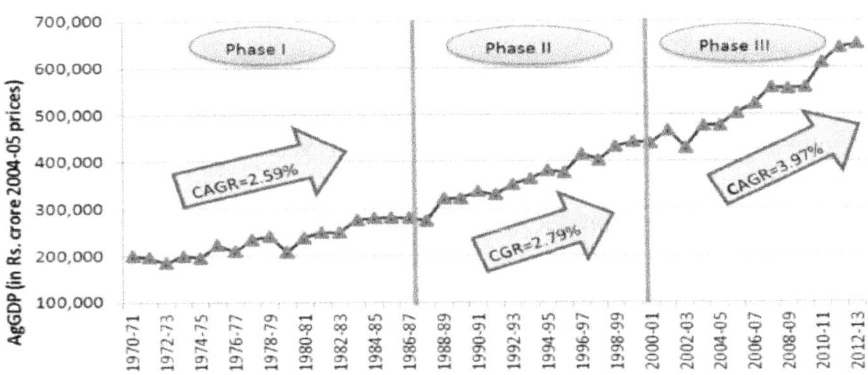

Output of food grain, driven mainly by wheat and rice, the main crops of the Green Revolution, increased almost 6 times between 1951 and 2017, oilseeds 6 times, cotton almost 11 times and milk more than 10 times. The production of pulses had traditionally remained stagnant, but it took off over the past three years due to the government 's special initiative, making the country more than self-sufficient in pulses. For the oilseed economy, where yields are still low and India is meeting 70% of its worldwide edible oil demand through imports, a similar approach / strategy is required.

AGRIPRENEURSHIP IN THE VALUE CHAIN

Poverty drains resources from families,Agripreneurship helps them build resources (Dobson and Markley, 2010). Agriculture may generate wealth along the value chain, by processing, adding value, and exporting, among others, processed or unprocessed products. There are several areas of agriculture within the supply chain that entrepreneurs can explore in both on-farm and off-farm operations. On-farm operations include production, processing (feed and seed processing), production agricultural products, and agro-service projects. Off-farm projects will include, among others, agri-tourism entrepreneurship (Uneze,2013) and other business areas such as transport, storage and packaging, workshops and service centers for the maintenance of agricultural implements in the value chain for agriculture (Birwa et al 2014). Post-harvest technology such as manufacturing, packaging, and storage can sustain food production and ensure against excessive food import reliance (Ado, 1997), even as it creates more job openings.

Strategies to increase growth in India's agriculture sector (Sachin Ghai, Journal of Economics and Sustainable Development, 2012) is managing the product's value chain to increase the value of the commodity at each point of the chain without increasing costs. The rise in income levels and health-consciousness make the customer very aware of what he eats. This leads to the tremendous opportunities for the agricultural sector to set up agro-business that will increase the value of the farmer's food produced. By improving the value chain productivity, transaction costs will decrease and thus reduce the producers' risk. The overall value added for agriculture produced in India is less than 2 per cent. Where, as in the case of other developing countries such as Malaysia, Brazil contributes more than 40 percent of the value of its agricultural products, one of the reasons for that is inadequate storage and distribution service facilities. The price of the commodity was very high due to the commission agents' so much intermediation. The demand for high-value food would increase due to the increasing income of the people. And thus the supply chain needs to be improved for perishable goods in order to provide the consumers with high quality items. Because of the convergence in the supply chain, the farmer would know better what to grow based on experiencing the market's real-time demand. For this kind of agribusiness, bankers in the country have to fund the entire value chain. In this phase, the bankers have to play the catalyst role.

As the world's second-largest agricultural economy, India must effectively harness its AFS to break the mysterious coexistence of high economic growth and deprivation (an infamous Indian Enigma). The food and agriculture system must be transformed so that the associated growth does not only mean increased production but also production plus, with equal focus on farmers' net earnings, physical and economic access to adequate quality food, safe drinking water, sanitation and hygiene, environmental productivity, environmental health, nutritional adequacy seeking a wholesome and integral approach – the cherished Sustainable Development Goals.

The forecast growth in production indicates that by 2030 the trend will be sustained, and supply will largely surpass demand, resulting in significant export surplus, except for edible oils. With regard to pulses, India will have surplus production of pulses by 2030, having already produced nearly 25 million tonnes during 2018-19. Therefore, the challenges in the future will be to achieve these gains with enhanced input productivity and equal economic and social access to adequate nutritious food, and to substantially enhance the 'cost-risk-return' structure of agriculture in the face of the rising uncertainty of climate change and markets (Kumar & Joshi, 2016).

Table 5. Supply-demand gap of food grain and high value commodities in 2030

Commodities	Supply	Demand	Gap (+/-)
Rice	122.1	122.4	-0.3
Wheat	128.8	114.6	14.2
Coarse cereals	64.2	47.2	17
Pulses*	23.7	26.6	-2.8
Food grain	338.8	310.8	28
Sugar	40.3	39.2	1.1
Edible oil	19.1	21.3	-2.1
Milks	179.4	170.4	8.8
Fish	11.9	11.1	0.8

Source: Kumar & Joshi (2016)* to be revised based on the 2018-19

AGRIPRENEURSHIP AND EMPLOYMENT GENERATION

Access to labor supply is a major problem in the agricultural sector. Since technology adoption in the Indian agriculture sector is much less so, most of the agricultural activities are still done by humans. Since the cost of labor in India is increasing with the introduction of MNREGA programmes (IMA India, 2009). Moreover, the main problem in the agricultural sector is becoming disguised jobs due to the small holdings of the land. Cooperative farming, i.e. the sharing of labor, should be encouraged to avoid this issue. This phenomenon in agriculture is not new, farmers can trade labor works in labor prices during the time of inflation. If the government promotes cooperative farming practices, the farmers themselves would conduct their farming activities with a high degree of reliance on outside labor.

The Indian government often implemented and followed a growth and development policy plan, which promoted entrepreneurship and self-employment, to promote and enhance economic growth and development. However, as regimes come and go, the programme was fraught with contradictions, insufficient articulations and regular policy truncations. Transitioning from paid jobs to self-employment has been a significant agendum in many developing nations ‘ economies. Individuals must move from paid jobs to self-employment to be economically self-sustaining (Palanivelu and Madhupriya, 2013).

India is in a strong position to promote this transformation by using agriculture as the launch pad since a substantially large number of the population are already active in the agricultural sector. If agribusiness is prepared to take up the challenge, the transformation does not pose serious problems. With the rise in unemployment and poverty in India and limited economic growth options in other sectors, there is a need to develop entrepreneurship in agriculture to create jobs, increase productivity and grow the economy because, for now, agriculture employs the highest amount of labor in the country and there are limited sustainable opportunities outside of agriculture. The application and practice of entrepreneurship in agriculture can produce a wide range of benefits such as increased productivity, developing new agribusiness projects that will lead to job creations, product innovations and service delivery, and wealth gains (Birwa et al 2014). Agripreneurship can contribute to social and economic development in the areas of reduction in poverty index, good nutrition and food security in the economy. In addition, It will contribute to a diversification of the bases of economy and wealth, generating jobs and entrepreneurship opportunities.

Anon (2016) and Ado (2017) noted that prohibiting the importation of certain agricultural products is an opportunity for additional investment by the private sector, but India only imports what it can produce in a comfortable and cheap manner, negatively impacting the country’s employment status. In the crop field, where the blends of products that can be produced are boundless, there are greater opportunities to be wasted. Investors can manufacture for specific markets, according to the study, simply by adjusting the shape of a product before selling, modifying the packaging and market labelling etc. thus Value addition can create jobs which are critical needs in tackling unemployment at this moment of economic recession.

ROLE OF AGRIPRENEUR IN ECONOMIC DEVELOPMENT

Agripreneurship plays a vital role in the growth and development of the national economy through the creation of entrepreneurship which increases the level of income and employment opportunities in both rural and urban areas (Bairwa&Kushwaha2012). Agripreneur helps smallholder farmers induce productivity gains and incorporate them into local, global, and foreign markets. It helps to reduce the cost of food, provide uncertainty and improve the country's rural and urban poor diets. The role of agripreneurship in the economic system also follows (Sah, 2009).

Table 4. Types of enterprises in agri-business in India

Types	Description
Farm Level Producers	At the individual family point, every family is to be treated as venture, to enhance the production by making best use of the technology, possessions and demand in the market.
Service Providers	For optimizing agriculture by every family business, there are diverse types of services requisite at the village level. It include the input borrowing and distribution, hiring of equipment like tractors, sprayers, seed drills, threshers, harvesters dryers and scientific services such as setting up of irrigation amenities, weed curb, plant security, yielding, threshing, conveyance, warehouse, etc. related opportunities exist in the livestock husbandry sector for providing breeding, immunization, disease diagnostic and treatment services.
Input Producers	There are many flourishing enterprises, which need critical inputs. a few such inputs which can be produced by the home entrepreneurs at the village level are biopesticides, soil amendments, biofertilizers, vermicompost, plants of diverse species of vegetables, fruits, ornamentals, root media for raising plants in pots, production of cattle feed concentrate, agricultural tools, irrigation accessories, mineral mixture and complete feed. There are good openings to support, fishery, sericulture and poultry as well, during sponsorship of critical service amenities in rural areas.
Processing and Marketing of Farm Producc	Well organized management of post-production processes requires higher level of knowledge as well as investment. Such venture can be handled by People's Organizations', either in the form of cooperatives, service joint stock companies or societies.

(Source: Baiwara&Lakra 2014)

ISSUES AND CHALLENGES FOR AGRIPRENEURS IN INDIA

Table 5

Issues	Challenges
Lack of Funds& Infrastructure(Small land holding)	Poor Entrepreneurial Culture among People
Risk	Poor Infrastructural Facilities
Low skill level among farmers	Problems in marketing of Agricultural Products
Marketing Problems and Competitions	Unresponsive Government Policies
Lack of technological Dissemination	Poor Technologies and Equipment
Legal formalities and regulations	Talent Migration from rural to Urban
Availability of Resources	High Costs of Physical Logistics
Lack of technical knowledge	

(Source: Baiwara&Lakra 2014)

SUMMARY

Agripreneurship shares many characteristics of 'generic' entrepreneurship, but also has its distinct characteristics due to the unique background of the agricultural sector. With better discipline in entrepreneurial and industrial education, entrepreneurs can of course benefit from the availability of large human capital. It shows that there is a wide scope for entrepreneurship in agriculture and that this potential can only be exploited through effective management of agri-elements that a person with risk-bearing ability and a search for the latest information in the agricultural sector can prove to be a successful agripreneur. The agriculture sector has a great potential to contribute to national income while providing direct employment and income to the society's numerically larger and vulnerable section. Agripreneurship is not only an opportunity but also a necessity in the agricultural sector to improve production and profitability

Agripreneurship is the need of hours to make agriculture a more desirable and profitable business. It is clear that there is a great scope for entrepreneurship in agriculture, and this potentiality can be exploited only by effective management of agri elements such as – soil, crop, water and consumer needs. A person with risk-bearing ability and a search for cutting-edge information in the agricultural sector can prove to be a good agripreneur. The agriculture sector has a great potential to contribute to national income while providing direct employment and income to the society's numerically larger and vulnerable segment. In addition to being an incentive, agripreneurship is also a requirement for improving production and profitability in agriculture and the related sector.

CONCLUSION

Agripreneurship is a must to make our economy attractive to agriculture and a lifeline. India's existing economic history demands diversification and a paradigm change from the agrarian economy to business opportunities. Our large population is based on agriculture, and therefore the development of the real sector such as agriculture should be supported in order to boost employment, reduce poverty and food insecurity. In agriculture, there is obviously huge scope for entrepreneurship that can be tapped. A significant section of the population would get direct jobs and income from agripreneurship. Agripreneurship outside offering opportunities is also a necessity to improving agricultural productivity and trigger the growth of allied sectors. India must make agriculture the centerpiece of industry to take off in the search for economic development and diversification. Government must create enabling environments that impact and concentrate on mitigating the constraining factors against the development of agriculture and entrepreneurship.

REFERENCES

Alex, L. (2011). *A Review and Analysis of Policies on Farmers' Entrepreneurship Development*. PELUM.

Bairwa, S. L., Kalia, A., Meena, L. K., Lakra, K., & Kushwaha, S. (2014b). Agribusiness Management Education: A Review on Employment Opportunities. *International Journal of Scientific and Research Publications*, *4*(2), 1–4.

Bairwa, S. L., & Kushwaha, S. (2012). Agro Industry scenario in India. In Agricultural research and Sustainable development in India. Bharti Publications.

Bairwa, S. L., Kushwaha, S., Meena, L. K., Lakra, K., & Kumar, P. (2014a). Agribusiness Potential of North Eastern States: A SWOT Analysis. In Agribusiness Potentials in India: experience from hill states. EBH Publishers.

Birwa, S. L., Kushwaha, S., Lark, K., & Mena, L. K. (2014). Agripreneurship Development as a Tool toUpliftment of Agriculture. *International Journal of Scientific and Research Publications, 4*(3).

Brockhaus, R. H., & Horwitz, P. S. (1986). The psychology of the entrepreneur. In D. L. Sexton & R. W. Smilor (Eds.), The art and science of entrepreneurship. Ballinger publishing company (pp. 25–48). Academic Press.

Carr, S., & Rollin, A. (2016). *An exploration of Agripreneurship Scope, Actors and Prospects*. Academic Press.

Chandra Shekhra, P. (2003) Third wave in Indian agriculture: introduction to Agriclinics and agribusiness centre scheme. *Manage Ext. Res. Rev.*, 10-20.

Dollinger, M. J. (2003). Entrepreneurship – Strategies and Resources. Pearson International Edition.

Drucker, P. (1985). *Innovation and Entrepreneurship*. Harper and Row. www.worlconnectors.nl

Global Agrisystem. (2010). *Evaluation Study of Agriclinics and agribusiness centre Scheme*. Global Agrisystem Private Limited.

Goyal, S. Z., & Obeng, F. K. (2004). Public-Private Sector partnerships for improved agriculture services delivery: How do we make them work? *Journal of Social Development in Africa*, *19*(2), 49–67.

Gray, C. (2002). Entrepreneurship, Resistance to change and Growth in Small Firms. *Journal of Small Business and Enterprise Development*, *9*(1), 61–72. doi:10.1108/14626000210419491

Jose, S. (2014). *Innovation in agriculture: A key process for sustainable development*. Inter-American Institute for Cooperation on Agriculture (IICA). Institutional position paper, 1-20.

Nagalakshmi, T., & Sudhakar, A. (2013). Agri-preneurs: A Case Study of Dharmapuri Farmers. *International Journal of Scientific Research*, *2*(8).

Pandey, G. (2013). Agripreneurship Education and Development: Need of the Day. *Asian Resonance*, *2*(4), 155–157.

Srivastava, U. K. (1989). Agro-processing industries: Potential, constraints and tasks ahead. *Indian Journal of Agricultural Economics*, *44*(3), 242–256.

Uche, C., & Familusi, L. C. (2018). The Adoption of agripreneurship as a mitigating measure to unemployment in Nigeria: A topical review. *Global Journal of Management and Business Research: G Interdisciplinary, 18*(2), 24 – 31.

Verma. (2018). Opportunities in agri-preneurship in India: Need, challenges and future prospects. *Rashtriya Krishi*, *13*(1), 69–72.

Chapter 5
Business Model of Palm Oil Smallholding in South Sumatra, Indonesia:
Challenges and Future Prospects

Ngadi Ngadi
https://orcid.org/0000-0002-5249-0860
Indonesian Institute of Sciences, Indonesia

ABSTRACT

One of the business models that has been developed in the palm oil plantations in Indonesia is the consolidation of the land of smallholders while professionally managed by cooperatives. While there are cooperatives that perform well, some are not so much resulting in the business cease. This chapter aims to analyze a smallholding palm oil business model that consolidates 367 smallholders with a land area of 734 ha in Srimulyo Village, Tungkal Jaya District, South Sumatra. The results of the analysis show that the cooperative in this village has performed well, and the palm oil land has had major productivity. Several factors are associated with the success of the palm oil smallholding business model, such as the trust from members, the cooperative initiatives, and the sustainability of the plantations. In the framework of developing sustainable plantations, this cooperative has received the RSPO certificate. At the same time, the smallholders have also been able to set up savings funds for replanting to be used during the time to rejuvenate plants.

DOI: 10.4018/978-1-6684-2349-3.ch005

INTRODUCTION

Inequality in rural-urban development in Indonesia brings an impact on the large youth workforce who relocate from villages to cities in search of work. The agricultural sector has become less attractive to a large proportion of the young workforce in Indonesia. A situation that can be observed from the aging workforce of the agricultural sector. Of the 38.22 million agricultural workers in August 2020, around 21.2% were 60 years old and over, while the proportion of young smallholders (aged 29 years and under) was only 17.3%, lower than the Service and the Manufacturing sectors, respectively 27.2% and 27.9% (BPS, 2020). White, (2012) showed that young people were uninterested in farming or rural futures.

To cope with this matter, there is a need to develop entrepreneurship in the rural areas to support the rural development and attract young workers to work in the rural areas, especially the agricultural sector. That being said, various business models have been developed in the rural areas, one of which is land consolidation of smallholders in the palm oil plantations while being managed by an institution formed by its members, namely cooperatives. The cooperative is then assigned to manage the plantation led by a cooperative chairman. The chairman is elected by the members at the annual membership meeting. The cooperative then form a committee in charge of carrying out the cooperative's activities.

In practice, the palm oil plantation business model differs depending on the agreement of the cooperative members. Some cooperatives are only established to collect and sell smallholders' crops, while the plantation management is still carried out by the individual smallholders. With this model, the smallholders' income is obtained according to the productivity of their respective land. The plantation management standards between smallholders can also differ. Some other cooperatives are established to manage the plantations from maintenance, harvesting, to sales. In this model, all plantation activities are carried out by the cooperatives. Cooperative members will receive income based on the results of their respective plantations, but the standard of plantation management is relatively the same. There are also cooperatives established to manage the plantations from maintenance to sales. The revenue sharing is done by summing the crop production on land managed by the cooperative divided by the number of members. In this case, every month the cooperative members will get the same income from the palm oil plantation.

This paper aims to examine the smallholders of the palm oil business model in South Sumatra, Indonesia. The data for the analysis is the research data acquired from the Research Center for Population-LIPI that was conducted in Musi Banyu Asin Regency, South Sumatra, in 2015. The data analysis is carried out using descriptive analysis. The business model in this area is a business built by 367 plasma smallholders. The smallholders established a business venture in the form of cooperatives to manage the consolidated land.

BACKGROUND

Various models of palm oil business management in Indonesia have their respective strengths and weaknesses. In the overall models, cooperative managers play an important role in maintaining and developing palm oil. While proper plantation management has an impact on the welfare and sustainability of the palm oil business. Several research results indicate the important role of cooperatives in palm oil development in Indonesia (Rist, Feintrenie, &Levang, 2010; Jelsma, Slingerland, Giller&Bijman, 2017). Various challenges are also faced by the palm oil plantation business model, including technological advances, maintaining relationships with palm oil buying companies, and maintaining the sustainability of productivity. In the long term, they also have to prepare for replanting plants since the productive age of palm oil is only up to 25 years. After reaching that age, palm oil needs to be replanted so that it continues to produce well.

The success of the business model can be recognized from the business performance which can be observed from various perspectives. Mutasowifin, (2002) mentions several dimensions in the assessment of cooperative performance such as employee productivity, interrelationship, business networking, and implementation of training programs, innovation, community concern, business relations and transactions, and the existence of members.

The assessment of the cooperative performance can be done in two ways, which are (a) discussing the organization as a unit of analysis and (b) discussing the perspective of social membership which considers members as a unit of analysis (Benos, Nikos, Martin, De Ruyter&Joost, 2018). Based on reviews of the studies that have been conducted, these researchers found that most of the research (58.04%) on the performance of cooperatives used a business financial appraisal (BFA) approach with the most reported metrics namely profitability, depths, liquidity, and efficiency ratios. Some other studies (30.07%) used a business efficiency appraisal (BWA) approach with the most reported metrics being technical and allocative efficiency. Others (14.00%), used the objective membership appraisal (OMA) approach where the most commonly reported metrics were prices paid and side selling.

Nkuranga& Wilcox (2013) measured the performance of cooperatives in Rywanda using the Cooperative Performance Index (CPI) technique. The five dimensions used in this measurement are (1) Legal Status and Cooperative Planning and Strategy, (2) Management Structure and Accounting System, (3): Production and Quality of Inputs, (4) Market Linkages and Business Relations and (5) Recruitment &

Member Retention Strategy. Each dimension consists of several questions. There are 73 questions in the formation of the index in the study. Based on this index, cooperatives in Rywanda are divided into early transition to growth cooperatives, mid-transition to growth cooperatives, and cooperative models.

In this study, the author used several variables to observe the cooperative work in palm oil plantations as a business model in rural areas. Some of these indicators are land productivity, member trust, management structure, market-linked, business relations, and work performance based on the awards received. The business relations in this study are the relations between plasma smallholders' cooperatives and nucleus companies, banks, other private institutions, and the government.

MAIN FOCUS OF THE CHAPTER

The Process of Becoming Palm Oil Smallholders

Before becoming palm oil smallholders, the residents of Srimulyo Village were food crop farmers who had been practicing the trade for 14 years. This is due to the status of Srimulyo Village, which was originally the location for the transmigration of food crops. Most of the residents of Srimulyo Village were the residents who moved from Java Island occurring in 1980. In 1994/1995 there had been a change in the type of commodity from food crops to palm oil. These changes were caused by: (a) frequent crop failures due to pests and diseases in food crops, and (b) the government's making Srimulyo Village a transmigration PIR village. Economic reasons were the main factor in changing agriculture from food crops to oil palm as the conversion of land functions from food crops to palm oil plantations had also taken place in other areas in Indonesia such as in Bungo District (Feintrenie, et al. 2010), Bengkulu Province (Astuti, Wibawa, &Ishak, 2011). The main purpose for the conversion of land use was economic, such as an increase in household income.

There was a long process before the change from food crops to palm oil plantations happened which could be detected from the initial phase of the village establishment to the transition to a palm oil project village. At the beginning of its establishment, the community life in Srimulyo Village was more prosperous than when they had lived in Java. The transmigrants received a living subsidy of 10 months which was sufficient for the daily life of the transmigrants. Furthermore, during the first two years, the crop fields at the location of the transmigration were still fertile. All types of plants cultivated by residents grew well without much disturbance from plant pests, including wild boar. Thus, smallholders enjoyed fairly good yields. However, various social facilities such as basic education were not yet properly available and the locations for children's education were still moving around in empty houses that had not yet been occupied by the transmigrants. The community's fairly prosperous economic conditions continued until 1984 due to two years' aid of food from the government.

However, the economic condition of the people in Srimulyo then drastically deteriorating from 1986 through 1993. The life of residents at that time became concerning, especially from the economic standpoint. The government's aid had been revoked, but crop failure often occurred as a result of the prolonged dry season and was attacked by wild boar pests. Many residents failed to harvest impacting their lives to become poor. This situation lasted for approximately seven years with many residents that were not tough forced to return to their areas of origin. The local community leaders stated that approximately 70 percent of the transmigrants in the village had returned to their areas of origin because they could no longer stand the increasingly difficult economic conditions. The land and houses they owned were sold inexpensively, which was important as means for transportation to go back home.

Transmigrants who chose to stay in the location were those who did not have other job opportunities at their birthplace. Without hope in their original village, since they no longer had a house and land, they were compelled to stay in the transmigration area. At that time, there was a lot of idle land in the transmigration area because the cost to cultivate the land was not proportional to the results obtained. The risk of crop failure was also very high due to drought and plant pests. Many residents chose to abandon food crop farming on their land to work as manual laborers outside the village, especially in Jambi Province. They usually went in groups by truck and then worked outside their area for about two months. Apart from being laborers in the agricultural sector, some residents also worked as masons as they had acquired the skills from their places of origin in Java.

Generally, the jobs outside the village were carried out by men so that only the wives and children who were not yet adult could stay at home. While the heads of the family worked outside the village, the mothers were in charge of taking care of the household and cultivating agricultural land, although with limited abilities. Much of the land, however, was idle and only overgrown with grass.

The conditions of poor transmigrants kept on even after the Sumatra road crossing the village was opened in 1987. With unfavorable economic conditions, Srimulyo Village was then made the village for the location of the palm oil PIR project. This change marked the beginning of the development of Srimulyo Village into a more developed village.

Business Model of Palm Oil Smallholding

The palm oil development project in Srimulyo Village has involved 367 participating smallholders with a land area of 734 ha so that each farmer owns 2 ha of oil palm land. To run the business, the smallholders in Srimulyo Village formed a joint business in the form of a cooperative which was given the name "KUD Bersama Makmur". The cooperative members were divided into 16 groups with each group consisting of 15 to 31 smallholders. Thus, along the line, this cooperative then manages 734 ha of oil palm plantations starting from weeding, eradicating pests, fertilizing, and transporting them to the palm oil mill. In this scheme, the palm oil smallholders remain as owners of the land and each month receives income from the plantation through the cooperative. Smallholders who are involved in the plantation management will receive wages from the cooperative, either as the cooperative managers or the part-time workers. Smallholders who do not participate in managing the plantation usually allocate their time to work outside the palm oil plantation sector. During its development, the cooperatives played an important role in the success of the palm oil plantations. This situation is in line with the research results of Jelsma et al. (2017) which showed that cooperatives were the key to the high and low productivity of palm oil.

Figure 1. The business model of palm oil smallholding in Srimulyo village
Source: KUD BersamaMakmur (processed data)

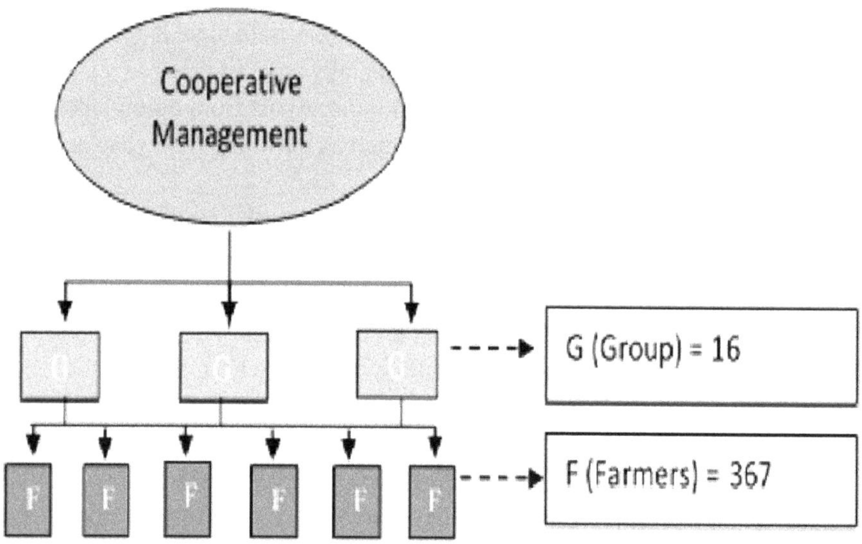

The plantation management model with this system has various benefits. First, the plantation management is implemented in the same standard, so that the quality and quantity of results will be almost similar. The cooperative also has personnel who are more professional in maintaining the quality of the palm oil products that are produced. Second, the smallholders have a high bargaining position in negotiating as partners with the local governments, the palm oil mills, or the nucleus companies. This bargaining position allows smallholders to obtain a competitive purchase price according to the decision of the provincial pricing team. Smallholders also find it easier to negotiate with partners from the financial institutions and the government, especially in obtaining subsidized fertilizers.

As a cooperative member, the palm oil smallholders receive income every month even though they are not involved in the plantation management. This income is obtained from the harvest of land which serves as the right of ownership of the members. If smallholders are involved in plantation activities such as fertilization and harvesting, they will receive wages based on the rules agreed upon in the cooperative. Thus, smallholders can receive income from the crops they own on the land as well as wages from the work done in the cooperative. Smallholders who do not work in the palm oil plantations managed by cooperatives can get additional income from working outside the plasma palm oil plantations. With the cooperative's plantation management, the smallholders can have free time to work outside the palm oil plantation sector.

The cooperative in Simulyo Village has a management structure consisting of a chairperson who is concurrently a member assisted by four heads of divisions, namely the head of the plantation, administration, and finance, enterprise and business, as well as human resources and organizations (Figure 2). In carrying out the duties, the head of the division is assisted by staff, for example, the head of the plantation sector has 2 staff, the head of the administration and finance division has 2 staff, and the head of the enterprise and business sector has one staff. The monitoring and controlling of the implementation of the management's daily duties are carried out by a supervisory body consisting of one chairman and two members. The Supervisory Body has to provide input to the management to carry out their duties properly. Meanwhile, the highest decision of the cooperative relied on the annual members' meeting.

The working period of the cooperative management is 5 years and after that, they can be re-elected. During this period, the cooperative management makes an annual accountability report which is then submitted at the annual members' meeting. The performance of the managers in administering the cooperative will become the determination on whether they can be trusted to continue to be the managers of the cooperative or whether they should be replaced. As of now, the cooperative management has continuously been trusted by the members. It can be seen from the fact that in almost 20 years since its establishment, cooperative management has never been replaced. While it can also lead to stagnation in various cooperative policies, this condition also reflects the size of members' trust in the existing management. This huge credence becomes the basic capital of the cooperative to develop more.

Figure 2. The organization of KUD BersamaMakmur in Srimulyo village
Source: KUD BersamaMakmur

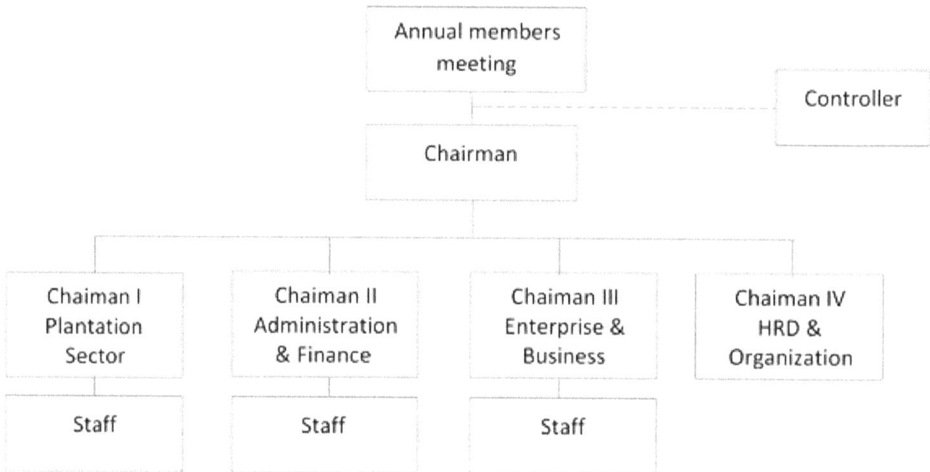

To manage the plantation, the cooperative recruits daily laborers and foremen who are given a certain wage. The various types of work that are usually done by the daily laborers include fertilizing, cleaning plants, harvesting, loading FFB into trucks, and weighing. The provision of fertilizers/pesticides is usually carried out by daily workers under the supervision of the foreman. Working hours for this work are between 7:00 a.m. to 12:00 p.m. with wages of IDR 85 thousand / day for daily workers and IDR 95 thousand for the foreman. Fertilizer is applied based on the results of leaf analysis taken out from several plant samples. After that, the appropriate fertilizer application is implemented to maintain the quantity and quality of the plantation production.

The harvesting of palm oil takes place every two weeks and smallholders will receive their plantation products once a month. Smallholders will then receive the net proceeds from the sale of palm oil, which is the total sales minus the plantation management costs. These costs consist of wages for harvesting, wages for loading and weighing, spewing, road repairs, farm maintenance, compulsory savings, and savings for replanting plants. With this management model, the productivity and quality of FFB from plasma smallholders will be the same because it is administered with the same standards. That being said, the smallholders' income will be calculated by the performance of the cooperative in managing the palm oil land.

Do the Business Model Success?

Many variables can be used to measure cooperative performance. Some of the variables used in this study are land productivity, member trust, management structure, market-linked, business relations, and work performance based on the awards received.

Land Productivity and Smallholders Income

The main indicators of the cooperative's performance are determined by the productivity and income of the smallholders. Various studies have shown that the productivity of the smallholders of the palm oil plantations is generally lower than that of the larger companies (Jelsma et al., 2017). Meanwhile, the productivity of plasma smallholders is generally greater than that of the independent smallholders (Vermeulen and Goad, 2006). Nationally, the productivity of palm oil plantations in 2015 was 3.63 tonnes of CPO / ha/year. This figure varies according to the manager, namely large community plantations of 3.15 tonnes/ha, state companies 3.80 tonnes/ha, national private plantations 3.93 tonnes/ha, and foreign private plantations 4.43 tonnes/ha (Directorate General of Estate Crops, 2016).

Based on the data from the cooperative Bersama Makmur, the productivity of plasma palm oil plantations in Srimulyo Village in 2015 was 27.76 tonnes of FFB / ha/year. If it is assumed that the yield of palm oil in Srimulyo Village is 16 percent[1], then the productivity of palm oil plantations is 4.44 tonnes of CPO / ha/ year. This figure shows that the palm oil plasma smallholders in Srimulyo Village possess high productivity. This indicates that the cooperative in Srimulyo Village has been performing well.

The productivity of palm oil fresh fruit bunches (FFB) in Srimulyo Village has relatively fluctuated from 2007-2015. This is mostly caused by some natural factors, especially the rainy and dry seasons. The long dry season will then have an impact on low land productivity in the following year because palm oil is a type of plant that requires a lot of water. The highest land productivity occurred in 2012, reaching 28.07 tonnes/ha. This was then followed by an increase in the average income from oil palm, which was 6.34 million / month. The lowest productivity occurred in 2007 at 22.46 tonnes/ha, resulting in the low income of the smallholders, with only IDR3.17 million / month. Apart from having influenced by land productivity, the average income of plasma smallholders is also influenced by the price of palm oil on the international market. In 2009, while land productivity was quite high (25.2 tonnes/ha), the average income of smallholders from oil palm was relatively low (IDR 3.8 million / month). This situation occurred because the price of palm oil in the international market was falling.

Figure 3. The average of palm oil productivity (ton/ha) dan palm oil income (million IDR/month) in Srimulyo Village 2007-2015
Source: KUD BersamaMakmur (data processed)

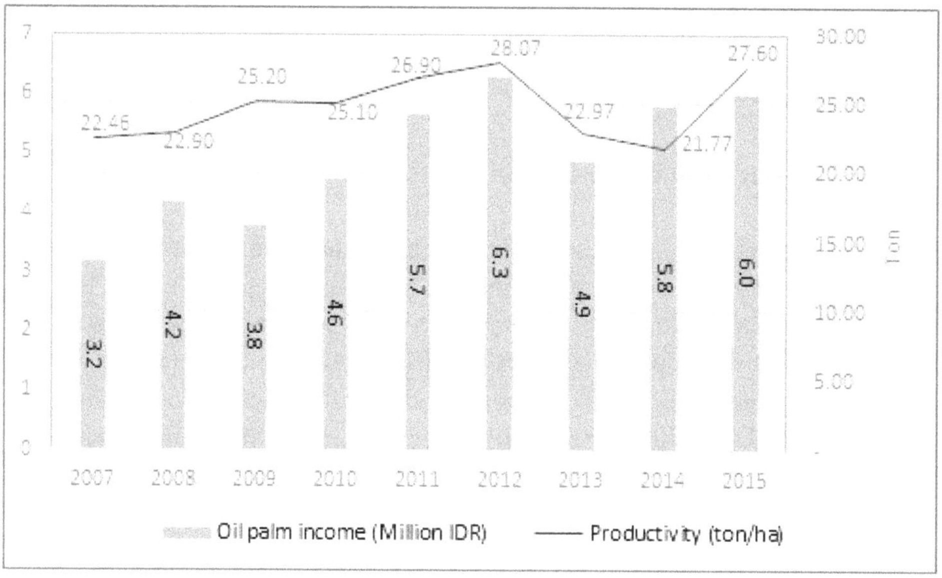

The land productivity and plasma smallholders' income fluctuations also took place between January-December. In 2015, the low productivity of the palm oil plantations came about between January-March. The plantation productivity hit its lowest point in January (2.4 tonnes/ha/month). After that, the productivity managed to slowly increase and reached its peak in July 2015. This means that in the middle of the year, the peak season for the palm oil harvest happened. This condition could arise because in the months of November-March the rainfall was fairly good, and the plants were able to thrive. With the provision of fertilizer and the sufficient water sources and long sunshine, palm oil could then produce optimally. The peak of palm oil productivity at that time was also followed by the high income of smallholders, reaching IDR 9.03 million / month (Figure 4). At that time the smallholders became prosperous, especially those who had side jobs outside of the palm oil plantations.

The low land productivity also went along with the low income for the residents of the palm oil plantations. In January the plantation productivity reached 2.4 tonnes / ha / month with a production value of IDR 3.15 million. While it was considered low, the income from these plantations was still higher than the minimum wage for the South Sumatra Province, which was IDR 1.97 million / month (Government of South Sumatra, 2015). This fact showed that the income of oil palm smallholders in Srimulyo Village was quite high so that most of Srimulyo's population at that time lived above the poverty line. The average income of plasma smallholders in one year was IDR 6.04 million / month, which indicated that the people in Srimulyo Village had a decent income. However, income fluctuations between months should serve as a guide for the farmers to come into a habit of saving during the peak season to cover shortages in the dry season.

Figure 4. The average of small holders income (IDR) and palm oil production (Kg), 2015
Source: KUD Bersama Makmur (processed data)

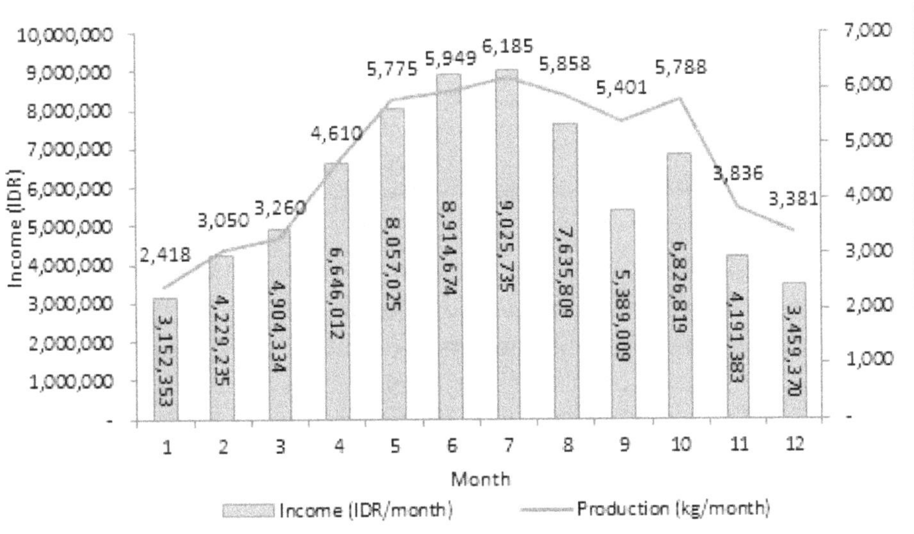

Village Development

The cooperative's performance has had a direct impact on the development of Srimulyo Village. The results of the Focus Group Discussion showed that the people's economy began to improve starting in 1994, ever since Srimulyo Village was converted into a PIR-Trans area for the palm oil plantations. Acting as the core company for this, was PT Goa Manurung Jaya which was domiciled in Sungai Lilin District. Since then, many residents in Srimulyo Village have stopped working outside of the region. They then started working as daily laborers at PT Manurung Jaya. The development of the palm oil plantations in Srimulyo Village became a turning point in the lives of the transmigrantswho began to enjoy improvements in their socio-economic lives. They no longer needed to go outside of their area for two months or so to work, because the same types of work started to become available close to their residential area.

Figure 5. The changing of the socioeconomic condition in Srimulyo village 1982-2016
Source: FGD with Management of KUD

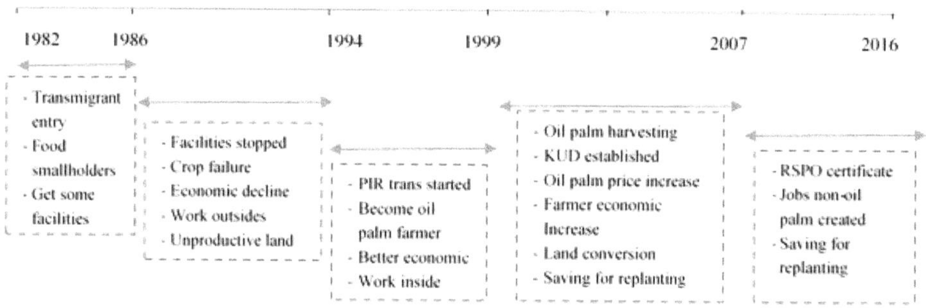

In 1999 several events were considered to be influential in the lives of the smallholders, such as the establishment of the BersamaMakmur Cooperative which managed the palm oil plasma smallholders in Srimulyo. 1999 was marked as the first year of the conversion of land for palm oil plantations from nucleus companies to smallholders. During this period, the economic crisis in Indonesia increased the profitability of export-oriented products, including palm oil (Syafa'at, Friyatno, Mardianto, &Suryadi, 2004). In 2003, smallholders had started saving in the form of idapertabun savings as means for the replanting preparation. There were originally only 30 participants in the idapertabun and 400 households in 2009. Starting in 2004, the economy of the community has then increased, seen from the increasing number of households owning motorbikes in Srimulyo village.

In 2007 the economy was growing again due to the high production and selling price of palm oil. During that year, many residents started owning their trucks and cars. In line with the improvement in the economic condition of the population, migrants from outside began to enter the village for some reasons such as being married to the residents or getting a job in Srimulyo Village. The high production of palm oil also brought about an impact on the increasing assets owned by Bersama Makmur cooperative. The growth of the cooperative's assets and activities in time led to an increase in the need for manpower in the cooperative, both as staff in the office and as supervisors. Outside the plantation sector, people's businesses have also begun to arise, such as stalls, shops, and the wood industry. In 2015 in Srimulyo Village, there were five units of the wood industry, 37 units of grocery/candy stalls, and 10 units of food/beverage shops.

The completion of facilities and infrastructure in the village has also enhanced due to the district road construction program and road repair programs from the local community. In 2009, electricity from PLN had begun to enter the village. In 2010 the village road was paved to support the mobility of the local population. In 2013, Srimulyo Village was separated from Bayung Lencir District and became a separate sub-district along with 13 other sub-districts. This separation was implemented to improve services to the community since the area of Bayung Lencir was considered as already very large.

Cooperative Achievements

The performance of the cooperative as an agricultural business model in Srimulyo Village can also be seen from the various achievements it has received. In the 20 years since its establishment (since 2000) this business model has been recognized for its success at the regional and national levels. In 2010, the palm oil plantation managed by this cooperative was presented with a Roundtable Sustainable Palm Oil (RSPO) certificate. The certificate holds a high standard and therefore reflects the good performance of the cooperative. With this certificate, the cooperative's palm oil products can be accepted in countries that require the RSPO certification.

Aside from that, BersamaMakmur cooperative also received several other achievements. Which include: (a) in 2003, a PIR-Trans oil palm partnership program certificate from PT. Hindoli; (b) the title of the outstanding cooperative in South Sumatra Province; (c) in 2005, the title of the best cooperative in MusiBanyuasin Regency; (d) outstanding producer cooperatives in South Sumatra Province; (e) in 2007, winner of various services cooperative in MusiBanyuasin Regency; (f) outstanding cooperative in South Sumatra Province; (g) 1st place winner of Tk Producer Cooperative. South Sumatra Province; and (h) in 2008 became cooperative with the national level achievement.

Driving Factors for Success Business

Maintaining the Trust of the Members

Trust is the extent to which one believes that others will not act to exploit one's vulnerabilities (Barney and Hansen, 1994; McAllister, 1995). Trust is the belief that someone will find what they are looking for in their exchange partner. Trust involves a person's willingness to behave in a certain way because of the belief that the partner will give what they hope for with an expectation that someone generally knows that the words, promises, or statements of others can be trusted (Barnes, 2003).

Trust is the basis for cooperative relationships (Loomist, 1959). Therefore, the efforts to maintain the trust of the cooperative members are imperative so that a cooperative can maintain its survival through the formation of mutually beneficial relationships that lead to the loyalty of the cooperative members. There are four essentials conditions to establish cooperative relationships. First, each individual must be committed to reaching their goals. Second, each individual must understand that they will not be able to reach the goal without the help of others. Third, each individual must realize that the other people are similarly dependent upon each other for their rewards. Fourth, each individual must know that all members are mutually interdependent (Loomist, 1959). The loyalty of the cooperative members is considered as the future inclination for the members to continue to use cooperative products or services (Ranaweera and Prabhu, 2003). Loyalty is also seen as the members' perception of how long the cooperative can last (Liu and Wu, 2007).

Maintaining the members' trust in cooperative management is a major consideration for palm oil plasma smallholders' cooperative in Srimulyo Village. Transparency and communication with members are keys that have always been undertaken by the cooperative managers. The annual members' meeting is a means for the administrators to gain input on various issues faced by the cooperative along with the solutions on how to overcome them. The existence of a supervisory board that always monitors the performance of the management is also a strategy to keep the cooperative continues to exist and earn the trust of its members. Members' trust in cooperative management has still been well preserved. This has been proven since the number of cooperative members has remained stable from its establishment until now.

The transparency of cooperative management is supported by its members' internet usage. Through the internet, members can directly access the prices set by the standard team for the FFB prices at the provincial level. This way, the members know exactly how much prices are in the market as appose to the prices in their cooperatives. The use of the internet also facilitates the communication and coordination between the management, the smallholders' group leaders, and the members. With the internet or telephone, the information conveyed by the management and the government can be quickly acknowledged by the members.

The involvement of the cooperative members in the palm oil business activities is an important factor in maintaining the members' trust. Through this involvement, members can immediately know as well as monitor the business trips they have along with the process of harvesting to transporting fresh fruit bunches to factories. The yields are usually weighted in each group and separated according to land ownership. After the weighing process is complete, the fresh fruit bunches are sent to the palm oil mill. The results of the scales in the group are then used as a guide in the distribution of the farm products. Through this scale, each member can monitor the distribution of the results to match their respective fields.

Cooperative Business Networking

Another success factor of the palm oil business model in Srimulyo Village is maintaining a harmonious relationship with the core company. Especially since the cooperative management and the relationship between the nucleus company and smallholders are the main factors that affect the productivity of palm oil (Gatto, Wollni, Asnawi&Matin, 2017; Svatoňova, David, & Abraham, 2015). Along with the harmonious relationship, the cooperative can get direct assistance from the company in performing the management duties and harvesting practices of smallholders so that the quality of palm oil products is maintained as well as meeting the standards that have been set.

Preserving the quality of the palm oil products is very important so that the RSPO certificate obtained can be held. To maintain the quality of these products, the core company places supervisors at the village level so that they can monitor the agricultural activities ranging from plant maintenance to harvesting. The palm oil fruit harvest has to have certain criteria that must be met so that the yield is according to standards. For the determination of the base price for the FFB, the core company determines the price based on the recommendation by the FFB pricing team at the provincial level. The smallholders can access the floor price, so they can refuse if the company price is lower than the base price. Through price transparency and strong cooperative leadership, conflicts between core companies and smallholders can be prevented (Feintrenie et al., 2010).

Various technologies have been employed by the cooperative as well to maintain the quality of oil palm products, namely: leaf analysis, making oil palm seeds and eradicating rats using owls. Leaf analysis is carried out to determine the type and volume of fertilizer that should be applied in the palm oil plantation area. In terms of providing fertilizer, the cooperative has also collaborated with fertilizer distributors in the regions to prevent the shortage of fertilizer from happening. The practice of making superior palm oil seeds has also been carried out by the cooperative to maintain the quality of the palm oil as well as to facilitate the outside community to get the quality seeds.

Even though there has been a mutually beneficial relationship between the plasma smallholders and the nucleus company, the cooperative's work has still limited to plant production activities on the farm. While this collaboration allows smallholders to obtain satisfactory results from the palm oil plantations, it seems that the cooperation relationship cannot be extended to processing activities from FFB to CPO since the activities are carried out extensively by the palm oil processing factories owned by the nucleus company. With this condition, smallholders do not have the opportunity to increase their income from activities in the factory. The involvement of the smallholders in the post-harvest activities seems necessary so that the transfer of knowledge can take place which in turn increases the income of the plasma smallholders.

Business Sustainability

Palm oil plantations do have advantages over seasonal crops, especially when it comes to the productive age of the plants. Palm oil can be productive up to the age of 25 years, while the annual crops must be changed during the season. The long productive life of the plants has been beneficial for the smallholders because they do not need to rejuvenate the plants until the plants reach the age of 25 years. During this productive life, the smallholders enjoy the yield from the palm oil, as long as the plants are maintained and supplied with nutrients according to their needs. According to this, the energy required for crop management is then more manageable than that of seasonal crops.

While so, the costs required to replant palm oil when it reaches its unproductive age are considerably high because of the 25 years accumulation. Therefore, smallholders need to plan to effectively replant their plants from an early age. This is necessary so that when they enter the replanting time, they are equipped with the necessary costs and labor. Without the proper planning, the smallholders can find the process troublesome especially during the first 5 years of growth. Therefore, it is estimated that they also need to employ other types of work outside the plantation as well as planting seasonal crops on the plantation area during the initial 5 years of growth. Smallholders also need to save, so they do not experience trouble with funds during replanting. The palm oil smallholders are usually greatly helped by replanting funds from the plantation fund management institution of 25 million rupiah/ha.

In terms of the preparations of replanting, the cooperative in Srimulyo Village coordinates a savings fund for the preparation of plasma plantation replanting which is named idapertabun. The collection of savings funds is in collaboration with the 1912 Bumiputera Life Insurance in Palembang. The smallholders as the owner of the land are the holders of the insurance policy and the funds will be disbursed at the time of replanting the plant. Idapertabun participants not only come from the cooperative members in Srimulyo Village but also other cooperatives around Srimulyo Village. In 2015 there were 857 Idapertabun participants with funds collected of approximately IDR 11.88 billion. With this replanting savings fund, the palm oil smallholders can be better prepared in doing the replanting when the plants reach the maximum age of 25 years and over. Through replanting, the continuity of the smallholders' work can be maintained and be even more productive since replanting uses new and more superior seeds.

Cooperative Initiatives

The role of smallholders and the cooperatives in the palm oil business model in Srimulyo Village is still limited to the activities in the plantation sector by relying on the assistance given by the nucleus companies. This makes the innovation initiative at the cooperative level is still not visible. This cooperative policy on palm oil plantations is prioritized to meet the quality and quantity demands of the company. Therefore, those plantation practices are carried out based on the company standards.

Several cooperative initiatives have been recognized to promote smallholders' businesses. One of the initiatives was the cooperative success in becoming an analyst during the plantation development of the KKPA PIR project at its inception. At that time, the cooperative faced a problem, which was a request from the core company to increase the loan value to 43.5 million per lot since the original loan value of 31.4 million was deemed insufficient. The government, in this case, Bank Mandiri as the executing bank, has set a credit ceiling for the development of IDR 31.400.000.-. However, for the development of the 297 plots of land, the core company could not take a credit ceiling from Bank Mandiri because the ceiling was deemed insufficient. They wanted to increase the credit limit to IDR 43.5 million / lot.

However, the cooperative as the institution for the smallholders rejected the credit limit that PT. Hindoli originally recognized and fought to find other solutions. After much deliberation with the community and consultation with various parties, the cooperative decided to take over the role of the core company to become an analyst (debt guarantor) for the development of a 594 ha plantation. The cooperative then applied itself for the intended credit and was approved with a credit ceiling of IDR 38.7 million per plot.

The change in role as an analyst has provided the cooperative with various benefits for the smallholders. First, the credit ceiling which before was a burden for the smallholders' debt has decreased from IDR 43.5 million as determined by the core company to be IDR 38.7 million, this brought the credit installment to be smaller. Second, the escrow account funds (50% of the proceeds from the sale of FFB before the 48 months of the conversion period) have then become the rights of the smallholders and the funds have been included in the reserve fund for the smallholders' credit installments. With the funds, when the credit payment is due, the smallholders can pay from IDR 2 million to IDR 6 million per lot.

This situation has implications for the timely implementation of conversion, which is when the plants are 48 months old. The profit for farmers has increased because after one year the Bank of South Sumatra has lowered interest rates from

20.5% per year to 16.75% per year. This is because the interest expense of 20.5% per year was admittedly seen as quite a burden by the smallholders, the cooperative was then trying to find a solution so that the interest expense could be reviewed by the Bank. Facilitated by the local government, the Bank reduced interest expense to 16.75% per year. With a fairly good production capacity and a lighter debt burden, the smallholders' credit could then be paid off in a period of 34 to 42 months. This is supported by a fairly good FFB price, a cut in credit installments of 40% of farmers' income, and a reserve for credit installments obtained from the escrow account funds before the conversion period. After the expiration of the smallholders' credit coverage, their economic conditions have improved so that the village economy has grown.

The other initiative was to make the cooperatives a place for palm oil education and training activities for people in need. Through this activity, the participants are directly involved in the plantation activities starting from maintenance, harvesting, to transportation to the palm oil mill. This activity will have a positive impact on the development of the cooperative itself. First, they will try to maintain the land productivity so that it is attractive to the trainees. Through this training, the cooperative managers are also required to improve their skills on the palm oil plantations so that they can teach others. Second, the cooperative is expected to maintain mutually beneficial relationships between the communities and the cooperative.

The Government Support

The support from both the local and national governments is an important factor in the development of the palm oil business model at the smallholder level. In the early stages of the development of palm oil, the government funding support was very high before then started to decrease (Pramudya, Otto, &Termeer, 2017). The financial support at the beginning of the development of palm oil with the PIR-Trans project was a key factor in the development of the palm oil plantation business. Without it, this business model would not have taken place since originally, Srimulyo Village was created to become a food crop farming village. As so, partnerships with banks also could not happen without a recommendation from the government.

Aside from the funding for the development of the plantation, the local government also provided the facilities through the development of the infrastructure at the village level and the construction of the trans-Sumatra road through Srimulyo Village. The construction of this highway increases the mobility of the local population which includes facilitating the transportation of palm oil products from the cooperative to the factory. The construction of village asphalt roads, health centers, and schools were also an important part of the development of Srimulyo Village as a palm oil plantation village.

Along the line, the biodiesel program made from palm oil has become the government's effort to increase the absorption of CPO in Indonesia. This program has a dual function, namely reducing dependence on imported fuel as well as minimizing the dependence on the foreign palm oil market. Through this program, part of the CPO can be utilized domestically, so that the palm oil smallholders are better protected from fluctuations in the CPO prices on the international market. This policy is also a solution to the prohibition of palm oil products in several countries because palm oil is a commodity that damages the environment.

SOLUTIONS AND RECOMMENDATIONS

The smallholders deal with various challenges in running a palm oil plantation business in Indonesia at the household, group, and international market levels. Since maintaining the trust of group members is a fundamental factor in the palm oil business, the cooperative management must then be transparent and hands-on on various elements of the cooperative members through the smallholders' group during the decision making. Information on prices, yields, and distribution of the remaining income from the business must be easily accessed by the cooperative members as owners of the palm oil land. At the same time, business innovation must be carried out continuously, so that the cooperative can continue to move forward and receive full support from members.

The development of the cooperatives cannot be separated from the roles of various stakeholders, especially the core companies, fertilizer factories, and local governments. Preserving a harmonious relationship with these stakeholders will guard and enhance the development of the cooperative. To do so, constructive communication with these stakeholders must be continuously promoted. The role of the cooperatives in the palm oil business cycle is still limited to activities on the land and has not been involved in post-harvest activities. However, to improve the smallholders' welfare, it is appropriate that smallholders are given a portion of shares for the post-harvest activities. With this model, the palm oil smallholders can get income from both the plantation products and the CPO.

In the meantime, the challenge from the international community regarding the palm oil business is the rejection of the CPO by several countries due to its projection as damaging to the environment. Therefore, sustainable plantation practices must be carried out by all palm oil business actors. The RSPO and ISPO certificates, which are the main indicators for the sustainability of oil palm plantations, need to be maintained by smallholders. On the other hand, the government must enforce various negotiations to ensure that the palm oil commodity in Indonesia is environmentally friendly. Besides, efforts to produce CPO-based fuel in Indonesia need to be intensified so that some of the palm oil can be absorbed in the domestic market.

FUTURE RESEARCH DIRECTIONS

The palm oil business model in this study is one of the models developed in South Sumatra, Indonesia. Apart from this model, there are still many other models that have been developing in the palm oil business. How to implement other models and ways to increase the success of the palm oil business model will need further investigation. Since the plantations have been managed by the cooperative, many smallholders do not participate in the plantation activities. With these circumstances, the smallholders can do other types of work outside the palm oil sector. Therefore, the research on the diversity of the sources of livelihoods for the smallholders in Srimulyo Village is important. One of the challenges in increasing the income of palm oil smallholders is that their activities are limited to activities in the garden. Therefore, it is imperative to develop a model so that the palm oil smallholders can be involved from the plantation activities to the post-harvest activities.

CONCLUSION

The oil palm plantation business model in Srimulyo Village was implemented by consolidating 734 ha of palm oil land owned by 367 farmers. This land combination was managed professionally by the cooperative. The empirical data shows that this business model has succeeded in maintaining the productivity of the palm oil plantations so that the smallholders' income can be above the poverty line. The success of plantation management has been one and the same from the trust of the members and cooperation with the core companies, banks, and the government. The change in Srimulyo Village from a food crop center village to a plantation village has altered the village's development from an underdeveloped village to a developed village. The socio-economic life of the community continues to improve in line with the high productivity of palm oil land which is the main source of income for most smallholders in Srimulyo Village.

To continue its existence, this cooperative business model must adapt to technological advances including digital technology. Since only the innovative business actors can survive in the increasingly fierce competition between business groups, this business model has not only faced challenges at the group level but at the international level as well. Concerning the international business competition, the government plays an important role in negotiating and providing alternative markets for palm oil products. In doing so, efforts to market palm oil in the form of finished products also need to be continued, so that palm oil can offer greater benefits for the progress of both the smallholders and the country.

The palm oil business that is managed by the cooperative allows benefits to its smallholder members. Some smallholders do not even need to participate in managing the farm while still enjoying the harvest every month. The available spare time is utilized by some of the smallholders' households to work in other sectors so that the household income can double. This condition raises the heterogeneity of the sources of livelihoods of the smallholders in Srimulyo Village. Which in turn, providing many other types of work outside of the agricultural sector that invigorate young people to stay in the rural areas.

REFERENCES

Astuti, U. P., Wahyu, W., & Andi, I. (2011). *Mempengaruhi Alih Fungsi Lahan Pangan menjadi Kelapa Sawit di Bengkulu : Kasus Petani di Desa Kungkai Baru*. Academic Press.

Barnes, A. (2003). *How to maintain the trust: A study in banking customer*. Academic Press.

Barney, J. B., & Hansen, M. H. (1994). Trustworthiness as a source of competitive advantage. *Strategic Management Journal*, *15*(S1), 175–190. doi:10.1002mj.4250150912

Benos, T., Kalogeras, N., Wetzels, M., De Ruyter, K., & Pennings, J. M. (2018). Harnessing a 'currency matrix'for performance measurement in cooperatives: A multi-phased study. *Sustainability*, *10*(12), 4536.

Directorate General of Estate Crops. (2010). *Tree Crop Estate Statistic of Indonesia: 2009-2011, Palm Oil. Directorate General of Estate Crops*. Ministry of Agriculture.

Directorate General of Estate Crops. (2016). *Tree Crop Estate Statistic of Indonesia: 2015-2027, Palm Oil. Directorate General of Estate Crops*. Ministry of Agriculture.

Directorate General of Estate Crops. (2019). *Tree Crop Estate Statistic of Indonesia: 2018-2020, Palm Oil. Directorate General of Estate Crops*. Ministry of Agriculture.

Feintrenie, L., Chong, W. K., & Levang, P. (2010). Why do farmers prefer oil palm? Lessons learnt from Bungo district, Indonesia. *Small-scale Forestry*, *9*(3), 379–396. doi:10.100711842-010-9122-2

Gatto, M., Wollni, M., Asnawi, R., & Qaim, M. (2017). Oil palm boom, contract farming, and rural economic development: Village-level evidence from Indonesia. *World Development*, *95*, 127–140. doi:10.1016/j.worlddev.2017.02.013

Government South Sumatra. (2015). Decree of the Governor of South Sumatra, Number: 838 / KPTS / Disnakertrans / 2015, concerning the minimum wage of South Sumatra Province in 2016

Jelsma, I., Slingerland, M., Giller, K. E., & Bijman, J. (2017). Collective action in a smallholder oil palm production system in Indonesia: The key to sustainable and inclusive smallholder palm oil? *Journal of Rural Studies*, *54*, 198–210. doi:10.1016/j.jrurstud.2017.06.005

Liu, T. C., & Wu, L. W. (2007). Customer retention and cross-buying in the banking industry: An integration of service attributes, satisfaction and trust. *Journal of Financial Services Marketing*, *12*(2), 132–145. doi:10.1057/palgrave.fsm.4760067

Loomis, J. L. (1959). Communication, the development of trust, and cooperative behavior. *Human Relations*, *12*(4), 305–315. doi:10.1177/001872675901200402

McAllister, D. J. (1995). Affect-and cognition-based trust as foundations for interpersonal cooperation in organizations. *Academy of Management Journal*, *38*(1), 24–59.

Mutasowifin, A. (2002). Penerapan Balanced scorecard sebagai tolok ukur penilaian pada badan usaha berbentuk koperasi [The application of the Balanced scorecard as a measure of assessment for cooperative business entities]. *Paramadina University Journal*, *1*(3), 245–26.

Nkuranga, T., & Wilcox, K. (2013). *Cooperative performance index field results and analysis. In USAID-enabling market integration through rural group empowerment*. EMIRGE.

Pramudya, E. P., Hospes, O., & Termeer, C. J. A. M. (2017). Governing the palm-oil sector through finance: The changing roles of the Indonesian State. *Bulletin of Indonesian Economic Studies*, *53*(1), 57–82.

Ranaweera, C., & Prabhu, J. (2003). The influence of satisfaction, trust and switching barriers on customer retention in a continuous purchasing setting. *International Journal of Service Industry Management*.

Rist, L., Feintrenie, L., & Levang, P. (2010). The livelihood impacts of oil palm: Smallholders in Indonesia. *Biodiversity and Conservation*, *19*(4), 1009–1024.

Svatoňová, T., Herák, D., & Kabutey, A. (2015). Financial profitability and sensitivity analysis of palm oil plantation in Indonesia. *Acta Universitatis Agriculturae et Silviculturae Mendilianae Brunensis*, *63*(4), 1365–1373.

Syafa'at, N., Friyatno, S., & Mardianto, S,. & Suryadi. (2004). Kinerja Nilai Tambah dan Produksi Sector Pertanian 2000-2003 [Value Added and Production Performance in the Agricultural Sector 2000-2003]. *Agricultural Policy Analysis*, *2*(1), 1–16.

Vermeulen, S., & Goad, N. (2006). *Towards better practice in smallholder palm oil production*. Iied.

White, B. (2012). Agriculture and the generation problem: Rural youth, employment and the future of farming. *IDS Bulletin*, *43*(6), 9–19.

World Bank. (2020). *Employment in agriculture (% of total employment) (modeled ILO estimate)*. https://data.worldbank.org/indicator/SL.AGR.EMPL.ZS

KEY TERMS AND DEFINITIONS

Business Sustainability: A strategy that prioritizes the long-term survival of a business while connected to the ecological, social, and cultural systems. Through this strategy, cooperatives can continue to exist and develop.

Cooperative: An association that works together to run a business as a family to improve the welfare of its members. In managing a cooperative, its members can freely enter and leave the business entity.

Free Daily Workers (Casual Workers): Workers who are recruited by the cooperative with a daily or piece-rate system. These workers are given wages according to the output produced.

Idapertabun: Savings fund of the cooperative members that is paid monthly as a support fund for replanting plants once the plants reach the age of 25 years and over.

Transmigrants: People who move from the island of Java-Bali to outside of the island of Java-Bali through what is called a transmigration program.

Trust: The trust of certain parties to others in conducting transaction relationships based on a belief that the person whom they trust has all their obligations properly conducted as expected.

ENDNOTE

1 The yield of high-quality oil palm is between 22.1-22.2 percent (Ministry of Industry, 2007). Currently, the yield of oil palm in Indonesia is still below 20 percent, which is between 15-18 percent (www.medanbisnisdaily.com).

Chapter 6
Cognitive Antecedents of the Agricultural Entrepreneurship Intentions of Indian University Students

Syed Abid Hussain
Aligarh Muslim University, Aligarh, India

Gayas Ahmad
https://orcid.org/0000-0003-3094-9869
Aligarh Muslim University, Aligarh, India

Adil Khan
https://orcid.org/0000-0002-1309-2472
O. P. Jindal University, India

Aamir Hassan
Aligarh Muslim University, Aligarh, India

Mohd Shamim
Aligarh Muslim University, Aligarh, India

ABSTRACT

The purpose of this chapter is to extend the research on determinants of entrepreneurial intentions in the agricultural industry by using the theoretical framework of determinants of entrepreneurial intentions and entrepreneurial event model. By employing the DEI and EEM, the researchers evaluate how perceived desirability, perceived feasibility, individual background, and triggering events can influence the attitude of an individual and in turn how entrepreneurial attitude can control entrepreneurial intentions. To achieve the objective, a questionnaire survey was held using the sample of 335 PhD, master, and bachelor students in commerce, business, and economics from an Indian central university. The data was analysed using a linear regression model. The findings advocate that perceived desirability, perceived feasibility, individual background, and triggering events are positively related to entrepreneurial attitude, and the entrepreneurial attitude positively and significantly influences entrepreneurial intentions in the agricultural sector.

DOI: 10.4018/978-1-6684-2349-3.ch006

INTRODUCTION

India is the second most populated country, of which a large amount of population is unemployed. As per ILO reports, there is an expected rise of 0.6 million in the number of unemployed persons in India in the year 2019. India ranks second worldwide in farm outputs. "The history of Agriculture in India dates back to Indus Valley Civilization era and even before that in some parts of southern India" (Brese & White, 1993). As per 2018, 50% of the Indian jobs were created by agricultural sector and has contributed 17-18% to the GDP of the country (Sundar, 2019). But interestingly as per the reports of CMIE, during 2018 there was an increase on employment among urban men by 500000 jobs, whereas the rural men lost 2.3 million jobs(Vyas, 2019). This process is dominated by the increasing number of unemployed youth. As we know the people living in rural areas are more inclined towards agricultural activities, hence it is necessary to supplement the human resource productivity in agricultural sector through agricultural entrepreneurship.

In agricultural sector the youth interest is decreasing year by year, in addition many of them who primarily depend on agricultural activities are mostly small scale farmers and are facing low productivity. So there is a need to involve creative, dynamic and young generation to develop this sector through entrepreneurial activities. Now what drives entrepreneurship is a debatable question (Shane &Venkataraman, 2000;Stevenson &Jarillo, 1990). Entrepreneurial intentions and cognitions have captured a pertinent place in research for setting up a new business venture (Krueger,1993;Krurger,2000;Bird &Schjoedt,2009). Shane (2007) and Fitz-Koch, Nordqvist, Carter, & Hunter, (2018) argues that there is the requirement of in-depth study so as to understand how and why Entrepreneurs identify opportunities and create new ventures.Surprisingly on EI only a few studies (Roy, Akhtar, & Das, 2017; Giacomin, Janssen, Pruett, Shinnar, Llopis, & Toney, 2011; Pathak & Wang, 2012;Kar, Subudhi, &Padhy 2017; Arafat, Saleem, Dwivedi, & Khan, 2018; Arafat, Saleem&Dwivedi, 2020) had been held in Indian setting.Moreover, the existing literature also shows that among the studies in the agricultural sector less attention has been paid towards new venture creation aspect (Arafat et al., 2018). The available literature on the subject of Entrepreneurial behaviour portraits that there are very few studies focusing on Agricultural Sector (McElwee 2006; Brunjes& Revilla 2013; Arafat et al., 2018). Number of studies were held among university students i.e. (Marques, Ferreira, Gomes, & Rodrigues, 2012

Nasip, Amirul, Sondoh&Tanakinjal, 2017) which have tried to investigate the effects of various psychological and behavioural factors but the focus was mainly on the generic entrepreneurship.

The gap between human resources and education problems is the main indicator of insufficiency in entrepreneurs, were colleges play a vital role in giving birth to the young agricultural entrepreneurs (Ridha & Wahyu, 2017). Given the role of agricultural sector in employment and economic development, there is a wide range of opportunity for research to promote entrepreneurship in agricultural sector. So the objective of the current study is to unfold and analyse how various precursory factors affect entrepreneurial activity in the agricultural sector. Based on the above statements we have taken two models in this study, the first one is an Entrepreneurial Event Model (EEM), which is a behavioural psychological model (Shapero&Sokal, 1982) and the second model is Determinants of Entrepreneurial Intentions (DEI), which is an economic-psychological model (Davidsson, 1995).

In the next section we provide a literature background on EEM and DEI models. In the remaining subsequent sections, we described our research methodology followed by analysis and discussion of study results and finally, the conclusions are drawn.

LITERATURE REVIEW

The procedure for the foundation of new businesses, progressing of standing ones and reorientation of national institutions is concomitant with the changes in economic system through entrepreneurship (Reynolds,2015). Moreover, entrepreneurship is considered as one of the main mechanism of economic development as it ensures employment, innovation and welfare effects (Acs et al.,2008). According to Gelderen, et al.(2008, p555) "understanding modern careers as well as entrepreneurship, knowledge of the factors that explain entrepreneurial intentions is indispensable." As per Krueger, Reilly, &Carsrud, 2000to form a novel business venture, intention is possibly the greatestprognosticator of entrepreneurship. Gelderen et al. (2008, p.555) also argued that "understanding modern careers aswell as entrepreneurship, knowledge of the factors that explain entrepreneurial intentionsis indispensable". Ajzen (1991, p. 181) has clearly defined entrepreneurial intention as "intentions are assumed to capture the motivational factors that influence a behaviour; they are indication of how hard people are willing to try, of how much of an effort they are planning to exert, in order to perform the behaviour". While choosing to perform a particular behaviour recognition of opportunities is a vital preliminary exercise. As the process of entrepreneurship takes place over the time, entrepreneurial intention (EI) can be regarded as an initial step in this process (Lee & Wong, 2004;Nguyen, 2017; Hassan, Saleem, Anwar &Hussain, 2020; Hassan, Anwar, Saleem, Islam &Hussain, 2021). So to understanding why an individual behave entrepreneurially it is important to analyse and understand how the development of entrepreneurial intention takes place and it is also essentially pertinent to understand the nature and characteristics of the precursors that may vary entrepreneurial intentions (Shane &Venkataraman, 2000). Hence the researchers have exhibited much attention to entrepreneurial intentions throughout the world. The prevailing literature comprises of several studies available that have discussed entrepreneurial intentions of the students with respect to various studies(Schlaegel & Koenig, 2014). Such that (Fitzsimmons & Douglas, 2011) studied the role of perceived feasibility and desirability in the formation of entrepreneurial intentions, (Nguyen, 2017) studied the association between entrepreneurial intentions of international students of business and their attitude towards entrepreneurship, (Franke&Luthje, 2004) investigatedthe association between education and entrepreneurial intentions of the students, and in similar manner(Ali, Tajddini, K., Rehman, K., Ali, & Ahmed, 2010) have discussed the external factors that might impact students entrepreneurial intentions.

With respect to this, existing literature has already explained the effect of entrepreneur's characteristics, over and above theentrance behaviours and the targeted objectives of those entrepreneurial behaviours (Autio, Kenney, Mustar, Siegel, & Wright, 2014). Consequently, thecore and basic issue for entrepreneurship research is to understand the effect of both internal and external environments on entrepreneurial process (welter, 2011; Stenholm, Acs, &Wuebker, 2013; Pindado & Sánchez, 2017). And the review of the literature currently in existencediscloses that the agricultural sector has not yet been regardedin the general entrepreneurship research (Alsos, Hogskolani, Internationella, &Ihh, 2011).Besides this, the agricultural sector has a different environmental and economic features which are difficult to ignore as they differentiate other economic activities from agricultural entrepreneurship(Pindado & Sánchez, 2017).

The level of business activity in an industry is directly linked to higher expected returns of entrepreneurship in comparison with employment (Yaseen, Asif, Saleem, Sadaf, &Israr. 2016)

So, the cognitive factors "desirability and feasibility" play a vital role as the determinants of intention influencing new start-ups (Krueger at el., 2000, Elfving, Brannback, &Carsrud, 2009). Davidsson (1995) proposed the EEM model based on behavioural dispositions and contingent factors that provides possibly good explanations for both environmental and dispositional factors. However, it appears that the researchers have underestimated the grandness of personal and situational factors which might help us to explain the ability and the motivation to start new business in a better way (Krueger, 2000; Elfving et al., 2009).

So in this research, we focussed on EEM and DEI to articulate the theoretical framework. The rationale behind is that both models give adequate explanation on transforming intention into action of starting new business (Schlaegel& Koenig, 2014). Furthermore, such characterisations have not yet been properly investigated in agricultural sector specially in a developing country like India.

ENTREPRENEURIAL EVENT MODEL

The EEM is one among the famous avenues informing number of theoretical frameworks business (Schlaegel& Koenig, 2014).In Shapero's Entrepreneurial Event Model (Shapero, 1975), it contemplates the venture creation as an outcome of the relationships among various contextual elements, which may perform via their impacts on the perceptions of an individual. According to Fitzsimmons and Douglas (2011) the presence of any of the two (i.e. perceived desirability or perceived feasibility is very vital to become an entrepreneur, if both are absent in an individual then he or she will be non-entrepreneur. In EEM, Triggering Events are mentioned with respect to "displacement" of inertia. Shapero and Sokol (1982) stated that the behaviour is directed by inertia unless something shifts that inertia and the shift in inertia can be positive or negative and has a potential to change one's life and could act as a major force for venture creation(Shapero&Sokol, 1982; Learned, 1992).Furthermore, themost commonly observed triggering event is job related displacement (Shapero&Sokol, 1982;Summers, 2000).

Shapero mention three elements that regulateentrepreneurial intention, viz. "Perceived desirability," "Perceived feasibility," and "Propensity readiness".Shapero emphasizes that to envisage the intention it is necessary to study these perceptions clearly. The desirability, feasibility and readiness are the basic requirements to act in a particular way.
The three constituents of Entrepreneurial Event Model are described as follows:

- Perceived desirability denotesa given behaviour under which people feels attraction towards entrepreneurship.
- Perceived feasibility refers to the degree to which people consider themselves personally capable of carrying out adefinitebehaviour.
- Perception of readiness refers tothe willingness of an individual to perform on decision. But we have not taken this component for the study.

Over their powerof changing individual's values system. Social and cultural factors determine the above three perceptions (Shapero, 1975).

Determinants of Entrepreneurial Intention

So as to explain intention to start entrepreneurial activity Davidsson (1995), introduced the notion of entrepreneurial conviction (EC). DEI model particularly focused on the explanatory power of entrepreneurial conviction along with attitude and individual background (Yaseen et al., 2016). Among the components of DEI model we have taken only two components i.e. individual background and entrepreneurial attitude in this study.

Next section of this study discusses the formulation of hypothesis for the conceptual model.

PERCEIVED DESIRABILITY (PD)

Entrepreneurial opportunities tend people to assess their competence to exploit these opportunities and if they find themselves capable, they cultivatean optimistic attitude towards behaviour (Ajzen, 1991). That is, when someone asks himself about the presence of business opportunities, what he is actually does is the evaluation of his confidence in the economic climate (Maula, Autio, &Arenius, 2003).Krueger (1993) described perceived desirability as "the degree to whichone finds the prospects of starting a business to be attractive; in essence, it reflects one's affection toward entrepreneurship".Krueger and his associates have recognized that the perceived desirability of an individual may create entrepreneurial opportunity and hence can be prudently characterized as the precursors of entrepreneurial intentions (Krueger &Carsud, 1993;Krueger &Brazeal, 1994; Krueger et al., 2000) and for becoming an entrepreneur personal desirability raises attitude towards entrepreneurship (Kolvereid, 1997).Therefore, it is the normative environment, which substantially influences the perception of belief that opportunity is attractive enough to be exploited (Yaseen et al., 2016).

Hence, the following hypothesis is formulated.

Hypothesis 1: Perceived desirability positively influences the entrepreneurial attitude.

PERCEIVED FEASIBILITY (PF)

While thinking tentatively for setting up of new business venture one needs to consider what actually he or she can achieve if succeeded, that is, the extent to which new venture creation is feasible to them. So the confidence in one's own ability to successfully execute refers to the perceived feasibility (Shapero&sokol, 1982). Starting a new business is not an accidental act rather it is an intentional act that involves frequentendeavours to exercise control over the process in order to reach the desired aftermaths (Gartner, 1985).Perceived feasibility perception is a higher order product of self-efficacy and collective efficacy to translate perceptions into revealed behaviour of starting business (Yaseen et al., 2016). The psychological literature has established the importance of self-efficacy and collective efficacy for entrepreneurial behaviour. An individual with high self-efficacy will develop morepositive attitude

of starting a new venture as the self-perceived dispositional tendency may be relevant to attitude towards entrepreneurship in particular (Zampetakis, Kafetsios, Dewett, &Moustakis. 2009). Thus, it was hypothesized that:
H2: Perceived feasibility is positively related to entrepreneurial attitude.

INDIVIDUAL BACKGROUND (IB)

As per Davidsson (1995) in the model of DEI, the relationship between the individual's personal background and intention is supplementary through entrepreneurial general attitude. According to Lüthje and Franke (2003), when an individual makes an economical assessment of choosing an entrepreneurial career path, contextual factors can be realized to assist or hamper entrepreneurial activities. Individual background can be experienced of producing entrepreneurial intention incidentally as a contextual factor (Kim & Hunter, 1993; Krueger, 2000;Yaseen, Saleem, Zahra, &Israr, 2018) and societal context of an individual, environmental and personal have a significant impact to the develop attitudinal behaviour positively (Robinson, Stimpson, Huefner, & Hunt, 1991; Luthje&Franke, 2003).
Given the above, the hypothesis tested in this study is:
H3: Individual background is positively related to entrepreneurial attitude.

TRIGGERING EVENTS (TE)

People are hesitant towards the change. The reason of such hesitancy is the lack of motivational force or the perceptions of task difficulty (Shapero&Sokol, 1982;Tushman&Romanelli, 1985).Triggering Events are mentioned with respect to "displacement" of inertia. Shapero and Sokol (1982) stated that the behaviour is directed by inertia unless something dislocates that inertia and the shift can be positive or negative and has a potential to change one's life and could act as a major force for venture creation (Shapero&Sokol, 1982; Learned, 1992). Furthermore, the most commonly observed triggering event is job related displacement (Shapero&Sokol, 1982; Summers, 2000). An individual may get triggered either in positive or negative way on happening of an event and can prompt an individual to become an entrepreneur, as entrepreneurial start-ups encompass incremental collective change or an immediate shift with swings in between the period of inactive and active changes. Hence we hypothesized;
H4: Triggering event is positively related to entrepreneurial attitude.

ENTREPRENEURIAL ATTITUDE (ATT)

The idea of establishing and starting a new business venture is intentional phenomena andthe attitude is the finest predictor of intention(Soomro, 2015).Scholarssimilar to Robinson et al. (1991) have argued that the study of attitude is a superior approach for the study of entrepreneurship than the personality characteristics. The attitude toward behaviour refers to the amount of extent an individual assesses somewhatfavourableor unfavourable. More the expectations and beliefs are present toward self-employment, the more will be favourable attitude toward entrepreneurship of an individual (Nguyen, 2017). In other words,the maximum optimisticthe perception of establishing a new business venture (see,e.g.Shapero&Sokol, 1982; Krueger et al., 2000; Segal Borgia, &Schoenfeld 2005; andGelderen& Jansen, 2006), the maximum favourable will be the attitude towards setting up a new business venture and, subsequently, the stronger the individual's entrepreneurial intention. So this is a very critical factor for an entrepreneur as the entrepreneurial intention of an individual is affected by an optimistic belief about entrepreneurship.With respect to Zampetakis, Anagnosti, &Rozakis, (2013)andDevi (2015), attitude toward the behaviour has upper hand in the formation of entrepreneurial intentions in agricultural sector.Besides,(Krueger et al., 2000;Franke&Luthje, 2004;Fayolle, 2005; Kickul, Wilson, Marlino, & Barbosa, 2008), confirmed a significant, positive relationshipamong one's attitude and entrepreneurial intentions.

Hence it was hypothesized that;

H5: Entrepreneurial attitude is positively related to entrepreneurial intention.

Figure 1.

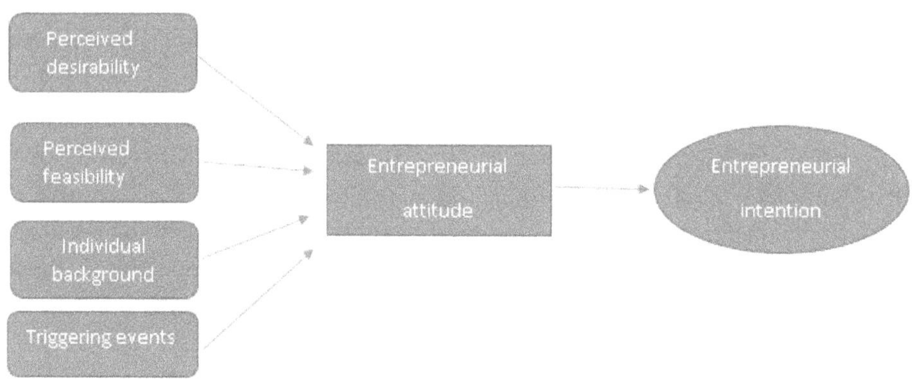

RESEARCH METHODOLOGY

This study was conducted with three objectives. First, we wished to assess the factors precursor to entrepreneurial attitude in the context of agricultural sector see figure 1. Second, we wanted to assess the association between entrepreneurial attitude and entrepreneurial intention in the context of Indian agricultural sectorand to know how strong it is? Third, we intended to show how entrepreneurial intention in agricultural sector exists among university students in an economicallyemerging country like India, where the livelihood of large population depends upon agricultural sector.

Table 1. Result of exploratory Factor Analysis

	Factor Loadings	Communality	Variance Explained (%)	Reliability (alpha)
Perceived Desirability				
PD1	.733	.537	61.291	0.67 (0.662)
PD2	Removed			
PD3	.787	.619		
PD4	.826	.682		
Perceived Feasibility				
PF1	811	.658	65.344	0.733
PF2	.835	.698		
PF3	.777	.604		
TE				
TE1	.880	.775	77.475	0.709
TE2	.880	.775		
Attitude				
AEA1	.807	.650	56.102	0.603
AEA2	.662	.438		
AEA3	.771	.595		
Intention				
AEI1	.851	.724	69.793	0.782
AEI2	.885	.784		
AEI3	.766	.586		
IB				
IB1	.783	.613	46.806	0.513(0.396)
IB2	.784	.615		
IB3	Removed			

Table 2. Methodological synthesis

Population of the study around 30000
Sample design convenient sampling
Information collection methodPersonal survey (EIQ)
Rate of response 67%
Size of sample500
Statistical tools Regression analysis LocationAligarh, UP, India

Table 3: Regression 1

Dependent Variable: Agricultural Entrepreneurial Attitude			
	Standardised Beta	**Significance**	**Model Fit**
PD→AEA	.178	.008	R Square =0.366 Durbin-Watson 1.824 F=32.421, p<0.000 VIF=1.214 to 1.177
PF→AEA	.207	.001	
IB→AEA	.290	<0.001	
TE→AEA	.209	<0.001	

Table 4: Regression 2

Dependent Variable: Agricultural Entrepreneurial Intention			
	Standardised Beta	**Significance**	**Model Fit**
AEA→AEI	.259	<0.001	R Square =0.067 Durbin-Watson 1.832 F=16.427, p<0.000

As the Undergraduates, postgraduate and PhD students probably have a good chance to search for employment after finishing their education, the sample is based on the same (Bosma, &Schutjens, 2007).In entrepreneurship research the convenient method of sampling has been regularly used(Kolvereid, 1996; Tkachev&Kolvereid, 1999; Krueger et al., 2000; Liñán& Chen, 2009; Raposo, &Paço, 2011; Fayolle&Gailly, 2015).

The questionnaire was adopted from (Yaseen et al., 2016). All the questions were formulated on a five-point Likert-scale, with 1 "completely disagree" and 5 "completely agree". Questionnaires were administrated to the students of commerce, economics and engineering. In the month of March and April 2019, the investigation was accomplished with the response rate of 67%. Finally, 335 questionnaires were usable as the incomplete or inconsistent questionnaires were rejected.

Using SPSS, the data was statistically analysed and interpreted after the collection of data. Regression analysis technique was used to measure the impact of (PD, PF, IB, TE) on EA and of EA on EI.

EXPLORATORY FACTOR ANALYSIS

Exploratory Factor Analysis (EFA) was conducted for each scale to check the uni-dimensionality of the scales. All factor loadings were above 0.70 except for one item of attitude, for which the factor loading was .662 (see table 1). Factor loading of 0.6 or above is acceptable if sample size is large (Hair, Anderson, Babin, & Black, 2010). All communalities were above 0.40 (Hair et. al., 2010). The variance explained in each case was also above the threshold value of 50 percent. To check the reliability of the scales Cronbach's alpha were also calculated. The values of Cronbach's alpha for each scale was 0.70, except for two scales. These two variables were IB and PD for which the Cronbach's alpha were 0.513 and 0.67 respectively. Some authors advocate that the values above 0.60 and sometimes above 0.50 are adequate to accept in exploratory studies. Overall, the reliability and validity of the instrument seems suitable for the further analysis.

RESULT OF REGRESSION

To test the hypotheses, the regression was conducted in two steps. In step 1 multiple regression was conducted between four independent variables (PD, PF, IB and TE) and one Independent variable (Attitude) see table 3. In the second regression attitude has been considered as the independent variable and Entrepreneurial Intention was used as the dependent variable. The result of first regression indicates that all the independent variables were significant in influencing the *Agricultural entrepreneurial attitude*(AEA).Therefore, all the hypotheses H1 (PD→AEA: beta=0.178, p=.008), H2 (PF→AEA: beta=0.207, p=0.001), H3 (IB→AEA: beta=0.290, p<0.001) H4 (TE→AEA: beta=0.209, p<0.001) were accepted. The model was significant (F=32.421, p<0.000) and the all independent variables could explain the 36.6% variation (R Square =0.366) in the dependent variable. The values of the residuals are independent as the Durbin-Watson statistics is close to two (Durbin-Watson= 1.824). Further, there was also not the issue of multi-collinearity as the all the VIF values were low (VIF=1.214 to 1.177).

In the second step, simple regression was conducted between attitude (independent variable) and entrepreneurial intention (dependent variable). The model was significant ($F=16.427$, $p<0.000$). The hypotheses H5 (AEA→AEI, beta=0.259, $p<0.001$) was also accepted. In this model, the Durbin-Watson was also close to two indicating that residuals are independent. All the relationships were checked through the curve estimation to establish the linearity.

DISCUSSION AND CONCLUSION

This study investigates the precursors of entrepreneurial behaviour in the agricultural sector of India. This research provides an empirical evidence supporting the relationship between cognitive factors and entrepreneurial intention through entrepreneurial attitude among university students of India. The results above demonstrate that three behaviour-psychological variables i.e. (perceived desirability, perceived feasibility and triggering event) and one economic-psychological variable i.e. (individual background) have a positive and significant influence on the attitude of entrepreneurs and entrepreneurial attitude have a significant and positive influence on entrepreneurial intention. The discussion of the findings is as follows:

First, the results have shown that perceived desirability has a significant impact on attitude towards entrepreneurship. As per the statement of Kolvereid (1997), for becoming an entrepreneur personal desirability raises attitude towards entrepreneurship. So the level of attraction towards starting a new business will have a positive direct impact on the entrepreneurial attitude. Hence, future entrepreneurs are required to have the sense of attractiveness towards the entrepreneurship, which may develop entrepreneurial attitude. The result indicates that perceived feasibility is positively related with entrepreneurial attitude. This finding confirms the results of the study carried out Zampetakis et al. (2009) and Izquierdo and Buelens (2011). So the confidence in one's own ability to successfully execute is the key entrepreneurial skill which leads to the improvement of entrepreneurial attitude in an individual. With regard to the individual background, the findings of this research has shown that individual background is positively associated with his entrepreneurial attitude. In other words, education, prior experience and family background helps an individual to assess entrepreneurial career favourable and make an economical assessment of choosing an entrepreneurial career path. Hence, this finding of our study is also in line with (Robinson et al., 1991; Davidsson 1995; Zapkau, Florian, Schwens, Steinmetz, &Kabst, 2015; Yaseen et al. 2016). As for as triggering event is concerned, the results depict that there is a significant impact of triggering event on the entrepreneurial attitude. In other words, the external force that motivates an individual to become active from the state of rest or inactiveness may develop an entrepreneurial attitude in an individual.

As anticipated, the findings indicate that the entrepreneurial attitude has a substantial influence on entrepreneurial intention. This research finding supported various previous researches on the relationship between entrepreneurial attitude and entrepreneurial intention (e.g. Davidsson (1995),Izquierdo and Buelens (2011);Zapkau et al., (2015);Liu, Xianyue, Lin, Zhao, & Zhao, 2019). The more the beliefs and expectations will be towards the self-employment the more will be the willingness of an individual in undertaking entrepreneurial activity.

As a conclusion, the basic purpose of the study was to understand the precursors of the agricultural entrepreneurship with the emphasis on the behavioural psychological factors and an economical psychological factor "individual background". The study tested a self-constructed conceptual model by using a sample of 345 students of a public university in India. The findings of the research provide various implications for the policy makers and parents of the educated youth. As through the agricultural entrepreneurship, issues of unemployment and other economic matters can be solved by the collaborative efforts of the policy makers who may encourage the educated youth and students to engage in agricultural entrepreneurship and also by the guardians who can play a big role by encouraging and equipping the potential youth for becoming successful entrepreneurs.

ACKNOWLEDGMENT

We thank Dr Mohd Yasir Arafat for his assistance and valuable suggestions and we would also like to show our gratefulness to Imran Saleem, professor, Aligarh Muslim University for sharing pearls of wisdom with us during the course of this research.

REFERENCES

Acs, Z. J., Desai, S., & Hessels, J. (2008). Entrepreneurship, economic development and institutions. *Small Business Economics*, *31*(3), 219–234. doi:10.100711187-008-9135-9

Ajzen, I. (1991). The theory of planned behavior. *Organizational Behavior and Human Decision Processes*, *50*(2), 179–211. doi:10.1016/0749-5978(91)90020-T

Ali, I., & Tajddini, K., urRehman, K., Ali, J. F., & Ahmed, I. (2010). University student's inclination of governance and its effects on entrepreneurial intentions: An empirical analysis. *International Journal of Trade. Economics and Finance*, *1*(1), 36.

Alsos, G. A., Hogskolani, J., Internationella, H., & Ihh, E. (2011). *The handbook of research onentrepreneurship in agriculture and rural development*. Edward Elgar Publishing Limited. doi:10.4337/9780857933249

Arafat, M. Y., Saleem, I., & Dwivedi, A. K. (2020). Understanding entrepreneurial intention among Indian youth aspiring for self-employment. *International Journal of Knowledge and Learning*, *13*(3), 185–200. doi:10.1504/IJKL.2020.109882

Arafat, M. Y., Saleem, I., Dwivedi, A. K., & Khan, A. (2020). Determinants of agricultural entrepreneurship: A GEM data based study. *The International Entrepreneurship and Management Journal*, *16*(1), 345–370. doi:10.100711365-018-0536-1

Autio, E., Kenney, M., Mustar, P., Siegel, D., & Wright, M. (2014). Entrepreneurial innovation: The importance of context. *Research Policy*, *43*(7), 1097–1108. doi:10.1016/j.respol.2014.01.015

Bird, B., &Schjoedt, L. (2017). Entrepreneurial behavior: Its nature, scope, recent research, and agenda for future research. *Revisiting the Entrepreneurial Mind*, 379-409.

Bosma, N., &Schutjens, V. (2007). Patterns of promising entrepreneurial activity in European regions. *Tijdschriftvooreconomische en socialegeografie, 98*(5), 675-686.

Brünjes, J., & Diez, J. R. (2013). 'Recession push' and 'prosperity pull' entrepreneurship in a rural developing context. *Entrepreneurship and Regional Development*, *25*(3-4), 251–271. doi:10.1080/08985626.2012.710267

Davidsson, P. (1995). *Re: Determinants Of Entrepreneurial Intentions.* Type To Research, W. I. E.

Devi, M. (2015). A study on the influencing factors for a literate youth to take up agricultural entrepreneurship. *International Journal of Management and Commerce Innovations*, *3*(1), 692–700.

Elfving, J., Brännback, M., & Carsrud, A. (2009). Toward a contextual model of entrepreneurial intentions. In *Understanding the entrepreneurial mind* (pp. 23–33). Springer. doi:10.1007/978-1-4419-0443-0_2

Fayolle, A. (2005). Evaluation of entrepreneurship education: Behaviour performing or intention increasing? *International Journal of Entrepreneurship and Small Business*, *2*(1), 89–98. doi:10.1504/IJESB.2005.006072

Fayolle, A., & Gailly, B. (2015). The impact of entrepreneurship education on entrepreneurial attitudes and intention: Hysteresis and persistence. *Journal of Small Business Management*, *53*(1), 75–93. doi:10.1111/jsbm.12065

Fitz-Koch, S., Nordqvist, M., Carter, S., & Hunter, E. (2018). Entrepreneurship in the agricultural sector: A literature review and future research opportunities. *Entrepreneurship Theory and Practice*, *42*(1), 129–166. doi:10.1177/1042258717732958

Fitzsimmons, J. R., & Douglas, E. J. (2011). Interaction between feasibility and desirability in the formation of entrepreneurial intentions. *Journal of Business Venturing*, *26*(4), 431–440. doi:10.1016/j.jbusvent.2010.01.001

Franke, N., &Lüthje, C. (2004). Entrepreneurial intentions of business students—A benchmarking study. *International Journal of Innovation and Technology Management, 1*(3), 269-288.

Gartner, W. B. (1985). A conceptual framework for describing the phenomenon of new venture creation. *Academy of Management Review*, *10*(4), 696–706. doi:10.5465/amr.1985.4279094

Giacomin, O., Janssen, F., Pruett, M., Shinnar, R. S., Llopis, F., & Toney, B. (2011). Entrepreneurial intentions, motivations and barriers: Differences among American, Asian and European students. *The International Entrepreneurship and Management Journal*, *7*(2), 219–238. doi:10.100711365-010-0155-y

Hair, J. F., Anderson, R. E., Babin, B. J., & Black, W. C. (2010). *Multivariate data analysis: A global perspective* (Vol. 7). Academic Press.

Hassan, A., Anwar, I., Saleem, I., Islam, K. B., & Hussain, S. A. (2021). Individual entrepreneurial orientation, entrepreneurship education and entrepreneurial intention: The mediating role of entrepreneurial motivations. *Industry and Higher Education*.

Hassan, A., Saleem, I., Anwar, I., & Hussain, S. A. (2020). Entrepreneurial intention of Indian university students: The role of opportunity recognition and entrepreneurship education. *Education + Training*, *62*(7/8), 843–861. doi:10.1108/ET-02-2020-0033

Izquierdo, E., & Buelens, M. (2011). Competing models of entrepreneurial intentions: The influence of entrepreneurial self-efficacy and attitudes. *International Journal of Entrepreneurship and Small Business*, *13*(1), 75–91. doi:10.1504/IJESB.2011.040417

Kar, B., Subudhi, R. N., & Padhy, R. (2017). Impact of self-efficacy and contextual variables on entrepreneurial intention. *Pertanika Journal of Social Science & Humanities*, *25*(3), 1121–1138.

Kickul, J., Wilson, F., Marlino, D., & Barbosa, S. D. (2008). Are misalignments of perceptions and self-efficacy causing gender gaps in entrepreneurial intentions among our nation's teens? *Journal of Small Business and Enterprise Development*, *15*(2), 321–335. doi:10.1108/14626000810871709

Kim, M. S., & Hunter, J. E. (1993). Relationships among attitudes, behavioral intentions, and behavior: A meta-analysis of past research, part 2. *Communication Research*, *20*(3), 331–364. doi:10.1177/009365093020003001

Kolvereid, L. (1996). Prediction of employment status choice intentions. *Entrepreneurship Theory and Practice*, *21*(1), 47–58. doi:10.1177/104225879602100104

Kolvereid, L. (1996). Prediction of employment status choice intentions. *Entrepreneurship Theory and Practice*, *21*(1), 47–58. doi:10.1177/104225879602100104

Krueger, N. (1993). The impact of prior entrepreneurial exposure on perceptions of new venture feasibility and desirability. *Entrepreneurship Theory and Practice*, *18*(1), 5–21. doi:10.1177/104225879301800101

Krueger, N. F. (2007). The cognitive infrastructure of opportunity emergence. In *Entrepreneurship* (pp. 185–206). Springer.

Krueger, N. F. Jr, & Brazeal, D. V. (1994). Entrepreneurial potential and potential entrepreneurs. *Entrepreneurship Theory and Practice*, *18*(3), 91–104. doi:10.1177/104225879401800307

Krueger, N. F. Jr, & Brazeal, D. V. (1994). Entrepreneurial potential and potential entrepreneurs. *Entrepreneurship Theory and Practice*, *18*(3), 91–104. doi:10.1177/104225879401800307

Krueger, N. F., & Carsrud, A. L. (1993). Entrepreneurial intentions: Applying the theory of planned behaviour. *Entrepreneurship and Regional Development*, *5*(4), 315–330. doi:10.1080/08985629300000020

Krueger, N. F. Jr, Reilly, M. D., & Carsrud, A. L. (2000). Competing models of entrepreneurial intentions. *Journal of Business Venturing*, *15*(5-6), 411–432. doi:10.1016/S0883-9026(98)00033-0

Learned, K. E. (1992). What happened before the organization? A model of organization formation. *Entrepreneurship Theory and Practice*, *17*(1), 39–48. doi:10.1177/104225879201700105

Lee, S. H., & Wong, P. K. (2004). An exploratory study of technopreneurial intentions: A career anchor perspective. *Journal of Business Venturing*, *19*(1), 7–28. doi:10.1016/S0883-9026(02)00112-X

Liñán, F., & Chen, Y. W. (2009). Development and cross–cultural application of a specific instrument to measure entrepreneurial intentions. *Entrepreneurship Theory and Practice*, *33*(3), 593–617. doi:10.1111/j.1540-6520.2009.00318.x

Lüthje, C., & Franke, N. (2003). The 'making' of an entrepreneur: Testing a model of entrepreneurial intent among engineering students at MIT. *Research Management*, *33*(2), 135–147.

Marques, C. S., Ferreira, J. J., Gomes, D. N., & Rodrigues, R. G. (2012). Entrepreneurship education: How psychological, demographic and behavioural factors predict the entrepreneurial intention. *Education + Training*, *54*(8/9), 657–672. doi:10.1108/00400911211274819

Maula, M., Autio, E., & Arenius, P. (2003). *What drives micro-angel investments? A large sample study of the factors explaining micro-angel investments.* Unpublished working paper.

McElwee, G. (2006). Farmers as entrepreneurs: Developing competitive skills. *Journal of Developmental Entrepreneurship*, *11*(03), 187–206. doi:10.1142/S1084946706000398

Nasip, S., Amirul, S. R., Sondoh, S. L. Jr, & Tanakinjal, G. H. (2017). Psychological characteristics and entrepreneurial intention: A study among university students in North Borneo, Malaysia. *Education + Training*, *59*(7/8), 825–840. doi:10.1108/ET-10-2015-0092

Nguyen, C. (2017). Entrepreneurial intention of international business students in Viet Nam: A survey of the country joining the Trans-Pacific Partnership. *Journal of Innovation and Entrepreneurship*, *6*(1), 1–13. doi:10.118613731-017-0066-z

Pathak, G. S., & Wang, M. (2012). Entrepreneurial self-efficacy and entrepreneurial intention in India: A comparative study of professional students. *International Journal of Psychology*, 47.

Pindado, E., & Sánchez, M. (2017). Researching the entrepreneurial behaviour of new and existing ventures in European agriculture. *Small Business Economics*, *49*(2), 421–444. doi:10.100711187-017-9837-y

Raposo, M., & Do Paço, A. (2011). Entrepreneurship education: Relationship between education and entrepreneurial activity. *Psicothema*, *23*(3), 453–457. PMID:21774900

Reynolds, P. D. (1995). Who starts new firms? Linear additive versus interaction based models. *Frontiers of Entrepreneurship Research*, 32-46.

Ridha, R. N., &Wahyu, B. P. (2017). Entrepreneurship intention in agricultural sector of young generation in Indonesia. *Asia Pacific Journal of Innovation and Entrepreneurship.*

Robinson, P. B., Stimpson, D. V., Huefner, J. C., & Hunt, H. K. (1991). An attitude approach to the prediction of entrepreneurship. *Entrepreneurship Theory and Practice, 15*(4), 13–32. doi:10.1177/104225879101500405

Roy, R., Akhtar, F., & Das, N. (2017). Entrepreneurial intention among science & technology students in India: Extending the theory of planned behavior. *The International Entrepreneurship and Management Journal, 13*(4), 1013–1041. doi:10.100711365-017-0434-y

Schlaegel, C., & Koenig, M. (2014). Determinants of entrepreneurial intent: A meta–analytic test and integration of competing models. *Entrepreneurship Theory and Practice, 38*(2), 291–332. doi:10.1111/etap.12087

Segal, G., Borgia, D., & Schoenfeld, J. (2005). The motivation to become an entrepreneur. *International Journal of Entrepreneurial Behaviour & Research, 11*(1), 42–57. doi:10.1108/13552550510580834

Shane, S. A. (2003). *A general theory of entrepreneurship: The individual-opportunity nexus*. Edward Elgar Publishing. doi:10.4337/9781781007990

Shapero, A. (1975). *The displaced, uncomfortable entrepreneur.* University of Illinois at Urbana-Champaign's Academy for Entrepreneurial Leadership Historical Research Reference in Entrepreneurship.

Shapero, A., & Sokol, L. (1982). *The social dimensions of entrepreneurship*. University of Illinois at Urbana-Champaign's Academy for Entrepreneurial Leadership Historical Research Reference in Entrepreneurship.

Soomro, B. A., & Shah, N. (2015). Developing attitudes and intentions among potential entrepreneurs. *Journal of Enterprise Information Management, 28*(2), 304–322. doi:10.1108/JEIM-07-2014-0070

Stenholm, P., Acs, Z. J., & Wuebker, R. (2013). Exploring country-level institutional arrangements on the rate and type of entrepreneurial activity. *Journal of Business Venturing, 28*(1), 176–193. doi:10.1016/j.jbusvent.2011.11.002

Stevenson, H. H., & Jarillo, C. J. (1990). A paradigm of entrepreneurship: Entrepreneurial management. *Strategic Management Journal, 11.*

Summers, D. F. (2000). *The formation of entrepreneurial intentions*. Garland Publishing.

Sundar, S. (2019). India Economic Survey 2018: Farmers Gain As Agriculture Mechanisation Speeds Up, But More R&D Needed. *Financial Express*. Https://Www.Financialexpress.Com/Budget/India-Economic-Survey-2018-For-Farmers-Agriculture-Gdp-Msp/1034266/

Tkachev, A., & Kolvereid, L. (1999). Self-employment intentions among Russian students. *Entrepreneurship and Regional Development, 11*(3), 269–280. doi:10.1080/089856299283209

Tushman, M. L., & Romanelli, E. (1985). Organizational evolution: A metamorphosis model of convergence and reorientation. *Research in Organizational Behavior*.

Van Gelderen, M., Brand, M., Van Praag, M., Bodewes, W., Poutsma, E., & Van Gils, A. (2008). Explaining entrepreneurial intentions by means of the theory of planned behaviour. *Career Development International, 13*(6), 538–559. doi:10.1108/13620430810901688

Van Gelderen, M., & Jansen, P. (2006). Autonomy as a start-up motive. *Journal of Small Business and Enterprise Development, 13*(1), 23–32. doi:10.1108/14626000610645289

Venkataraman, S., & Shane, S. (2000). The promise of entrepreneurship as a field of research. *Academy of Management Review, 25*(1), 217–226.

Venkataraman, S., & Shane, S. (2000). The promise of entrepreneurship as a field of research. *Academy of Management Review, 25*(1), 217–226.

Vyas, M. (2019). 11 Million Jobs Lost In 2018 - Cmie. *Cmie*. Https://Www.Cmie.Com/Kommon/Bin/Sr.Php?Kall=Warticle&Dt=2019-01-0809:28:37&Msec=666

Welter, F. (2011). Contextualizing Entrepreneurship—Conceptual Challenges And Ways Forward. *Entrepreneurship Theory and Practice, 35*(1), 165–184. doi:10.1111/j.1540-6520.2010.00427.x

Yaseen, A., Saleem, M. A., Zahra, S., & Israr, M. (2018). Precursory effects on entrepreneurial behaviour in the agri-food industry. *Journal of Entrepreneurship in Emerging Economies*.

Zampetakis, L. A., Anagnosti, A., & Rozakis, S. (2013, September). *Understanding entrepreneurial intentions of students in agriculture and related sciences*. Poster session presented at the meeting of the EEAE 2014 Congress" Agri-Food and Rural Innovations for Healthier Societies," Ljubljana, Slovenia.

Zampetakis, L. A., Kafetsios, K., Bouranta, N., Dewett, T., & Moustakis, V. S. (2009). On the relationship between emotional intelligence and entrepreneurial attitudes and intentions. *International Journal of Entrepreneurial Behaviour & Research*, *15*(6), 595–618. doi:10.1108/13552550910995452

Zapkau, F. B., Schwens, C., Steinmetz, H., & Kabst, R. (2015). Disentangling the effect of prior entrepreneurial exposure on entrepreneurial intention. *Journal of Business Research*, *68*(3), 639–653. doi:10.1016/j.jbusres.2014.08.007

Chapter 7
The Implications of the New Geography Framework of Urban Agro Ecology on Urban Planning

José G. Vargas-Hernández
https://orcid.org/0000-0003-0938-4197
Postgraduate and Research Department, Tecnológico Mario Molina, Zapopan, Mexico

ABSTRACT

This chapter has the objective to analyze the implications that the new geography framework of urban agro ecology has on urban planning. It departs from the assumption that the new geography is a theoretical framework for the for the analysis of the economic, social, political, ecological, technological, research, and science based on the interrelationships between urban agro ecology and urban planning. The methodology is based in a constructive analysis of the reviewed theoretical and empirical literature to infer a model based on the construct of the new geography. Finally, it is concluded that urban planning of local governments can formulate and implement strategies based on the new geography framework in urban agro ecology to proving incentives in new urban developments and to benefit disadvantaged communities.

DOI: 10.4018/978-1-6684-2349-3.ch007

INTRODUCTION: THE NEW GEOGRAPHY FRAMEWORK

The new geography is a new theoretical and analytical framework that adds new layers of meaning of urban agroecology and security in the food system based on the interrelationships between economic, social, political, ecological, technological, and public health factors threaten human survival—the analysis of policy formation and consolidation relationships for policy action of the food system. Sustainable food systems can be approached from other interdisciplinary research beyond agro ecology such as economics, environmental sciences, ethics, sociology, etc. Theoretical and practical analysis of local food strategies is relevant to developing a more sustainable food system vision.

Agroecology is an integrative discipline that includes other elements and principles from agronomy, ecology, sociology and economics (Dalgaard, Hutchings, & Porter, (2003). Urban agroecology has been analysed from multidisciplinary perspectives and meanings, giving urban forms while reconnecting urban green areas, vacant spaces and food growing, supporting new geographical and sociopolitical configurations. Food ascetics (Osti, 2006), a new geography of food (Cook, 2008), ethical foodscapes (Morgan, 2009), and alternative food networks (Harris, 2009) are analytical streams emerging from the critical geography of urban agroecological systems.

The critical geography of urban agroecology practices is beyond the urban food production to analysing the embedded economic-social and environmental dynamics from a transdisciplinary dialogue to propose alternatives of policymaking (Marcuse, 2009; Moulaert, MacCallum, &Mehmood, 2013). Agroecology tends to become more transdisciplinary knowledge and practice. The critical geography of urban agroecology and agroforestry are alternative paradigms more suitable for an urban context beyond market-driven sustainability (Gliessman, 2012).

Agroecological forests using organic fertilisers protect water springs that provide clean water for irrigation. Urban agroforestry and intercropping are viable agroecological practices used to improve higher soil fertility and better moisture. An agenda for a critical geography of urban agroecology initiatives must analyse the cultural and political identity meanings, their values, objectives and claims of practitioners in the emerging projects in current times.

In the new geography of food security, sustainability and food security concerns are potential elements for urban innovations in the complexity of socioeconomic and environmental dynamics. New geography adds new layers of meaning of security in the food system based on the interrelationships between economic, social, political, ecological and public health factors that threaten human survival. The new geography of food security has to design strategies and policies to deal with inequalities, exclusion and poverty in urban areas, addressing variations across the different socioeconomic groups where available food is expensive (UN Habitat 2010; FAO 2011; Holt-Giménez2008).

Agroecology practice, movements, and scientific perspectives are discursive assessments of the emerging interactions in the context of yield maximisation models to elucidate a responsive agricultural approach. In this sense, urban agroecology practices and activities explicitly reject the corporate food system (McClintock, 2010; McClintock & Simpson, 2017; White, 2011).

Blended agroecology and indigenous knowledge systems have developed new technologies and approaches to conserve natural resources and enhance food security. The holistic way indigenous peoples see agriculture considers agroecology in the nucleus at the centre of all aspects of community life. A holistic approach to urban sustainability assumes the notions of healthiness and freshness of a sustainable and resilient food economy identified as the main objective of urban food strategies in localisation, which is a major contribution to environmental and community health.

From a holistic perspective, the re-localisation strategy of a sustainable food system produces important repercussions, benefits and outcomes (Renting, Marseden, & Banks, 2003; Sonnino 2013). The academic analysis on local food re-localisation has the challenge of developing a relational approach to the theoretical and practical discussion on the food system to study the alternativeness of initiatives concerning the conventional dynamics shaping the global food system (Allen, FitzSimmons, Goodman, & Warner, 2003; Goodman 2004; Sonnino and Marsden 2006).

Agroecological identities are identified as a practice, movement, and science, recognised in different forms and assessed in different contexts. The different agro ecological perspectives spark the synergies among all the participants and stakeholders. Agroecology perspectives are presented as a farming practice, social movement, scientific knowledge, etc. Humanity needs new agroecological forms of farming that are more locally biodiverse, sustainable, and socially just.

This study can contribute to the development of the economic, social and political urban agroecology framework. The biological diversity in agroecological systems based on long-standing principles brings socioeconomic, environmental sustainability and ecological benefits. As an alternative to industrial input-intensive agricultural production, agroecology avoids the economic, social and environmental consequences.

The principles of agroecological economics can be applied to the utilisation of agricultural, natural resources to meet the needs of sustainable development and thus to create an agroecological, economic regionalisation. The regionalisation of thc local food system is more concerned with food security and has many quality attributes such as agricultural and labour practices and production methods. Promotion of local and urban food production as a wider strategy for sustainability can be a means to create employment for urban residents and decrease energy costs.

Urban agroecology has constraints and potentials for overcoming the traditional industrial agroecology practices reflecting on the procedural capabilities of socio-environmental forms of unjust distribution of benefits that unfold from urban food production. Agroecology, as a scientific discipline to study living organisms in the context of their interrelationships with land use and agriculture, emerged over a century ago as the basis of sustainable agriculture.

The development of urban agroecology farming is difficult to implement since there are barriers and challenges to face, such as access to urban land as a valuable commodity, the required level of soil fertility, water availability for food production, pollution of industrial areas, the high costs of inputs, etc. Urban agroecological farming can supply low-carbon food overcoming the need for oil-derived inputs in food production—the agroecological system functions based on a set of ecological processes.

THE ECONOMIC IMPLICATIONS

The new economy and culture of producers and consumers from communities of alternative food networks of sustainable local food systems. Urban agroecology can become one of the most dynamic economic sectors by adapting its products to changing urban demands and conditions and diversifying its functions. The agroecology sector can promote solidarity economy activities and inclusive employment and labour conditions. Urban agroecological, social economy enterprises committed to economic and social values and utilising urban vacant land supported by public policies and using social networks is a driving force of sustainable urban development.

The agro-food system is undergoing unprecedented changes in the urban realm, rising concerns in times of economic crisis and related social and environmental problems (Cohen and Garret, 2010; Partalidou and Anthopoulou, 2016). Urban agroecology is conceived as an economic development strategy for promoting healthy and nutritious food while improving residents' economic growth and income (Lovell, 2010). Urban agroecology contributes to local socioeconomic development, food security, social inclusion and poverty alleviation, the greening of the urban areas and the productive reuse of wastes.

Some cities, such as Rotterdam, have adopted an urban agroecology strategy to recycle human waste as nutrient sovereignty and asset for urban cultivation and food production. Human biodegradable wastes recycling, such as surplus or residual food, human and animal excrements, rainwater, tree-and lawn cuttings, etc., can become an essential source of nutrients for urban food cultivation and production.

Agroecology as sustainable food production has a different meaning than organic agroecology because it is a solution for agricultural development aligned with the work of nature and its complementarities between soil, plants, animals, etc., reducing dependency on chemical inputs and fossil energies. Food production based on urban agroecology generates local knowledge, strengthens economic viability, promotes social justice, and nurtures urban communities' identity and culture.

The agroecological production gives more dignity to smallholder production. When most agroecological production is for the self-consumption of the family itself, the use of land is reduced only to produce what is needed, and the exceeding land is taken out. The agroecological farm system is multifunctional and multidimensional based on smallholder and cooperative farms labour-intensive. Local governments can develop meaningful democratic mechanisms to sustain long term and mutually respectful relationships to enhance participative perspectives from diverse backgrounds of urban agroecology and food justice organisations, communities and participants (Cohen and Reynolds, 2015).

Local governments are integrating urban agroecology in their urban sustainability plans and zoning regulations, considering the benefits from this greening strategy. Urban zoning is relevant in urban agroecology as a regulated activity to know the allowed growing vegetables and plants, raising animals, processing, distribution, commerce, selling, cultural issues, and consuming healthy food. Local governments support community farming and gardening as a strategy for urban revitalisation despite the restrictive zoning inhibiting urban agriculture (Bartling, 2012, Brown & Carter, 2003; Vitiello& Brinkley, 2013).

Updated inventories based on the zoning regulations provide valuable and useful information for stakeholders involved in urban agroecology to know which areas are banned or excluded of agroecological production as primary land use, depending on the priorities over other development initiatives.

Agroecological cooperatives linked to local government management could facilitate capacity building and joint development. The formation and development of cooperatives for the production, creation of distribution and commercialisationcentres of agroecological products to supply service organisations such as hospitals and schools require incentives from local governments. Urban agroecology organisations can benefit from access to vacant urban land for their agroecological projects. Urban agroecological productive reorganisation in smaller and more cooperative farms in urban vacant land and public spaces, giving in usufruct to residents who may have the right to usufruct the use of the soil and take advantage of its production.

Urban agroecology production systems and processes are heterogeneous in urban green areas and gardens and crop mixes, production and supply levels to the needed neighbourhoods and communities. Urban agroecology impacts positively upon the cleaning and greening of the urban spaces by turning into green urban areas free of housing and impacting the micro-climates.

To leverage its economic power of producers, connectivity's of urban agroecology to coordinate the supply of products focuses on infrastructural development aimed at food processing facilities, support distribution and marketing through the development of alternative models of retail outlets and wholesale farmers' markets. Funding in infrastructure and increasing the amount of land available for urban agroecology strategy should be targeted to benefit the more disadvantaged groups and counter urban displacements. Removal of restrictions on urban agroecology as a strategy enables limited participation of residents to have access to land.

An agro ecology-based farming system should be economically efficient and socially inclusive to increase crop yields in a sustainable environment. Small farms may be conducive to the agroecological model development, more diverse in resource conservations, and more efficient in inputs- outputs production. The urban agro-system is a high input consumption from the ecological system. There is evidence favouring small farmers to have more efficient use of land, natural resources, biodiversity, water, etc., in terms of converting inputs into products while allowing the small farmers to be employed and remain with their families. Crop auxiliaries increase agricultural productivity and decrease pesticides' health and financial costs (Östman, Ekbom, &Bengtsson. 2003;Östman, Ekbom, &Bengtsson. 2003;Weisenburger 1993).

The movement for agroecology addresses issues such as land access and uses, food distribution and food justice for the most vulnerable groups, such as the indigenous. Land access for agroecological urban food growing, production and allocation strategies are disconnected from food consumption patterns

The challenge is to scale up the agro ecology-based production programs by creating and increasing incentives within a policy framework. Agroecology addresses these challenges among an increasing range of scientific experts and international agencies (McIntyre et al., 2009). Government-supported policies expanding agroecological export-oriented production must be based on the scale of economies and standardisation focusing on economic competitiveness in global markets.

Food producers integrated into global markets have led the global food system into a deep crisis. Food producers and sovereignty organisations work at community, local, regional, national and transnational levels to develop practical and politically agroecology. The globalisation process of the food systems combined with the green revolution causes urban food insecurity. The industrial agrion system has negative impacts on the wellbeing and autonomous development of agricultural producers and farmworkers. Sustainable agroecological systems are more autonomous closing cycles, although they depend upon a network of ecological processes and functions.

SOCIAL IMPLICATIONS

National food policies were changed by international policies in neoliberal trade agreements that freed commodity prices, eliminating the food reserves, increasing the import food dependency, losing the food sovereignty and destroying rural communities and ruining the life of small farmer producers and peasants, turning them into urban refugees in a large metropolis. Regressive and neoliberal agendas, potential inclusionary and exclusionary dynamics, and social inequality reproduction are becoming a reality through urban agroecological projects.

The social benefits of urban agroecology are limited in reducing inequities by resourcing organisations and displacing lower-income households. Unfortunately, urban agroecology projects may contribute to perpetuating existing inequities, marginalisation and displacement of urban communities by benefitting already privileged ones.

Agroecology production of healthy foods challenges neoliberal industrial agriculture based on agribusiness and agro exports. Urban inhabitants are subjugated to the industrial agro-food system subject to controversial ethical views of production, distribution, marketing and consumption. Emerging urban agroecological practices embedded in consumer ethical views are relevant to explain the relationships between urban agroecological food producers and market consumption.

The agro-industrial model is dominated by a few large corporations controlling the access to the market and the distribution channels through the intermediaries between farmers and consumers controlling prices with a high margin of profits. Agribusiness corporations structure the global agriculture system dependent on toxic inputs and influence local governments to commodify food, degrading biodiversity, changing the climate, wasting and polluting the soil and water. The export-led agroecology production promoted by these large multinational corporations leads to environmental degradation, low wages, hunger, poverty, and migration. Consumers usually have to purchase agroecological goods at an unfairly high price.

Transnational agro-food corporations increased their power by implementing industrial agroecology, the adoption of monocultures and biological homogenisation, which have brought consequences such as the loss of farmer knowledge, growing impoverishment and inequality, etc. Mergers and acquisitions between seed and agrichemical transnational agribusiness have reduced agroecological technologies, dismantled commodity trade and marketing, reduced the power of the market and been subject to price discrimination.

Agroecology strengthens food identities and cultures and may alleviate hunger and malnutrition. Growing epidemics of malnutrition is linked to poverty (Darmon, Darmon, Maillot, Drewnowski, 2005;Tanumihardjo, Anderson, Kaufer-Horowitz, Bode, Emenaker, Haqq, Satia, Silver,&Stadler, 2007)due to the high costs of nutrient-dense foods and the low-cost availability of energy-dense foods from industrial agriculture (Schmidhuber and Shetty 2003; Darmon et al. 2005). Agroecological practices improve nutrition by focusing on cereal crops. Human nutrition requires more diverse agroecological systems to ensure a diversified nutrient-rich diet.

Most urban agroecological practices, at the level of grassroots or institutional, are sustained on the notion of a sustainable city. Urban agroecology is the cultivation of organic food within the city more than suburban and peri-urban areas (Golden, 2013). Urban agroecology cultivation is practised at a variety of types of land tenures, scales, levels of public access and locations, from small to large-scale projects, such as community green areas and gardens, urban farms, greenhouses on industrial land, rooftops etc. (Hodgson, 2012; Mukherji, & Morales, 2010). Greenhouse agroecological methods function as natural dehumidifiers, and potential returns are higher crop yields. Urban agroecology has become a significant source of fresh and healthy produce and food for urban and suburban populations (Companioni, Rodriguez, & Carrion, 1997).

The spread of agroecology has been due to the social process and social movement dynamics contributing to farming practices, resulting in resilience to climate change and other benefits such as food security. Agroecological farming practices can be aligned to government-subsidised food programs by offering better prices for good quality agroecologist-based production.

Women and young people practice urban agroecology in diverse contexts with different meanings and expressions. Women play a relevant role in promoting specific agroecology policies to enable agroecological production and consumption.

POLITICAL, POLICY AND GOVERNANCE IMPLICATIONS

The framework of urban political ecology can be used to analyse the roles of humans in urban agroecological foods and other urban green spaces. The cultural, political ecology analysis considers that the goals of urban agroecology initiatives are linked with the contextual economic and sociopolitical arrangements, the economic structure, governance, food regimes, etc. The intensive food regimes of industrialised countries demarcate agricultural land for rural agricultural production systems and mass urban consumption spaces (Marsden and Sonnino 2012, 428) to enhance production for a growing urban population.

New localism is emerging to address the complex dynamics of urban food system strategies and the relationships between actors, spaces and governance scales. Local governments attempt to raise the profile of local food systems and coordinate policy agendas on the issue through new governance mechanisms. The governance mechanisms of a non-profit, local food entity should be represented and integrated by stakeholders of different economic, social, environmental, etc., sectors to analyse, design, implement and institutionalise policy issues.

Some social movements have claimed radical political agroecology as a contested co-option. Social movements claim that agroecology starts from the bottom-up model development, demanding public policy support for practice by communities. Social movements for agroecology have different backgrounds, ideologies, knowledge, values and practices, occurring at different spaces, places and times. In a globalised context, agroecological social movements are committed to marginal from the bottom-up, emergent, decentralised and heterogeneous political activities. Local governments are responsible for their urban agroecology outreach, accessible, culturally responsive, and targeted to benefit disadvantaged communities.

The political peasant agroecology is advocated for resistance by groups and social movements that challenge unequal power relations in the food system. Social movements and citizens develop the capacity to resist the cooptation to scaling up agroecology. The Agro ecology-food sovereignty movement has connected and developed the political dimensions of urban agriculture, farming and gardening. Agroecology is linked to food sovereignty conceived as the right of citizens to control food practice and policy.

Unfortunately, the agroecology movement has been coopted to fine-tune the goals of the industrial food system with names such as organic food, green production and consumption, ecological intensification and sustainability, climate-smart agroecology, etc. The food system based on agroecology is growing as a transnational social movement towards the building and defence of a fairer and sustainable environmental development. Environmental and social movements have proposed different frameworks aimed at fostering and managing natural resources and economic development, such as agroecology (Wezel and Soldat 2009), ecological design (Todd 2005), agroforestry (Nair 1993), and appropriate technology (Pursell 1993).

There is a link between urban agroecology and food sovereignty. The advancement of agroecology addresses the realities of food sovereignty struggles. The urban agroecology movement has clear intentions to promote better conditions of food accessibility to low and middle-income groups with effects on food sovereignty of the producers and consumers. Food sovereignty is the right of each nation or state to maintain, develop and enhance their capacity to produce sustainable basic food crops while promoting their cultural diversity and identity. Urban agroecology builds community capacity to improve the neighbourhoods by replacing vacant or neglected and lots of urban lands into community gardens and contributing to (Armstrong, 2000).

Agroecology is a political umbrella for agriculture and aquaculture. Agroecology has forms of agriculture and aquaculture. In this sense, agroecology is the political term for various forms of agriculture and aquaculture practised.

Local governments have a relevant role in facilitating the access to infrastructure, natural resources, supply of water and production inputs for urban agroecology. Local governments are less convinced to fund transdisciplinary agroecology and diversified farming than conventional agriculture. The provisioning of spaces by local governments for community urban agroecology is part of the expansion and protection of green spaces.

Local governments may facilitate the development of agroecological processing and packaging units and assist urban agroecological groups in setting up quality green labels, distribution and marketing strategies. Urban agroecology enhances organic crop production and food security using renewable and organic fertilisers in urban environments. Still, it is being challenged because of the nutrient requirements and plant disease incidences.

Local government and city investments in urban agroecology must be targeted to benefit historically disadvantaged communities from diverse backgrounds and to confront the displacement threats by the increased availability of land. Local governments may invest in urban agroecology infrastructure subject to the needs of disadvantaged households in locations. Some local governments tertiary water and incentivise Graywater for urban agroecology. Other local governments allow temporal access to fire hydrants for urban agroecological use, at least temporarily (Hagey, Rice, & Flournoy, 2012).

Urban agroecological initiatives are contextually related to potential exclusionary and inclusionary dynamics and socio-environmental injustice in promoting community engagement and participation, social equity, poverty alleviation, etc. Urban agroecological projects to achieve broader goals than healthy food production, improving environmental sustainability, social justice, developing communities and neighbourhoods require support from local governments, networks of non-profit organisations and practitioners and communities themselves. Agroecology empowers producers to farm with nature to enable local communities to improve agroecological use of resources and practices to control food production and achieve food sovereignty.

Beneficiaries of community-building of urban agroecology projects are the propertied class making the affordable communities more attractive to the newcomers rather than the neighbourhoods of disadvantaged groups, increasing the living costs and leading to processes of gentrification in cities with growing populations (Cadji&Alkon, 2014; Safransky, 2014; Walker, 2015). Urban agroecological initiatives engage communities to take advantage of their ecological knowledge in growing practices.

The activities of urban agroecology should be included in the local government budget. The local government should allocate resources to support its program on infrastructure development, capacity building and training, urban agroecology, etc.

Infrastructure for local processing, distribution, the establishment of green labels and marketing of fresh urban agroecological grown food has not been well developed in city farmers' markets.

When the local government supports regulations and measures that entail the ownership type of land property to be used, public spaces are better to contribute for more social purposes and benefits. Government policy can be designed and implemented in programs targeting agroecology to enhance nutritional security and food justice strategies. Food justice organisations have potential input and information on urban agroecological policies and programming to support local governments and cities, taking advantage of mutually respectful relationships.

Urban agroecology requires a food sovereignty policy to back it, and without urban agroecology, there is no food sovereignty. Urban agroecology and farming contribute to food systems change, food security, environmental sustainability, health, wellbeing, community development, etc. The benefits have been associated with food justice to benefit disadvantaged urban communities.

Urban agroecology enhances food justice by reducing food insecurity of disadvantaged communities experiencing barriers to full participation, long-term land ownership, funding and political support. Urban agroecology in advantaged communities has limited impacts on food justice in specific issues such as funding, leadership and representation, and insecure land tenure must have been experienced when associated with poor and diverse households (Lawson, 2005).

The practice of urban agroecology cultivation and growing food is dependent on public access and the type of land tenure. One barrier to achieving food justice in urban agroecology is residents who do not have homeownership and lack land tenure for farming and gardening space (Kuebler&Rugh, 2013). Practitioners of urban agroecology in private, business, or church-owned property can grow food in yards, rooftops, vertical farming within a controlled-environment building (Despommier, 2010).

Urban agroecology increased food self-determination based on more communal property ownership and land management forms and engaged in political initiatives for food system changes (Levkoe, 2011; Staeheli, Mitchell, & Gibson, 2002; Travaline&Hunold, 2010). Urban agro ecologists farm and garden on vacant privately owned properties clandestine with rights-of-way, without permission from the landowner (Crane, Viswanathan, & Whitelaw, 2013).

Urban agroecology is a strategy related to food justice (Reynolds, 2015). Learning and self-empowerment as the basis of urban agroecology systems are significant constitutive pillars to achieve urban food justice. Gender empowerment must be prioritised, avoiding discrimination against women in agroecological participatory working groups and networks in knowledge sharing initiatives. Urban agroecology should not be considered a white space for practices, knowing the dangers in this conception associated with white food culture (Cohen & Reynolds, 2016). Capabilities of political engagement and empowerment are the building blocks for urban food justice supported by the urban food movement challenging neoliberal urbanism.

In most agroecology systems, economic basis and environmental features are based on energy inputs from non-renewable resources and chemical pesticides and fertilisers. Urban agroecology focuses on the economic viability of non-renewable resources such as soil, land, water, etc., to produce food and achieving food security through self-provisioning and marketing flows and supply. Urban agroecology may be considered a strategy among other required strategies for seeking food justice, including poverty alleviation.

The local governments demand urban agroecology practitioners conduct their increasing use in vacant properties despite food justice organisations' calls (Baker, 2017). Local governments can identify urban agroecology sites on public owned property, converting some of the lands to community farming and gardening uses (Public Health Law and Policy, 2009). Local governments and cities can also acquire privately-owned vacant land. Urban agroecology has economic, social and environmental benefits for communities and cities by developing creative ways to fund land acquisition and development.

Permanent land tenure is relevant for access to urban agroecology practitioners. The amount of food required by the population leads to the land required for cultivation to support the production, which is subtracted from the total land. The remaining is available for conservation. Farmers and gardeners' practitioners of urban agroecology experience the destruction of their cultivations when the vacant lots of their own or community land are used to get higher profits (Schmelzkopf, 2002).

Many organisations and local governments that manage and run urban agroecology programs require the necessary resources to remain viable, facing significant challenges to providing the expected benefits by matching of needs of urban neighbourhoods and communities. These benefits of urban agroecological programs have to be distributed equitably and effectively between urban communities despite the pressures to maximise returns and reduce costs on public expenditure. Some urban agroecological programs adopt a missionary and patronising position and tone of bringing good food to everybody (Guthman, 2008).

Urban agroecology is a strategy in urban contexts to improve healthy food security and nutritional status. Urban agroecology provides nutrient-dense food directly contributing to food security, increases income and employment (FAO 2008; Dubbeling, Campbell, Hoekstra, &Veenhuizen, 2009; Zezza and Tasciotti 2010; De Zeeuw et al. 2011). Urban agroecology in food cultivation can be a strategy used to pursue the political values of justice and self-determination of communities (Reynolds & Cohen, 2016). Representation of urban agroecology organisations has dominant cultural values and leadership regarding procedural justice (Hislop, 2014).

Urban agroecology, meaning characterisation, is exposed to the various forms of agroecological projects, conflicts and challenges in specific contexts. The goals of urban agroecological farmers and producers are widely varied from growing food for their consumption, for internal use in community groups, for selling their produce at stores, farmer markets, restaurants, cafeterias, etc. (Taylor & Lovell, 2014). Urban agroecological projects play a critical role in urban ecological security (Hodson and Marvin, 2009; Whitehead, 2013).

Urban agroecology is also the source of struggles and conflicts between practitioners. An increasing number of participants involved in urban agroecology intensifies competition for scarce land and water, resulting in conflicts, contestations and struggles. Competition for resources in different geographic areas such as land, soil, water, labour, energy, public and private financial support, technology, etc., have impact urban agroecological activities.

Urban agroecological projects raise some policy challenges framed by the theoretical approaches of political economy in terms of structure-agency claims for land, allocation and distribution mechanisms, externalities on the environment. There are critical challenges to agroecology policies that may dissuade agroecological production, such as the pressure of industrialised export-oriented agriculture that favours large farms at the expense of small farmers.

Projects of urban agroecology operations require resources to meet their needs, goals and expectations of farm and garden viability and equity among all the involved farmers, gardeners, funders and supporters, local governments and all the stakeholders. Disparities of urban agroecological systems in access to material and financial resources make them unequal and constrain the efforts to achieve the goals of some farms and gardens to a fair distribution of benefits.

In most cities, urban agroecology is a temporary use of the land being left vacant and not permanently protected when the income made from cultivations is less than the same land is dedicated to other economic activity with higher profits. Urban agroecologicalorganisations taking advantage of funding opportunities and garnering policy support may identify significant race, poverty and class-based disparities in their lack of diverse representation for urban agroecological operations (Hislop, 2014).

Urban agroecology projects led by residents have been public unfunded. Funding allocation to agroecological research has increased in the last few years. Allocation of financial support to agroecological-related research programs has been very limited and reflective of the presence of agroecology on farms.

Agroecology management of urban and peri-urban agroecology has several challenges: creating an institutional and regulatory framework for procedures for the governance of urban areas, such as cooperation relationships between local government and communities.

ENVIRONMENTAL IMPLICATIONS

The industrial, agricultural and food system has caused social and environmental problems giving rise to social and environmental movements and the implementation of policies and strategies of land use and agroecology management practices. The agroecology industry has an opportunity to expand even in populated areas in small family farms and a variety of complex ways and economic-socio-cultural links to cities. However, if the agroecological production is more market-oriented, it may be increased depending on the prices aiming to increase the revenue. An overproduction will tend to lower the prices.

Agroecology is framed as a response to the environmental degradation caused by mechanised forms of the productive industrial orientation of agriculture. Agroecology plays a relevant role in adapting the most vulnerable groups of people interacting with plants, animals and the environment, enhancing the environmental resilience of biodiversity, and interacting with agroecological practices by applying innovative solutions.

Sustainable urban agroecology is a way to achieve a resilient food sovereignty system, self-sufficiency in food production and security to solve hunger. Agroecological biodiversity improves resilience to climate change cushioning its negative impacts. Improving the structure of energy use and environmental inputs also improves the productivity the urban farming, ecology and sustainable development.

Urban agroecology can be controlled by environmental conditions in conjunction with green urban infrastructure or open space playing important roles in the urban food system, such as the vegetable and community gardens, rooftop agroecology, etc. Some local governments have prioritised already agroecology practices to increase urban economic growth and employment and decrease migration and urban decline.

Urban agroecology may improve urban environments and the deterioration if not planned and practised in an environmentally friendly way (Mok, Williamson, Grove, Burry, Barker, & Hamilton, 2014;Duží, Tóth, Bihuňová, &Stojanov, 2014). Urban agroecology is viable and environmentally friendly. Urban environment place people in a position of dependence concerning food provisioning and leaving a place for urban agroecology to regain the urban food system. Currently, more people are engaged in urban agroecology systems in cities worldwide, some of them as market producers in urban farming and gardens (FAO/WB 2008; Leeuwen, Nijkamp, &Vaz, 2010). A sustainable urban agroecological farming model improves the environment and provides food security, employment and income, contributes to making a more livable and pleasant urban environment increasing the population's standard of living.

KNOWLEDGE AND RESEARCH IMPLICATIONS

The agroecological revolution is epistemological, technical, and socially capable of restoring self-reliance and securing healthy food production while regenerating agrobiodiversity, conserving natural resources, and empowering agroecologicalorganisations. Agroecology as a science combines local knowledge needs and practical experience of food farmers and producers with the modern scientific insight of ecology as a starting point for agroecological research and development. Strengthening agricultural knowledge, science and technology (AKST) towards agroecology contributes to addressing environmental issues increasing productivity.

Researchers, farmer producers and indigenous communities should share experiences and practical knowledge systems through participatory learning processes, which may be used for agroecological science research and education. Agroecological systems have a participatory and inclusive nature with a community orientation capability to support networking and sharing techniques. Creation of multi-organisational collaborative structures and partnerships to develop social networks aimed to integrate agroecological knowledge, innovation and studies through co-learning and technology transfers.

Agroecological innovations are depending changes in policies, economic and political institutions, and research and development. The agroecological process requires active participation and involvement of the farmers and urban communities in continuous innovation and dissemination of technology and resources, laying the foundation for their empowerment. Improved technology makes agroecology more profitable and reduces pressure to expand and intensify production to crop more food in smaller land areas.

More vulnerable groups use urban agroecology to learn and share cultural and culinary knowledge and build intergenerational connections (Airriess& Clawson, 1994; Hondagneu-Sotelo, 2004; Meek, Bradley, Ferguson, Hoey, Morales, Rosset, &Tarlau, 2017;

Saldivar-Tanaka &Krasny, 2004; White, 2011). Farmers have complained that local urban agroecology organisations usually have devalued their agroecological knowledge and practices that are not Eurocentric (Alkon& Mares, 2012).

Knowledge exchange between stakeholders on agroecological management practices contributes to building research resources and networks with agroecological management implications in cultural and environmental urban settings to develop agroecological production techniques further. Building capacity in agroecological research, education and extension requires incentives, investments in infrastructure and budget allocations to support researchers, farmers and producers, consumers, communities, organisations, etc.

Knowledge-based agroecological policy fosters the engagement and participation of stakeholders to deliver support for agroecological practices. Participatory agroecological approaches have positive outcomes despite some adverse environmental conditions. Developing sustainable agroecological technologies should use local knowledge. Agroecological innovators and early adopters usually do not have the funding to continue experimenting with better techniques and practices. Local knowledge systems and agroecology enhance biodiversity and respond to environmental stress while improving soil quality, nutrients, pests, and water management.

Training and development programs in urban agroecology provide participants with green-collar skills in horticulture and landscaping (Pinderhughes, 2007). However, little support exists to fund training programs in agroecology practices and activities. Farming knowledge sharing is a key pillar of agroecology should be the mainstream of agricultural training. However, these claims must be reviewed to prove the ability of the strategy of urban agroecology to contribute to food systems (Allen, 2008; Reynolds, 2015; Tornaghi, 2014).

Building an effective agroecological food system requires support and investments from the state, collective and private farming organisations, social organisations and networks. Localisedagroecological networks are relevant for interactive knowledge creation, sharing and dissemination between organised farmers and organisations.

The agroecology process of research institutionalisation and implementation process is taking place already around the world involving academic, independent, governmental institutions, private and community-owned farms. Agroecology requires to be supported by networks to create, develop, disseminate and share knowledge through participatory processes. The spread of agroecological innovations is through horizontal networks between researchers, food producers.

Scientific research findings combined with the peasant experience in agroecology and farming experiences contribute to the development of agroecology for sustainable and healthy food systems. Agroecology has evolved to an approach that applies ecological principles to the entire food system.

URBAN PLANNING

Urban planning and policymaking should provide the framework for developing the economic, social and ecological benefits of urban agroecological. Urban planning models may become a trap for further urban agroecological development, such as the low-density garden city model revisited as agrarian urbanism (Duany, 2011; Vander SchansandWiskerke, 2012) because this model does not contribute to urban agroecology intensification to solve the problem for everyone but the privileged due to the increasing rate of density (Viljoen, 2005). Urban planning models can be trapped by agroecological urbanism that represents a privilege for a few and not the solution for the living of everyone (Duany, 2011; Van der Schans and Wiskerke, 2012).

An integrated and transdisciplinary concept of urban agroecology democratise urban sustainable planning processes and support urban agroecological production. Urban agroecology active practitioners and participants gain skills in food democracy and the intersections with economic, social and environmental issues, becoming more aware of power complexities (Levkoe, 2011; Barron, 2016).

Urban agroecology used as a mechanism for food democracy can resolve the structural causes of food injustice in disadvantaged communities despite the movements toward structural change and social justice. Urban planning should provide policy options for reducing poverty and hunger through agroecological knowledge, science and technology, improving the health of urban livelihoods and facilitating economically productive and efficient, socially equitable and environmentally sustainable development.

Urban sustainable planning and policy management in the issue of urban food security incorporates urban agroecology. Sustainable urban planning initiatives begin to envision agroecological food production in cities. Agroecology supported by public policy requires different strategies developed for participants and stakeholders to support food sovereignty.

A public policy in agroecological food self-sufficiency and food security intends to continue the growth-oriented economic model with the fair use of natural resources. Agroecological food-centric urban sustainable planning is proposed to overcome problems of urban agro-food cultivation, production, storage, distribution and marketing (Duany 2011).

The urban planning of some cities promotes urban agroecology programs using a racial lens targeted to serve low-income people and change the conditions of disadvantaged communities. A representative sustainable urban planning agroecology advisory board can offer strategic direction to make decisions on programming, budgeting, funding, and other decisions. Urban planning implements urban agroecology policy using racial equity or equity lens to impact initiatives, programming and budgeting. Urban planning has to remove structural barriers to grant access to land for diverse urban agroecology programs by offering publicly owned land and investment funds.

Public and private sector investments for agro scientific and technological research is small for agroecology. Research in agroecology projects requires investments to scale up the already proven successful results to impact the environmental wellbeing, food security and income of all participants. Investments in applied agroecological research and development, innovation, education and technology transfer may support the farmers and producers, academics, scientists and economists to develop an innovative environment of knowledge-sharing and innovation.

New urban agroecology businesses are more action-oriented projects based on research and urban green innovations dealing with sustainable urban food production and consumption, sustainable provision of urban and peri-urban land use for food production, etc.

Sustainable urban planning must pay more attention to urban agroecology policies and programs to support the goals of food justice. Urban planning can be oriented toward urban agroecology and more explicitly toward food justice. Sustainable urban planning should include urban agroecology codes and plans to be developed and practised (Pothukuchi& Kaufman, 2000).

Urban planning should enhance the power of urban agroecology projects to spur gentrification. Urban agroecology is practised on vacant, underused, and publicly owned land or land owned privately. Non-profit entities are considered in urban planning as parcels that may be used for farming and gardening (Jaramillo, 2016). Urban agroecology planning must involve and enhance food justice in long-range efforts, helping disadvantaged residents, providing access to land and resources, promoting neighbourhoodstabilisation and livability, offering incentives, reducing fees and taxes, removing legal barriers, etc.

Local governments design and implement a strategy to reduce property taxes, utility fees, water and garbage fees to promote urban agroecology activities for food justice beneficial to disadvantaged communities. The incentive of low property tax rates to promote urban agroecology can be used by real estate investors and attract new residents.

Local governments tend to allocate agroecology land spaces and funds that benefit the upper and mid Local governments tend to allocate agroecology land spaces and funds that benefit the upper and middle classes and not the most disadvantaged groups.

Urban agroecology planning must address the food justice principle to offer potential social benefits by increasing access to healthy food, community development and skills building. Most urban farmers lack ecological horticultural skills leading to low and unproductive crop density or diversity. Fragmented activities of urban agroecology are included in urban planning strategies connected to sustainable development strategies and local food systems.

Long-term urban planning strategies can prioritise urban agroecology to develop relationships with food justice organisations and participants by targeting investments to benefit disadvantaged communities, increasing the availability of land and confronting gentrification and displacement processes.

The process of gentrification is powerful and far beyond urban agroecology. Urban agroecology projects are related to disadvantaged community improvements, but they are not a predictor of gentrification. Cities have to tackle the tendency of urban agroecology to make contributions to gentrification and investments requiring inputs from food justice-oriented organisations to benefit disadvantaged communities. Some disadvantaged communities are sometimes suspicious of the implications of a new form of paternalism and the receiving benefits from urban agroecology, impeding gentrification (Hern, 2016).

Municipal and local governments urban planning for urban agroecology does not necessarily are considering and doing food justice by examining who gains and who loses (Flyvbjerg, 2002) and not more explicitly benefitting disadvantaged communities. Sustainable urban planning should recognise potentials for urban agroecology sustainability, food justice, etc. (Neuner, Kelly, & Raja, 2011). The sustainable urban planning of urban agroecology can connect strategies to social equity and justice (Cohen and Reynolds, 2014).

Sustainable planning of urban agroecology on local governments has included public allotments in spaces and backyards (Lawson, 2005; Taylor & Lovell, 2014). Usually, disadvantaged communities have many limitations in resources and energies to get involved in urban agroecological practices, depending on contexts and individuals, although the interest among residents may be high (Bowens 2015).

Sustainable urban long-term planning should design and implement urban agroecological strategies to support food justice based on developing mutually respectful relationships with food justice organisations and urban agroecology participants. A sustainable urban long-term planning strategy may support the urban food justice movement by promoting the benefits to more disadvantaged urban communities as a priority.

Urban agroecological projects require cultural and political identity meanings and intervention strategies of residents and communities inscribed within the land appropriation for the food justice movement (Block, Chavez, & Allen, 2012;Gottlieb & Joshi, 2010; Tornaghi, 2011, 2012; Wekerle, 2004). Urban agroecology converged urban and agrarian movements and struggles for food communing justice and control over social production.

An urban food strategy aims to develop an integrated and cross-sectoral food policy linking agroecology, economic growth, public health, environmental sustainability, community development, education and cultural development, urban planning on land use and tourism, and waste management.

For Lang and Barling (2012, 322), urban food strategies are "the only food system to be secure is that which is sustainable, and the route to food security is by addressing sustainability". The movement towards a more comprehensive urban health-focused food system must reconcile land-use planning, urban and farming possibilities, environmental protection and affordability, etc.

Future developments on the land as the basis for sustainable urban agroecology planning must consider the food justice relationships and their needs and constraints between organisations, institutions and communities. Urban agroecology planning can be a strategy to incentivise affordable housing with opportunities to halt the displacement of economic and socially disadvantaged communities. Urban planning should support urban agroecological efforts to provide affordable housing and prevent displacements and other undesirable outcomes.

Local governments develop and implement long-range, comprehensive sustainable urban agroecology plans to involve stakeholders and make them accountable. These long-range and comprehensive sustainable planning efforts make more explicit connections between urban agroecology, social equity and food justice. The local government can use urban agroecology planning as a strategy to impact environmental justice-related program decisions, policy and funding proposals to overcome social inequalities and disparities of disadvantaged communities (Zapata, 2017).

Urban agroecology planning must explicitly focus on fostering food security and food justice supported by equity lens policy and programming as a strategy to benefit disadvantaged low-income communities and neighbourhoods to conduct better outreach. Urban agroecology planning concerning food justice has been subject to critical analysis in the scholarly literature (Reynolds & Cohen, 2016; Tornaghi, 2014).

Cities develop urban planning to target local action supporting large scale improvements for food system changes. Sustainable urban planning for urban agroecology fosters food justice favouring disadvantaged consumers by increasing their wellbeing through access to healthy food, capacity building, social and community development changes. Sustainable urban agroecology planning can be structured and used to foster more equity through food justice explicitly.

Urban food justice addresses concerns resulting from planning and implementing urban agroecology strategies and policies by removing regulations, reducing utility fees and taxes, offering funding incentives, programming support, land, infrastructure, etc. Urban agroecological strategies aim at conserving natural resources, increasing production and should target deliberately the poor.

Local food system planning should be integrated into relating planning of sustainable urban development, urban transportation, reduction of climate warming planning, etc. Local food systems planning and policymaking have evolved significantly since 2000 (Cohen, Reynolds, Wakefield, Yeudall, Taron, Reynolds, & Reynolds, 2014;Pothukuchi, 2010; Cohen, & Reynolds, 2015).

Policymaking on urban agroecology and food security should be established considering all the stakeholders involved in safe and sustainable food production and consumption. It can be achieved by accepting and enhancing access to urban land use, creating and enabling the policy environment for productivity and economic efficiency and reducing environmental and health risks.

Agroecological tourism planning promoted by local government in partnership with producers aimed to provide support and assistance in infrastructure development in open urban green spaces, business planning, etc.

The concept of agroecological parks is applied in several peri-urban areas (Fanfani, 2013;Roth, Frixen, Tobisch, &Scholle, 2015; IPR, 2015). Planning new agroecological parks in the urban and peri-urban areas can be supported by community participative approaches (Roth et al., 2015; IPR, 2015; Fanfani, 2013). MAgroecological parks and agro-urban parks focus on maintaining and preserving urban agroecological land devoted to urban food gardening production's basic role in contributing to healthy nutrition and food security (Simon-rojo, Recasens, Callau, Duží, Eiter, Hernandez-Jimenez., Laviscio, Lohrberg, Pickard, Scazzosi, &Vejre, 2015;Parham, 2015; Fanfani, 2013).

Strategies of sustainable urban planning to support urban agroecology includes creating incentives for programming and funding policy environment and public land (Butler, 2012). Sustainable urban planning in cities experiencing growth and development may incentivise urban agroecological spaces and practices in remaining vacant land in areas around affordable housing without competing for other land uses.

Sustainable urban agroecology strategies may value food justice explicitly by not benefiting the propertied class rather than disadvantaged communities. Some local governments have set planning goals and strategies in the sustainable city plan to create a policy environment to support urban agroecology and food production (Hodgson, 2012). Planning and development strategies targeting urban agroecological and natural resources are aimed to reduce utility fees, funding, infrastructure investments, subsidies and incentives to disadvantaged communities (Cohen and Reynolds 2014).

CONCLUSION

Urban agroecology is the integrative study of the ecology of the entire food systems, encompassing ecological, economic and social dimensions based on the new geography when related to urban planning. Encouraging sustainable agriculture includes promoting agroecological urban farming, favouring access to natural resources and improving food security.

Urban agroecology is on the agenda of many individuals, farming producers, environmentalists, city urban planners, community and advocacy groups, food justice. A comprehensive plan on agroecology should be supported by the new geography framework knowledge and analysis to build a dialogue framed by institutional arrangements and new social movement networks to enable and advocate a policy on collective self-organisation and action.

Practising agroecology is one part of the struggle involving the collective action in urban planning through the participation in the political context of food sovereignty social movements. Agroecology can be scale-up supported by action and collective self-organisation, which challenge corporate control of food systems.

Agroecological practices related to the achievements of science and social movements emerge as a response to social and ecological concerns associated with the industrial agricultural production model. The agroecological movement based on societal actors may create and mobilise using social capital through networks of farmers to overcome difficulties and promote innovation of alternative production forms. The peasant and indigenous-based agroecological movements argue that farmers need to access and control biodiversity, land, water and incentives to produce food to meet the growing demand of their communities.

A sustainable urban agroecological system must be just and fair, supported by economic, social and political structures that reduce disparities of inequality and precariousness, aligned resources, needs to city goals and expectations, and technical and financial assistance that enable farmers to produce healthy food.

Urban agroecology as a strategy has significant benefits to individuals, communities and neighbourhoods, by improving food access and reducing food insecurity. Cities and local governments can formulate and implement strategies of urban planning based on the new geography network to benefit disadvantaged communities to have long-term access to land, acquiring vacant properties and offering underused public properties to be used for urban agroecology and proving incentives in new urban developments based on the new geography.

An agroecological practices report should identify the weak management and social indicators in the new geography to improve the overall performance of farming. A network of urban agroecological advocates and practitioners has released a set of indicators to assess the effectiveness of urban agroecological projects addressing environmental concerns such as improved soil, water, and air quality, storm-water management, water conservation and increased biodiversity, which are the basis of urban planning. Other indicators are based on social objectives, equality, marginalised populations, social cohesion, and collaboration.

REFERENCES

Abati, A., Skarulis, M. C., Shawker, T., & Solomon, D. (1995). Ultrasound-guided fine-needle aspiration of parathyroid lesions: A morphological and immunocytochemical approach. *Human Pathology*, *26*(3), 338–343.

Airriess, C. A., & Clawson, D. L. (1994). Vietnamese market gardens in New Orleans. *Geographical Review*, 16–31.

Alkon, A., & Mares, T. (2012). Food sovereignty in US food movements: Radical visions and neoliberal constraints. *Agriculture and Human Values, 29*(3), 347-359. doi:10.1007/s10460-012-9356-z

Allen, P., FitzSimmons, M., Goodman, M., & Warner, K. (2003). Shifting plates in the agrifood landscape: The tectonics of alternative agrifood initiatives in California. *Journal of Rural Studies, 19*(1), 61–75.

Armstrong, D. (2000). A survey of community gardens in upstate New York: Implications for health promotion and community development. *Health & Place, 6*(4), 319–327.

Barron, J. (2017). Community gardening: Cultivating subjectivities, space, and justice. *Local Environment, 22*(9), 1142–1158.

Bartling, H. (2012). A chicken ain'tnothin'but a bird: Local food production and the politics of land-use change. *Local Environment, 17*(1), 23–34.

Block, D. R., Chávez, N., Allen, E., & Ramirez, D. (2012). Food sovereignty, urban food access, and food activism: Contemplating the connections through examples from Chicago. *Agriculture and Human Values, 29*(2), 203–215.

Bowen, S., & Mutersbaugh, T. (2014). Local or localized? Exploring the contributions of Franco-Mediterranean agrifood theory to alternative food research. *Agriculture and Human Values, 31*(2), 201–213.

Bowen, S., & Mutersbaugh, T. (2014). Local or localized?Exploring the contributions of Franco-Mediterranean agrifood theory to alternative food research. *Agriculture and Human Values, 31*(2), 201–213.

Bowens, N. (2015). *The color of food*. New Society.

Brown, K. H., Carter, A., & Bailkey, M. (2003). *Urban agriculture and community food security in the United States: Farming from the city center to the urban fringe*. Community Food Security Coalition.

Butler, W. H. (2012). Welcoming animals back to the city: Navigating the tensions of urban livestock through municipal ordinances. *Journal of Agriculture, Food Systems, and Community Development, 2*(2), 193–215.

Cadji, J., &Alkon, A. (2014). One day, the white people are going to want these houses again. *Incomplete streets: Processes, practices and possibilities*, 154-175.

Clark, K. H., & Nicholas, K. A. (2013). Introducing urban food forestry: A multifunctional approach to increase food security and provide ecosystem services. *Landscape Ecology*, *28*(9), 1649–1669.

Cohen, M. J., & Garrett, J. L. (2010). The food price crisis and urban food (in) security. *Environment and Urbanization*, *22*(2), 467–482.

Cohen, N., & Reynolds, K. (2014). Urban agriculture policy making in New York's "New Political Spaces" strategizing for a participatory and representative system. *Journal of Planning Education and Research*, *34*(2), 221–234.

Companioni, N., Rodríguez, A., Carrión, M., Alonso, M., Ojeda, R. M., & Peña, E. (1997). La Agricultura Urbana en Cuba: suparticipaciónen la seguridadalimentaria. *Proceedings III encuentronacional de agriculturaorganica*, 9-13.

Cook, I. (2008). Geographies of food: Mixing. *Progress in Human Geography*, *32*(6), 821–833.

Crane, A., Viswanathan, L., & Whitelaw, G. (2013). Sustainability through intervention: A case study of guerrilla gardening in Kingston, Ontario. *Local Environment*, *18*(1), 71–90.

Dalgaard, T. H., & Porter, J. R. (2003). Agroecología: escalamiento e interdisciplinariedad. *Agricultura, ecosistemas y medioambiente*, *100*(1), 39-51.

Darmon, N., Darmon, M., Maillot, M., & Drewnowski, A. (2005). A nutrient density standard for vegetables and fruits: Nutrients per calorie and nutrients per unit cost. *Journal of the American Dietetic Association*, *105*(12), 1881–1887.

De Zeeuw, H., Van Veenhuizen, R., & Dubbeling, M. (2011). The role of urban agriculture in building resilient cities in developing countries. *The Journal of Agricultural Science*, *149*(S1), 153–163.

Despommier, D. (2010). *The vertical farm: feeding the world in the 21st century*. Macmillan.

Duany, A. (2011). *Garden cities: theory & practice of agrarian urbanism*. Duany Plater-Zyberk& Company.

Duží, B., Tóth, A., Bihuňová, M., & Stojanov, R. (2014): Challenges of urban agriculture. Highlights on Czech and Slovak experience. In Current Challenges of Central Europe: Society and Environment (pp. 82–107). Charles University.

Fanfani, D. (2013). Local development and "agri-urban" domain: Agricultural park as promotion of an "active ruralship". *Planum: The Journal of Urbanism*, *27*(2), 38–47.

FAO. (2008). *Climate change and food security: a framework document.* Food and Agro ecology Organization of the United Nations. http://www.fao.org/forestry/15538-079b31d45081fe9c3dbc6ff34de4807e4.pdf

FAO. (2011). *Food, agro ecology and cities: Challenges of food and nutrition security, agro ecology and ecosystem management in an urbanizing world.* http://www.fao.org/fileadmin/templates/FCIT/PDF/FoodAgriCities_Oct2011.pdf

FAO/WB. (2008). *Urban agriculture for sustainable poverty alleviation and food security.* Food and Agriculture Organization of the United Nations. http://www.fao.org/fileadmin/templates/FCIT/PDF/UPA_-WBpaper-Final_October_2008.pdf

Flyvbjerg, B. (2002). Bringing power to planning research: One researcher's praxis story. *Journal of Planning Education and Research, 21*(4), 353–366.

Gibb, N., & Wittman, H. (2013). Parallel alternatives: Chinese-Canadian farmers and the Metro Vancouver local food movement. *Local Environment, 18*(1), 1–19.

Gliessman, S. (2013). Agroecology: Growing the roots of resistance. *Agroecology and Sustainable Food Systems, 37*(1), 19–31.

Golden, S. (2013). *Urban agro ecology impacts: Social, health, and economic: An annotated bibliography.* https://asi.ucdavis.edu/programs/sarep/publications/food-and-society/uaannotatedbiblio-2013.pdf

Goodman, D. (2004). Rural Europe redux? Reflections on alternative agro-food networks and paradigm change. *Sociologia Ruralis, 44*(1), 3–16.

Gottlieb, R., & Joshi, A. (2010). *Food justice.* MIT Press.

Guthman, J. (2008). Bringing good food to others: Investigating the subjects of alternative food practice. *Cultural Geographies, 15*(4), 431–447.

Hagey, A., Rice, S., & Flournoy, R. (2012). Growing urban agriculture: Equitable strategies and policies for improving access to healthy food and revitalizing communities. Report by PolicyLink, Oakland, CA.

Harris, E. (2009). Neoliberal subjectivities or a politics of the possible?Reading for difference in alternative food networks. *Area, 41*(1), 55–63.

Hern, M. (2016). *What a City is for: Remaking the Politics of Displacement.* MIT Press.

Hislop, R. S. (2014). *Reaping equity: A survey of Food Justice Organizations in the USA.* University of California.

Hodgson, K. (2012). *Planning for food access and community-based food systems: A national scan and evaluation of local comprehensive and sustainability plans.* Report by American Planning Association.

Holdsworth, B. (2005). *Continuous Productive Urban Landscapes: designing urban agriculture for sustainable cities.* Academic Press.

Holt-Giménez, E. (2008). Out of AGRA: The green revolution returns to Africa. *Development, 51*(4), 464–471.

Hondagneu-Sotelo, P. (2014). *Paradise transplanted: Migration and the making of California gardens.* University of California Press.

Horst, M., McClintock, N., & Hoey, L. (2017). The intersection of planning, urban agriculture, and food justice: A review of the literature. *Journal of the American Planning Association, 83*(3), 277–295.

IPR. (2015). *Cooperative planning for periurban park at Vltava and Berounka River.* Available at: http://en.iprpraha/clanek/1409/cooperative-planning-for-periurban-park-at-vltava-and-berounka-river

Jaramillo, C. (2016). *Urban agriculture leaders ask for citywide commitments to garden preservation and creation.* Academic Press.

Kuebler, M., & Rugh, J. S. (2013). New evidence on racial and ethnic disparities in homeownership in the United States from 2001 to 2010. *Social Science Research, 42*(5), 1357–1374.

Lang, T., & Barling, D. (2012). Food security and food sustainability: Reformulating the debate. *The Geographical Journal, 178*(4), 313–326.

Lawson, L. J. (2005). *City bountiful: A century of community gardening in America.* Univ of California Press.

Levkoe, C. Z. (2011). Towards a transformative food politics. *Local Environment, 16*(7), 687–705.

Lovell, S. T. (2010). Multifunctional urban agriculture for sustainable land use planning in the United States. *Sustainability, 2*(8), 2499–2522.

Marcuse, P. (2009). From critical urban theory to the right to the city. *City, 13*(2-3), 185–197.

Marsden, T., & Sonnino, R. (2012). Human health and wellbeing and the sustainability of urban–regional food systems. *Current Opinion in Environmental Sustainability, 4*(4), 427–430.

McClintock, N. (2010). Why farm the city? Theorizing urban agriculture through a lens of metabolic rift. *Cambridge Journal of Regions, Economy and Society*, *3*(2), 191–207.

McIntyre. (Eds.). (2009). *The International Assessment of Agricultural Knowledge, Science and Technology for Development: Summary for Decision-Makers Report of the Global Report*. Island Press.

Mehmood, A., & Parra, C. (2013). Social innovation in an unsustainable world. In *The international handbook on social innovation: Collective action, social learning and transdisciplinary research* (pp. 53–66). Edward Elgar Publishing.

Mok, H. F., Williamson, V. G., Grove, J. R., Burry, K., Barker, S. F., & Hamilton, A. J. (2014). Strawberry fields forever? Urban agriculture in developed countries: A review. *Agronomy for Sustainable Development*, *34*(1), 21–43.

Morgan, K. (2009). *Feeding the city: The challenge of urban food planning*. Academic Press.

Mukherji, N., & Morales, A. (2010). *Zoning for urban agro ecology: Zoning practice*. American Planning Association.

Nair, P. K. R. (1993). *An introduction to agroforestry*. Kluwer Academic Publishers.

Neuner, K., Kelly, S., & Raja, S. (2011). *Planning to eat? Innovative local government plans and policies to build healthy food systems in the United States*. Academic Press.

Östman, Ö., Ekbom, B., & Bengtsson, J. (2003). Yield increase attributable to aphid predation by ground-living polyphagous natural enemies in spring barley in Sweden. *Ecological Economics*, *45*(1), 149–158.

Parham, S. (2015). The productive periphery: food space and urbanism on the edge. *Localizing urban food strategies. Farming cities and performing rurality*, 118-130.

Partalidou M and Anthopoulou T. (2016) Urban Allotment Gardens During Precarious Times: From Motives to Lived Experiences. *SociologiaRuralis*.

Pinderhughes, R. (2007). *Green collar jobs: An analysis of the capacity of green businesses to provide high quality jobs for men and women with barriers to employment*. The City of Berkeley Office of Energy and Sustainable Development. Retrieved from https://community-wealth.org/sites/clone.community-wealth.org/files/downloads/paper-pinderhughes-schecter.pdf

Piotti, G. (2007). [Book Review] Osti, Giorgio: Nuoviasceti: Consumatori, imprese e istituzioni di fronteallacrisiambientale (Bologna: Il Mulino, 2006). Economic Sociology: The European Electronic Newsletter, 21-23.

Pothukuchi, K. (2009). Community and regional food planning: Building institutional support in the United States. *International Planning Studies*, *14*(4), 349–367.

Public Health Law and Policy. (2009). *Establishing land use protections for community gardens*. Author.

Pursell, C. (1993). The rise and fall of the appropriate technology movement in the United States, 1965-1985. *Technology and Culture*, *34*(3), 629–637.

Renting, H., Marsden, T. K., & Banks, J. (2003). Understanding alternative food networks: Exploring the role of short food supply chains in rural development. *Environment & Planning A*, *35*(3), 393–411.

Reynolds, K. (2015). Disparity despite diversity: Social injustice in New York City's urban agriculture system. *Antipode*, *47*(1), 240–259.

Reynolds, K., & Cohen, N. (2016). *Beyond the kale: Urban agriculture and social justice activism in New York City* (Vol. 28). University of Georgia Press.

Reynolds, K., & Cohen, N. (2016). *Beyond the kale: Urban agriculture and social justice activism in New York City* (Vol. 28). University of Georgia Press.

Roth, M., Frixen, M., Tobisch, C., & Scholle, T. (2015). Finding spaces for urban food production–matching spatial and stakeholder analysis with urban agriculture approaches in the urban renewal area of Dortmund-Hörde. Academic Press.

Safransky, S. (2014). Greening the urban frontier: Race, property, and resettlement in Detroit. *Geoforum*, *56*, 237–248.

Saldivar-Tanaka, L., & Krasny, M. E. (2004). Culturing community development, neighborhood open space, and civic agriculture: The case of Latino community gardens in New York City. *Agriculture and Human Values*, *21*(4), 399–412.

Schmelzkopf, K. (2002). Incommensurability, land use, and the right to space: Community gardens in New York City1. *Urban Geography*, *23*(4), 323–343.

Schmidhuber, J., & Shetty, P. (2003). *The nutrition transition to 2030: why developing countries are likely to bear the major burden*. http://www.fao.org/fileadmin/templates/esa/Global_persepctives/Long_term_papers/JSPStransition.pdf

Simon-Rojo, M., Recanes, X., Cllau, S., Duzi, B., Eiter, S., Hernández-Jiménez, V., ... Vejre, H. (2016). *From urban food gardening to urban farming*. Academic Press.

Sonnino, R. (2013). Local foodscapes: Place and power in the agri-food system. *Acta Agriculturae Scandinavica, Section B–Soil & Plant Science, 63*(sup1), 2-7.

Sonnino, R., & Marsden, T. (2006). Beyond the divide: Rethinking relationships between alternative and conventional food networks in Europe. *Journal of Economic Geography*, *6*(2), 181–199.

Staeheli, L. A., Mitchell, D., & Gibson, K. (2002). Conflicting rights to the city in New York's community gardens. *GeoJournal*, *58*(2), 197–205.

Tanumihardjo, S. A., Anderson, C., Kaufer-Horwitz, M., Bode, L., Emenaker, N. J., Haqq, A. M., ... Stadler, D. D. (2007). Poverty, obesity, and malnutrition: An international perspective recognizing the paradox. *Journal of the American Dietetic Association*, *107*(11), 1966–1972.

Taylor, J. R., & Lovell, S. T. (2014). Urban home food gardens in the Global North: Research traditions and future directions. *Agriculture and Human Values*, *31*(2), 285–305.

Todd, N. J. (2012). *A safe and sustainable world: The promise of ecological design.* Island Press.

Torghani, C. (2014). Critical geography of urban agriculture. *Progress in Human Geography*, *38*(4), 551–567.

Tornaghi, C. (2011). *Steal this veg*. The Red Pepper June.

Tornaghi, C. (2012). *Edible public space. Experimenting with a socio-environmentally just urbanism*. Territorio.

Travaline, K., & Hunold, C. (2010). Urban agriculture and ecological citizenship in Philadelphia. *Local Environment*, *15*(6), 581–590.

UN-Habitat. (2010). The State of African Cities 2010: Governance, inequalities and urban land markets. Nairobi: United Nations Human Settlements Programme (UN-Habitat).

van der Schans, J. W., & Wiskerke, J. S. (2012). Urban agriculture in developed economies. In *Sustainable food planning: Evolving theory and practice* (pp. 245–258). Wageningen Academic Publishers.

Van Leeuwen, E., Nijkamp, P., & de Noronha Vaz, T. (2010). The multifunctional use of urban greenspace. *International Journal of Agricultural Sustainability*, *8*(1-2), 20–25.

Vitiello, D., & Brinkley, C. (2014). The hidden history of food system planning. *Journal of Planning History*, *13*(2), 91–112.

Walker, S. (2016). Urban agriculture and the sustainability fix in Vancouver and Detroit. *Urban Geography*, *37*(2), 163–182.

Wezel, A., & Soldat, V. (2009). A quantitative and qualitative historical analysis of the scientific discipline of agroecology. *International Journal of Agricultural Sustainability*, *7*(1), 3–18.

Whitehead, M. (2013). Neoliberal urban environmentalism and the adaptive city: Towards a critical urban theory and climate change. *Urban Studies (Edinburgh, Scotland)*, *50*(7), 1348–1367.

Zapata, M. (2017). *Appendix 1: Equity lens for the Portland State University diversity, equity, and inclusion action plan*. Portland State University.

Zezza, A., & Tasciotti, L. (2010). Urban agriculture, poverty, and food security: Empirical evidence from a sample of developing countries. *Food Policy*, *35*(4), 265–273.

Chapter 8
A Comprehensive Review of Agricultural Policies in India

Amaan Ullah
Aligarh Muslim University, Aligarh, India

Abukasim Idrisi
Aligarh Muslim University, Aligarh, India

Mohd Mohsin Khan
Aligarh Muslim University, Aligarh, India

K. M. Bahrul Islam
Indian Institute of Management, Kashipur, India

ABSTRACT

India has been an agrarian economy since ancient times; despite the diminishing value added of agricultural activities to total domestic production or GDP, the agriculture sector has remained the largest employer and thus proved to be the driver of growth and poverty reduction. Moreover, in the recent period of a pandemic where every economic activity came to a halt and showed a negative growth rate, agricultural activities, on the other hand, grew positively, employing job losers in these challenging times. However, agriculture in India faces lots of obstacles due to its limitations in policy formulation and implementation. Therefore, this chapter aims to provide an outline of Indian agriculture growth – reviewing its agricultural policy reforms and observing the fundamental concerns that have shaped the expansion of the agriculture sector.

DOI: 10.4018/978-1-6684-2349-3.ch008

INTRODUCTION

In earlier years of independence, food security and self-sufficiency were the main objectives of the agricultural and trade policies. For India, it was necessary to achieve self-reliance through the policies seeking to achieve food security. Therefore, various policies were adopted over time, which were exclusively meant to achieve the objectives of pursuing 'Food Security in India.

In the 1950-60s, the agricultural policies aimed to abolish intermediary land lordship through the imposition of land ceiling acts. In the mid-1960s, the Green Revolution outstretched crop production through superior crop technologies and seed ranges. Moreover, in the year 1960, two key bodies, i.e., the "Food Corporation of India (FCI)" and Agricultural Prices Commission, later called as "Commission for Agricultural Costs and Prices (CACP)", played vital roles in moving the distribution and prices of rice and wheat. In the 1980s, with the widespread use of technology in primary crop production, the government also introduced schemes to increase agriculture subsidies(Anand Singh & Krishna, 1994). In the 1990s, policy reforms related to delicensing and deregulations were adopted throughout the economy, but the agriculture sector was left untouched (Selvaraj, 1993).

Therefore, the chapter will outline the scenario of agriculture since independence, providing, in brief, the main policy reforms during the period, further listing in short some of the most influential revolutions in agriculture and explaining in detail their contributions. In further section chapter draws a comparison of selected countries showcasing the agriculture sector's role in terms of its contribution to total Gross Domestic Product (GDP), Employment and Exports of these economies. The chapter also evaluates policy formation and implementation – describing the relations of central government with respective state governments - how the state governments are responsible for the actual implementation of policies formed at the center and how the state government and central government coordinates with the other related departments and agencies to implement these policies effectively. The chapter also tabulates different departments, agencies, and ministries operating for agricultural policy formulation and implementation. The chapter had further described the significant components of agriculture and food policy in India. The chapter then analyses the policies and reviews them to highlight key limitations and impacts of India's overall policies related to agriculture and allied activities.

Finally, the chapter concludes by giving some key suggestions and recommendations.

Table 1. Indian Agricultural scenario and policy reforms

Approximate Years	Major policy Initiatives
1950-60s	● Reforms abolishing intermediary landlord ship and, ● Imposition of Land Ceiling Acts.
1960-80s	● Green Revolution ● Foundation of FCI and CACP
1980s	● Increased subsidies and support for agriculture.
1990s	● Economic Liberalization ● Increase in input subsidies ● Targeting the beneficiaries of Public Distribution System (PDS) ● Some ease in the trade protection of crops like sugar, cotton, edible oil, wheat and rice etc.

Source:(Blyn, 2016; India, 2016)

INDIAN AGRICULTURE SCENARIO: TREND AND GROWTH

The Green revolution (1960), The White Revolution (1970), The Gene Revolution (2000), and the recent developments in the production of pulses, fruits and vegetables have pushed the agriculture sector of India to the next level. India is the world's biggest producer of pulses and the world's second-largest producer of fruits and vegetables. Moreover, steadily growing agricultural exports are transforming India from a food insufficiency to a significant exporter of agro-food products(Gulati et al., 2018). Export of agriculture and allied products include rice, cotton, fish and fish products, oilcake, vegetable extracts, meat & meat products and other products(OECD/ICRIER, 2018)

Employment remains the most vital beneficiary of agricultural activities in the selected economies. India being the agrarian economy, saw the most significant contribution of 56 per cent from agricultural activities to employment, while the US being the capital economy, have a minor contribution of about 2.5 per cent from agricultural activities to employment. Agricultural activities have contributed 18 per cent to the GDP of the Indian economy and 1.4 per cent to the GDP of the US economy. Most of the agricultural products are being exported from Brazil, amounting to about 34 per cent of its total exports, whereas China being a minor exporter of agricultural produce.

Figure 1. Role of agriculture: a comparison of different economies
[Note: an average of the years 2000-2018 are considered for all countries, figures are rounded off]
Source: Author's representation from the OECD statistical database and national data."

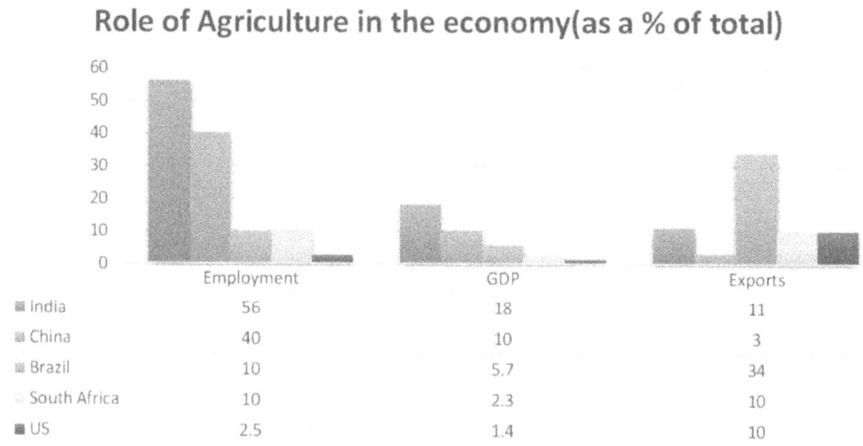

AGRICULTURAL POLICY PLANNING AND IMPLEMENTATION: CENTRE AND STATE RELATIONS

Indian Constitution enlists agriculture as a subject in the state list; the central government, however, plays a crucial role in considering agriculture as a matter of national importance. In line with the other policy initiatives, evolving and executing national policy and coordinating with the efforts of individual states are the co-operating functions of the government at the centre. Moreover, the state governments also delegate and devolve some of their authorities relating to agriculture to the local governments called *Panchayats*. Therefore, the administration of agriculture in India is multifaceted and comprises ministries at the state and central level, agencies and other institution and the administration at the local level.

The Role of Central Government

Through its Ministry of Agriculture and Farmers Welfare (MAFW), the central government issues broad guidelines for agricultural policies. Agencies at the central level administer and provide technical and financial support through Central Scheme (CS), Centrally Sponsored Scheme (CSS) and Additional Central Assistance (ACA) from time to time. However, the central government's priorities and approaches vary from state to state depending on the need, possibilities, and political relations between the centre and states. Depending on the issues, the formation and execution of agriculture policies require coordination among various ministries, departments and agencies. Occasionally temporary advisory committees are also set up to examine particular issues and provide recommendations.

In India, states differ in their natural resources, economic development and demography. The central government addresses the disparities and imbalances by identifying problems and opportunities in particular states and providing specific grants and aids to such states. For instance, the central government had provided its large share in various shared schemes in eight north-eastern states. (i.e., Arunachal Pradesh, Assam, Manipur, Meghalaya, Mizoram, Nagaland, Tripura and Sikkim)

The Role of State Government

The respective state governments have the primary role in implementing and administering agricultural policies through their different agencies and departments. State governments have their ministries and departments related to agriculture that cooperates with the centres' Department of Agriculture, Co-operation and Farmers' Welfare (DACFW) and Department of Food and Public Distribution (DFPD) to implement the central schemes and policies. However, many state governments employ specific policies in accordance to their primacies and funds availability. The state government's mutual and delegated efforts with the local administration are crucial as the policies had to be delivered in the local areas such as gram, tehsil, and districts.

MECHANISM OF AGRICULTURE AND FOOD POLICY IN INDIA

The agriculture policy in India consists of five significant heads; they comprise:

- *Dealing with the prices and marketing channels*: the pricing policy is concerned with three essential components, namely, minimum support price (MSP), buffer stock operations and the Public Distribution System (PDS). The main aim of the pricing policy is to assure farmers, reasonable and remunerative prices for their agricultural produce so that they can be protected against sharp decline or variations in prices of agricultural produce. The central government also administers marketing channels to procure agricultural produce through various agencies. These agencies include; "Food Corporation of India (FCI)", "Cotton Corporation of India (CCI)", "Jute Corporation of India (JCI)", "Central Warehousing Corporation (CWC)" and "Tobacco Board". The state governments also appoint state agencies to coordinate with central agencies and departments to implement price support schemes effectively.

- *Ensuring the availability of various farm inputs at government-subsidized prices*: agricultural subsidies are of two types, namely, input subsidies and investment subsidies. Investment subsidies are meant for infrastructural support such as developing irrigation systems, constructing rainwater harvesting systems and acquiring other farm equipment. Input subsidies are provided for subsidizing agrochemical fertilizers, water and electricity for irrigation, etc. Input subsidies are also meant for providing seeds, herbicides, and pesticides at a subsidized rate.
- *Providing Research and Development services;* The Central Government through the Indian Council of Agricultural Research (ICAR) reforms agricultural research and education in India. The government has funded many institutions through ICAR, which resulted in establishing the National Agricultural Research System with its vast network of research institutes and state agricultural universities.
- *Governing border transactions through trade policy*; international trade is regulated by the Ministry of Commerce and Industry through its Department of Commerce. The "Directorate General of Foreign Trade (DGFT)" formulates and implements India's Foreign Trade Policy; however, the government established a particular institution to regulate the exports of agro-products named the "Agricultural and Processed Food Products Export Development Authority (APEDA)". The APEDA has the special authority and responsibility to develop and promote exports of many agricultural products.
- *Environmental measures*; government at the centre and state coordinate through various missions and action plans of climate change on sustainable development. "Department of Agriculture Cooperation and Farmers Welfare (DACFW)" regulates the activities seeking sustainable agriculture, rain-fed area development and water conservation, water use efficiency, emphasizing soil health management.

Table 2. Institutions for the formulation and execution of agricultural policies in India

Issues	Central Ministries	Other Executive bodies.
Prices	● Ministry of Agriculture and Farmers' Welfare ● Ministry of Commerce and Industry	● Commission for Agricultural Costs and Prices (CACP) ● Other state-level departments
Inputs and production	● Ministry of Agriculture and Farmers' Welfare ● Ministry of Water Resources, River Development and Ganga Rejuvenation ● Ministry of Food Processing Industries ● Ministry of Power ● Ministry of Chemical and Fertilizers ● Ministry of New and Renewable Energy ● Ministry of Environment, Forest, Climate Change.	● Central Water Commission ● State-level counterparts.
Credit	● Ministry of Finance, and RBI	● National Bank for Agriculture and Development (NABARD)
Marketing, procurement	● Ministry of Agriculture and Farmers' Welfare ● Ministry of Food Processing Industries ● Ministry of Consumers Affairs, Food and Public Distribution ● Ministry of Textiles, Agricultural and Processed Food Products	● Food Corporation of India (FCI) ● Cotton Corporation of India (CCI) ● Jute Corporation of India (JCI) ● NAFED ● APEDA ● Central Warehousing Corporation (CWC) ● National Dairy Development Board (NDDB)
Public distribution	● Ministry of Agriculture and Farmers' Welfare ● Ministry of Consumer Affairs, Food and Public ● Distribution ● Ministry of Human Resource Development ● Ministry of Women and Child Development	● Food Corporation of India (FCI) ● NAFED ● Central Warehousing Corporation (CWC) ● State-level counterparts to center's Departments and Institutions
Trade	● Ministry of Commerce and Industry	● Agricultural and Processed Food Products Export Development Authority (APEDA) ● National Agricultural Cooperative Marketing Federation of India Ltd. (NAFED) ● Commodity boards ● Agro-export zones (AEZ) ● Food Safety and Standards Authority of India (FSSAI)
Research, education, extension	● Ministry of Agriculture and Farmers' Welfare	● Indian Council of Agricultural Research (ICAR) ● Veterinary Council of India ● Indian Council of Forest Research ● Central and deemed agricultural universities ● Indian Institute of Management (IIM) ● Agribusiness management institutes ● State-level counterparts, e.g., State Agricultural Universities (SAU), *Krishi Vigyan Kendras* (KVK, s)

Source: Agricultural Policies in India, OCED/ICRIERT 2018"

LIMITATIONS AND SHORTCOMINGS

The agricultural sector in India faces severe issues of structural development such as dependence of agricultural activities on moods of monsoon, which is inconsistent and uncertain; lack of infrastructure for the marketing of perishable commodities; high cost of critical farm inputs such as hybrid seeds, agrochemical fertilizers; lack of market guarantee; low and deteriorating returns per unit area; inadequate government support.(Mall et al., 2006; Sandeepa & Sarala, 2020)

One of the most contentious issues in India about the subsidies and other policy benefits is that these subsidies do not find their way to the farmers but are siphoned away beside the path. Moreover, the long processes of policy formulation and implementation that involve sharing of information, views, and orders among many institutions pose a challenge in the effective and efficient implementation of policy norms to the ground level.

CONCLUSION

The whole world is moving towards technological advancements in the agro-food sector; in contrast, the Indian agro-food sector faces severe challenges in line with availing multiple opportunities. Any policy's actual benefit and welfare can be achieved through its proper implementation, which lacks in the Indian economy. The proper policy implementation will define India's success in achieving food security, enhancing the quality of life of marginal landholders, sustainable productivity growth and an efficient agriculture system(Sharma & Sharma, 2018; Shukla & Dwivedi, 2015).

The long chain of authority and responsibility under various ministries, departments and regulating agencies is a challenge for effective and efficient policy reforms implementation. To overcome this challenge, the powerful and regulatory functions of these institutions should be delegated in proper form.

Moreover, farmers' literacy and skill development should be taken up as a priority for making them understand their rights and benefits that are particularly available to them.

Finally, less restrictive and more stable agricultural trade policies are needed to realize the full potential of farm inputs and products throughout the count.

REFERENCES

Anand Singh, K., & Krishna, K. (1994). Agricultural entrepreneurship: The concept and evidence. *The Journal of Entrepreneurship*, *3*(1), 97–111. doi:10.1177/097135579400300106

Blyn, G. (2016). *Agricultural trends in India, 1891-1947*. University of Pennsylvania Pre.

Government of India. (2016). Agricultural statistics at a Glance 2016. Ministry of Agriculture and Farmers Welfare, Department of Agriculture.

Gulati, A., Sharma, P., Samantara, A., & Terway, P. (2018). *Agriculture extension system in India: Review of current status, trends and the way forward*. Academic Press.

Mall, R. K., Singh, R., Gupta, A., Srinivasan, G., & Rathore, L. S. (2006). Impact of climate change on Indian agriculture: A review. *Climatic Change*, *78*(2), 445–478. doi:10.100710584-005-9042-x

OECD/ICRIER. (2018). Agricultural policies in India. In *OECD Food and Agricultural Reviews* (p. 296). OECD Publishing Paris.

Sandeepa, M., & Sarala, K. S. (2020). Opportunities and challenges of agriprenuership in India–Some reviews. *PalArch's Journal of Archaeology of Egypt/Egyptology, 17*(7), 9541–9554.

Selvaraj, K. N. (1993). Impact of government expenditure on agriculture and performance of agricultural sector in India. *Bangladesh Journal of Agricultural Economics*, *16*(2), 1–13.

Sharma, S., & Sharma, S. (2018). *Agri-entrepreneurial innovations for rural prosperity and sustainable development*. Academic Press.

Shukla, K. H., & Dwivedi, N. U. (2015). Sustainable development in agricultural sector in India. *The Business & Management Review*, *5*(4), 220.

Chapter 9
Optimization of Energy in Smart Farms Using a Genetic Algorithm

Brahim Lejdel
https://orcid.org/0000-0002-7193-5204
University Echahid Hama Lakhdar, El-Oued, Algeria.

ABSTRACT

The need for energy in the agriculture field is increasing as a result of increasing the productivity of these fields. The appearance of smart grid and IoT (internet of objects) enabled farmers to control, manage, and optimize the energy consumption. Agriculture will continue to rely on energy to increase its productivity in line with increasing population and great demand. In this chapter, the authors present an integrated model between SmartFarms, the smart-grid, and optimization methods. In this way, smart forms can participate actively and benefit from the energy market. In this chapter, they consider the electrical energy that is directly used in activities ranging from field processes such as irrigation of land. Energy is also indirectly consumed in synthetic additives notably fertilizers, pesticides, and herbicides. In addition, the authors can consider the electrical energy, which is used for powering some agricultural machines.

INTRODUCTION

Actually, the need for energy in the agriculture field is increasing as a result, increasing the productivity of these fields. The appearance of Smart Grid and IoT (internet of objects) enabled farmers to control, manage and optimize the energy

DOI: 10.4018/978-1-6684-2349-3.ch009

consumption. On the other sides, agriculture will continue to rely on energy to increase its productivity in line with increasing population and great demand. In this paper, we present an integrated model between SmartFarms, the Smart-grid and optimization methods. The objective of this model is to focus on ways, which can improve the decision support systems in Smart farms to enable optimal decision making in terms of energy consumption and crop quantity. In this way, SmartFarms can participate actively and benefit from the energy market. Also, it can support the reliable operation of Smart grids while minimizing their expenditures on energy and increasing farm productivity.

In the last decade, the energy used in agriculture is a recent issue of research (Hussain &Azlina, 2010). The flames consume increasingly the energy, which is used directly in farm activities such as fertilization, irrigation and indirectly in different forms such as liquid fuels to power vehicles and electricity for other processes. The rise in energy used has been a result of the intensification of agricultural production systems due to the need to match food production with increasing demand with increasing populations (MacLeod & Moller, 2006). The optimization of energy is an important subject in this decade. In this paper, we achieve the optimization of energy consumption using the genetic algorithm.

In this paper, we consider the electrical energy that is directly used in activities ranging from field processes such as irrigation of land. Energy is also indirectly consumed in synthetic additives, notably fertilizers, pesticides, and herbicides. In addition, we can consider the electrical energy, which is used for powering some agricultural machines.

SYSTEM ARCHITECTURE

In our system, we will use the Farm agent to model the smart farm. The farm agent uses information gathered from sensors to optimize energy consumption. It can also use satellite positioning data, remote sensing, and GIs technology to increase the precision of the information. The farm agent can connect with many other agents as irrigation agent, vehicle agent and crop agent with control, respectively, the irrigation, engine, and crop of the farm. Figure 1 presents an overview of our proposed system. Thus, in our system, we have three principal components, which discuss below.

Figure 1. Overview of our system

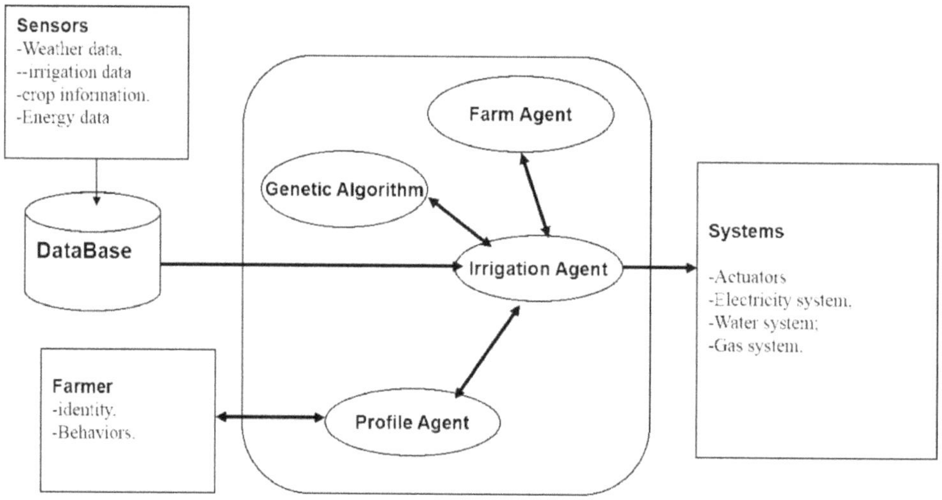

Sensors

Sensors can collect information about weather, irrigation, crop, and energy according to many field conditions. It can use map technology for increasing the precision of information as GPS and GIS. An example is Crop data, which is collected about crop factors such as disease infestation, nutrient, water stress and soil factors such as fertility moisture, electrical conductivity and environmental conditions such as air temperature (Wang, Zhang, & Wang, 2006).

Decision Support and Precision system

In agriculture, we generally use GIS database types such as ArcGIS, PostGIS or MapInfo to increase the precision and consistency of geographic information (Runquist, Zhang, & Taylor, 2001). The majority of GIS applications provides methods for performing much analysis as interpolation of data. This analysis consists of converting each point data in the ground into spatial data in Gis. The information obtained from sensors is converted into a format adequate in Gis with the specific parameters of interest. This information is stored in geographical databases. Thus, the GIs contains an interface which provides platforms for exporting and importing the data for execution issue in actuators.

Actuators

For making action in agriculture, we will use some actuators as machines, irrigation, and Fertilization system. The implementation of actuators requires, in general, location data, which are provided by GPS devices, and GIS databases. The actuators also have software to allow loading of maps and hardware to implement variable interventions basing on the maps. Examples include; agricultural machines powered using real-time kinematics, lightbar guidance systems where drivers are guided by means of horizontal display of lights.

Genetic Algorithm

Genetic algorithms are developed in Holland, (1975)to imitate the phenomena adaptation of living beings. They are an optimization technique based on the concepts of natural selection and genetics. It searches for an optimal solution among a large number of candidate solutions within a reasonable time (the process of evolution takes place in parallel). Each of these solutions contains a set of parameters that completely describe the solution. This set of parameters can then be considered as the "genome" of the individual, with each parameter comprising of one or more "chromosomes". They allow a population of solutions to converge step by step toward the optimal solution. To do this, they will use a selection mechanism for the population of individuals (potential solutions). The selected individuals will be crossed with each other (crossover), and some will be mutating by avoiding, whenever possible, local optima. Genetic Algorithms are used primarily to treat both problems (DeJong, &Sarma (1993).

1. The search space is large, or the problem has many parameters to be optimized simultaneously.
2. The problem can not be easily described by a precise mathematical model.

We will combine Multi-Agent Systems with Genetic Algorithms (SMA-GA) for permitting the agent to choose the optimal actions. Therefore, our proposed model is based on the three following points.

1. Farm Agent is a software agent who can manage, control the local optimization process and exchange relevant information with neighbouring Farm Agents.
2. Genetic patrimony, which is transformed between agents, is used as inputs to the genetic algorithm. This genetic patrimony represents values of the consumption energy of different systems that are collected by the sensor systems.

3. Genetic Algorithms is used to find the optimal solution for the current configuration; this is composed of the two objective functions, the energy consumption and the quality and quantity of the crop.

IMPLEMENTATION

In this section, we will introduce the different elements of implementation.

Chromosome Structure

To apply the genetic algorithm, we should define the genes and the chromosome structure. Its identifier and a set of values of different systems can be applied to perform the optimal energy consumption and increase the quantity of the crop, which can characterize the gene. We use multiple forms for coding the genes. Firstly, we use the strings for coding the identifiers, and then we use the real number for coding the values of the different systems. Figure 2 presents the structure of the gene.

Figure 2. Structure gene of room

ID-Farm	I	200	F	100	M	150	L	70

ID-Farm : Farm identify,

I: Irrigation system and its value,

F : Fertilization system and its value,

M: machinery system and its value,

L: light System ands its value.

Genetic Algorithm Steps

Initialization

The initialization operator determines how each chromosome is initialized to participate in the population of the genetic algorithm. Here, the chromosome is filled with the genetic material from which all new solutions will evolve. In this work, we will use the Steady State to initial the generation process and select the population of the genetic algorithm for the next generation. First, Steady State creates a population of individuals by cloning the initial chromosomes. Then, at each generation during evolution, it creates a temporary population of individuals, adds these to the previous population and then removes the worst individuals in order that the current population is returned to its original size. This strategy means that the newly generated offspring may or may not remain within the new population, depending upon how they measure up against the existing members of the population.

Crossover

The crossover operator defines the procedure for generating a child from two-parent chromosomes. The crossover operator produces new individuals as offspring, which share some features taken from each parent. The probability of crossover determines how often crossover will occur at each generation. In this approach, we will use the single point crossover strategy was adopted for all experiments. In this paper, the results for all experiments presented were generated using a crossover percentage of 50%, which is to say that at each generation, 50% of the new population was generated by splicing two parts of each chromosome's parents together to make other chromosomes.
Figure 3 shows the crossover operator.

Figure 3. Crossover operator

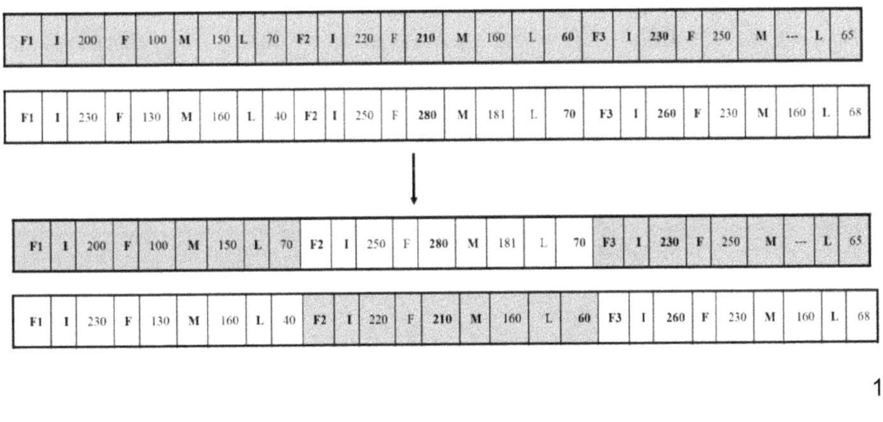

Mutation

The mutation operator is important. It defines the procedure for mutating the chromosome. Mutation, when applied to a child, randomly alters a gene with a small probability. It provides a small amount of random search that facilitates convergence at the global optimum. The probability of mutation determines how much of each genome's genetic material is altered, or mutated. If the mutation is performed, part of the chromosome is changed. The mutation should not occur too often as this would be detrimental to the search exercise. In this work, the results presented here were generated using a 1% mutation probability, which was determined experimentally, using a single case of vector HVAC-L system. We present a random mutation in figure 4.

Figure 4. Mutation operator

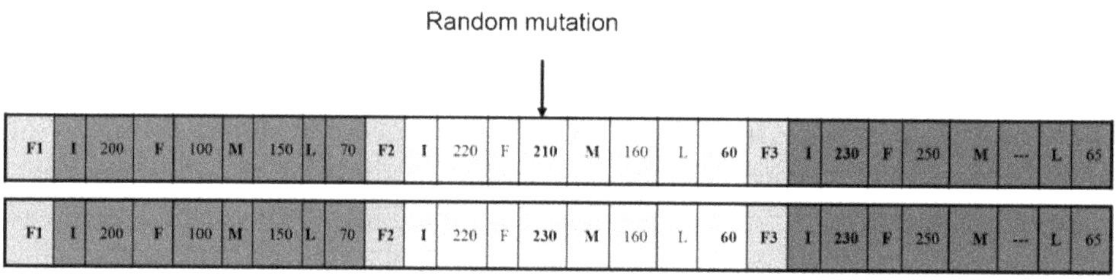

EVALUATION OF SOLUTIONS

We can say that the success of any discrete optimization problem rests upon its objective function, the purpose of which is to provide a measure for any given solution that represents its relative quality. In our resolution method of consumption energy problem in Farms, the objective function used here works by calculating and summing the energy associated with the Irrigation, the Fertilization and the machine system (IFM system) and light systems.

Thus, we have mainly *f(E)*, which permits us to evaluate the performance and efficiency of the proposed approach. Each Farm agent calculates the sum of these two energies.

$$f(E) = E_{IFM} + E_{L} \tag{1}$$

E_{IFM} and E_L represent the consumption energy of the IFM system and the lighting system, respectively.

RESULTS AND DISCUSSION

In this section, we present a case study that illustrates how to design the genetic algorithm of our system. We use Java (https://www.java.com/fr/) to implement the different steps of the genetic algorithm like a crossover operator, mutation operator, and the evaluation function. Thus, in experimentation, we use four farms as a case study. The experimentation aims to optimize the energy consumption of these farms. Firstly, each farm uses the sensors to learn the IFM-L data, which can be used as input in the genetic algorithms. The farmers can introduce their preferences in the profile via a graphic interface, as the energy consumption preferences. The system runs the genetic algorithm to find the optimal values of IFM-L system, which allow to optimize the energy consumption and increasing the quantity of crop of farms. Our genetic algorithm performs 200 generations. We use 1% for mutation. Figure 5 shows the evolution of the fitness function.

Figure 5. Evolution of fitness

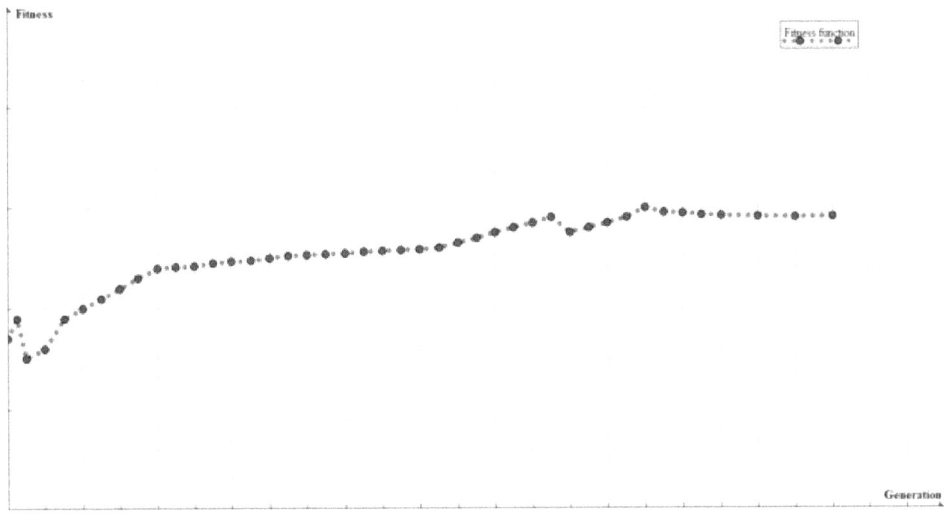

For comparing our approach with other approaches, energy consumption is solved using both Genetic Algorithms (GA) and Linear Programming (LP). In table 1, we introduce the quantity of the crop according to each crop type as potato, maize, tomatoes, garlic, and onions.

Table 1. Comparison of two approaches, GA and Pl in term of energy consumption and quantity of crop

N° product	Crops	Energy consumption *100 Kwh/ha		Quantity *1000 Kg/ha	
		GA	PL	GA	PL
1	Potato	33	35	50.00	42.8
2	Maize	13.20	15.2	30	29.61
3	Tomatoes	25.23	28.25	35.2	30.5
4	Garlic	20.5	24.22	23	20.1
5	Onions	22.36	26.35	25	22.8

In Figure 6, we present the quantity of crop obtained using both approach GA or Pl.

Figure 6. Comparison of the results obtained by GA and LP

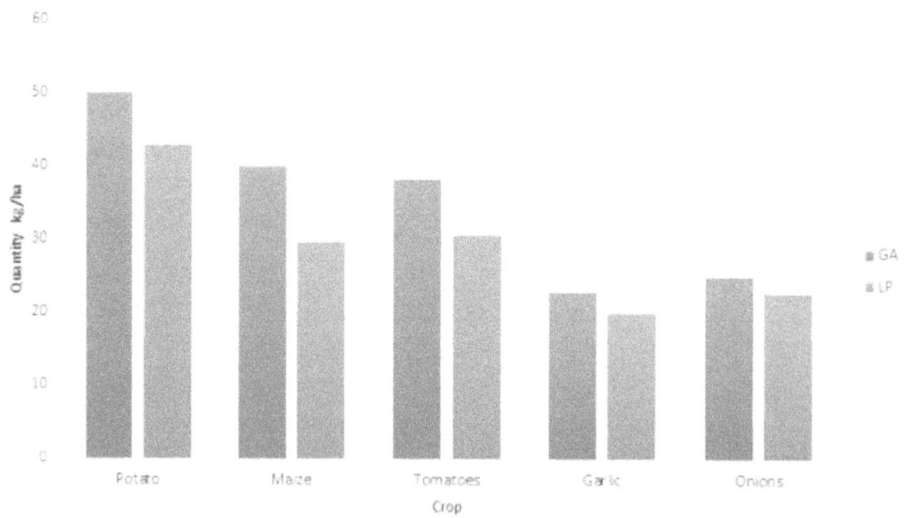

In Figure 7, we present the energy consumption using GA and Lp.

Figure 7. Energy consumption using GA and LP

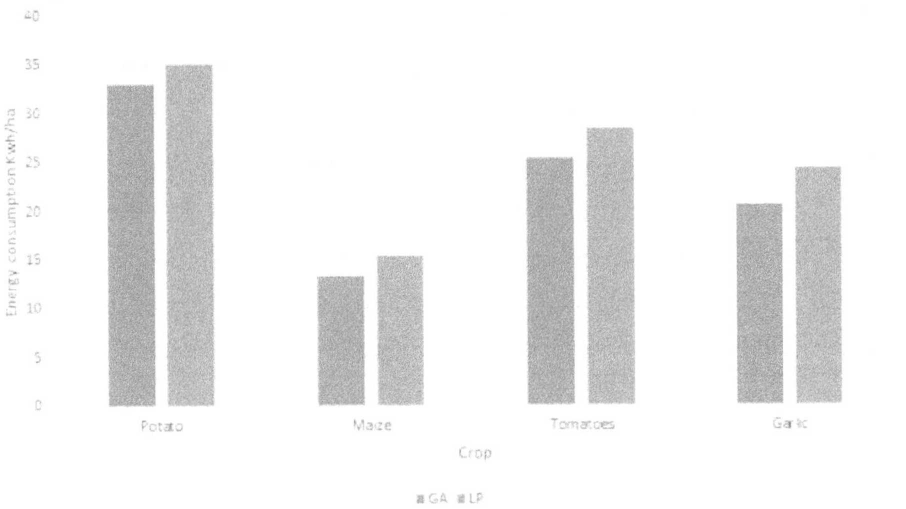

We conclude that Genetic Algorithms is an effective optimization tool for energy consumption and can be used for more complex systems of energy consumption.

CONCLUSION

We proposed in this paper a system that can optimize the energy consumption in the farms based on a Genetic Algorithm. The objective is to maximize crop quantity in farms with minimizing the energy consumed by the system of irrigations and machines. The results obtained from the GA model are compared with those obtained from the Linear Programming model. This study opens many future works:

- We can use Swarm optimization for optimizing energy consumption.
- We can study only the optimization of irrigation systems because it consumes more energy compared with the other systems.

REFERENCES

Bekhet, H. A., & Abdullah, A. (2010). Energy use in agriculture sector: Input-output analysis. *International Business Research*, *3*(3), 111. doi:10.5539/ibr.v3n3p111

DeJong, K., & Sarma, J. (1993). Generation Gaps Revisited. In *Foundations of Genetic Algorithms 2*. Morgan-Kaufmann Publishers.

Holland, J. (1975). *Adaptation in Natural and Artificial Systems*. University of Michigan Press.

MacLeod, C. J., & Moller, H. (2006). Intensification and diversification of New Zealand agriculture since 1960: An evaluation of current indicators of land use change. *Agriculture, Ecosystems & Environment*, *115*(1-4), 201–218. doi:10.1016/j.agee.2006.01.003

Runquist, S., Zhang, N., & Taylor, R. K. (2001). Development of a field-level geographic information system. *Computers and Electronics in Agriculture*, *31*(2), 201–209. doi:10.1016/S0168-1699(00)00155-1

Wang, N., Zhang, N., & Wang, M. (2006). Wireless sensors in agriculture and food industry - Recent development and future perspective. *Computers and Electronics in Agriculture*, *50*(1), 1–14. doi:10.1016/j.compag.2005.09.003

Chapter 10
Women Entrepreneurs and Agricultural Start-Ups:
Cognitive and Social Capital Perspective

Anjila Saleem
Aligarh Muslim University, Aligarh, India

Javed Ali
Aligarh Muslim University, Aligarh, India

Mohd Yasir Arafat
Aligarh Muslim University, Aligarh, India

ABSTRACT

Numerous nations hail the agriculture sector as a critical source of wealth creation, and past researches have shown the importance of entrepreneurship in the agriculture industry. However, there is a substantial difference in men and women's rates of taking entrepreneurial initiatives. Prior research has overlooked the significance of entrepreneurial inclination in creating agricultural start-ups from the perspective of gender. The primary objective of this study is to investigate the factors that influence women entrepreneurs working in the business endeavours of the agriculture sector. Using logistic regression, the study looked at a representative interview of 581 samples with individuals (18–65 years of age) from GEM countries. This model demonstrates the connection between the variables' qualities reliant on the data and the determinants. The chapter suggests that policymakers consider the consequences of promoting women's entrepreneurship in the agricultural industry and evolve the policies accordingly.

DOI: 10.4018/978-1-6684-2349-3.ch010

BACKGROUND

Several countries expect agriculture to be a significant revenue-producing source (Schmidhuber & Tubiello, 2007; Dubé, Pingali, & Webb, 2012). Consequently, experts are focusing on the agriculture industry since it can aid in producing jobs and economic prosperity (Arafat, Ali, Dwivedi, & Saleem, 2020a).Furthermore, scholars have acknowledged the importance of entrepreneurship's role in fostering innovation, competitiveness, expansion, and development (Acs, Braunerhjelm, Audretsch, & Carlsson, 2009; Arafat, Khan, Saleem, Khan, & Khan, 2020b;Khan, Arafat, Raushan, & Saleem, 2020; Hassan, Saleem, Anwar, & Hussain, 2020).The research earlier (Yaseen, Saleem, Zahra, & Israr, 2018; Pindado & Sánchez, 2017) has proven the importance of entrepreneurship in agriculture. (Lans, Seuneke, & Klerkx 2017; Grande 2011;.; Lans, Blok, & Wesselink, 2014; Barnes, Hansson, Manevska-Tasevska, Shrestha & Thomson, 2015). Despite the variations between the various activities of entrepreneurship drivers, these researchers looked at entrepreneurship in the agriculture business. A recent study found significant disparities in women's entrepreneurship rates compared to males (Arafat, Saleem, & Dwivedi, 2020c; Arafat, Saleem, Dwivedi, & Khan, 2020d), but it failed to give a clear perspective, especially in terms of gender (Kelley, Singer, & Herrington, 2015) regardless of whether previous research has found that the drivers of entrepreneurship are identical for men and women and that variations in the rate of entrepreneurship may be accounted for other entrepreneurial characteristics (Langowitz, & Minniti, 2007; Brush, 1990, 1992). Entrepreneurial intentions (EI) are recognised as prospective variables among other attributes of entrepreneurship (Arafat et al., 2017b; Khan, Arafat, Raushan, Khan, Dwivedi, Khan, & Saleem, 2019b; Hassan, Anwar, Saleem, Islam, & Hussain, 2021), making it even more critical for agricultural entrepreneurship research business based on gender (Kelley et al., 2015; Langowitz, & Minniti, 2007; Pindado & Sánchez, 2017).

The primary purpose of this research is to look at the factors of EI for women in the agriculture business to see what obstacles they face, (Ramos-Rodrguez, Medina-Garrido, & Ruiz-Navarro, 2012; Pindado & Sánchez, 2017; Arafat & Saleem, 2017b) attempted to assume sectoral entrepreneurship since sector-specific studies better understand the venture formation process. In addition, the study looks at entrepreneurship in the agriculture business from a gender viewpoint. The factors of entrepreneurial ambitions were chosen based on the findings of earlier research, which is focused on agriculture and other industries. The findings of this study describe the factors that influence agricultural entrepreneurship, with a focus on women. These findings might help explain the entrepreneurial phenomena and formulate strategies to encourage women to start businesses in the agriculture industry.

HYPOTHESES DEVELOPMENT AND THEORETICAL BACKGROUND

After accounting for other variables, recent studies show that women's entrepreneurship rates differ from men due to attitudinal and social characteristics (Wagner, 2007; Anwar, Saleem, Islam, Thoudam, & Khan, 2020). Further, on the other hand, prior research has shown that the drivers of entrepreneurial activity are the same for men and women (Langowitz & Minniti, 2005; Khan, Arafat, Raushan, Saleem, Khan, & Khan, 2019a, 2019c). As a result, by addressing external entrepreneurial elements, this dispute may be elucidated. Previous research has focused on entrepreneurship in the same socio-cultural context and environment of women and men, as these factors might influence entrepreneurship (Minniti & Nardone, 2007; Afandi, Kermani, & Mammadov, 2017). When it comes to gender, research can provide more specific entrepreneurial behaviour of women. As a result, the disparity in EI determinants between men and women may be attributable to industry difficulties or a sectoral split. As a result, a study of a particular industry might add to the literature and the policy-making process.

Previous agricultural entrepreneurship research (Pindado & Sánchez, 2017; Lans et al. 2014; Grande 2011; Lans et al. 2017; Barnes et al. 2015) has focused on its venture performance (for example, Barnes et al. 2015; Senger, Borges, & Machado, 2017; Lans et al. 2014; Grande, 2011; Ali, 2016). Without question, assessing the performance and profitability of an established entrepreneurial business is vital, but this is more a function of management competence than the founding process itself.

This study aims to find out more about the process of starting a new business in the agriculture industry. Evaluating the critical difficulties linked to the entrepreneur and its surroundings is possible based on the determinants measured in previous research. As a result, it has been discovered that the common characteristics of women and men influence the choice to establish a business. Income level, age, and education are the most important of these common characteristics, and they have a significant influence on the choice of the business to be commenced. Furthermore, these factors are uncontroversial and may produce similar outcomes in the agricultural industry.

Also, previous research has found that several other factors have a different influence on women and men in addition to shared features. As a result, numerous research suggests that perceptual features may influence entrepreneurial intent (Anwar & Saleem, 2018). Preliminary findings show women's perceptions of hostile surroundings, with fewer possibilities and higher impediments (Zhao, Seibert, & Hills, 2005). Furthermore, fear of failure is identified as a member of the most critical roadblocks to women entrepreneurs among the perceptual qualities (Wagner, 2007). In recent research, lack of confidence has been identified as another perceptual characteristic that hinders women entrepreneurs, and location dependency is also an essential factor in becoming a woman entrepreneur (Sofer & Saada, 2017). Despite the paucity of research that explores these factors of entrepreneurial aspirations in the same sector, we believe that business formation in the agriculture industry may be generalised.

In terms of social characteristics, previous research has found a strong link between entrepreneurial activity and community concerns, which might be attributed to an individual's social capital (Dahl & Sorenson, 2009; Freytag & Thurik, 2010) Kim and Kang, 2014). As a result, such factors as the existence of networks and the perception of entrepreneurial activity must be considered. Recent studies, notably in the agricultural business, have highlighted how networks impact entrepreneurship (Pindado & Sánchez, 2017).

Female entrepreneurship and the elements that impact it in the industry of agriculture are investigated in this paper, based on the theoretical framework of previous studies focusing on various other sectors such as the sector of agriculture (Pindado & Sánchez, 2017) and sector of tourism (Ramos Rodrguez et al., 2012; Lado-Sestayo & Vivel-Bua, 2018), among others. From the standpoint of essential factors considered as determinants, representative drivers of this amplitude are presented, and prior research findings on the influence of variables in the environment and culture. This study looks at similar elements found in earlier studies about factors of women's and men's entrepreneurship, such as education, income, age, and additional characteristics of individuals linked to Factors affecting perception and social environment factors. Because no previous research has considered the same conceptual framework and sector, it is critical to validate the drivers of agriculture, looking at entrepreneurship from a gender perspective. In this light, variations in perceived differences may be prompted by differences in external factors, and it is thus essential to study them to minimise prejudice. The findings of this research add to the current body of knowledge on female entrepreneurs and aid in the promotion of Women's agricultural entrepreneurship.

SOCIO-DEMOGRAPHIC FACTORS

Earlier research, especially those that include socio-demographic factors (Smallbone & Welter, 2001; Evans & Jovanovic, 1989), has suggested that an individual's earnings are favourably connected to entrepreneurial decisions since cash availability reduces financial restrictions. Similarly, (Carter, Williams, & Reynolds, 1997; Carter & Rosa, 1998) have concluded that the financial crisis has harmed women's entrepreneurship. Furthermore, Taylor (1996) and (Blanchflower & Oswald, 1998) have demonstrated the significance of employment status, and there is widespread consensus that working people are more interested in starting a business. This has been used by both men and

women in different nations (Minniti, Bygrave, & Autio, 2005; Acs, Arenius, Hay, & Minniti, 2005). Finally, no clear evidence of a link between entrepreneurship and education has been found (Blanchflower, 2004). However, literacy is a prerequisite for men and women who want to establish a business (Reynolds, Bygrave, Autio, & Arenius, 2004). Furthermore, some data suggest that female entrepreneurs have a greater level of education in certain industrialised nations than male entrepreneurs. In other occupations, women have a considerably higher education than men (Cowling & Taylor, 2001). Upon this foundation of the study, we propose:
H1: Income, job status, education, and age have all been significant in affecting women's tendency to establish new companies across all nations in our sample.

Entrepreneurial Cognition

Mental processes, such as attitudes and motivation, are considered cognitive elements (Krueger, 2003; Saleem, Arafat, Balhareth, Hussain, & Dwivedi, 2021). Individuals employ these cognitive processes to receive information and execute diverse tasks, such as decision-making and problem-solving. According to (Mitchell et al., 2002 p. 97), perceptions of entrepreneurship or cognitions are "the knowledge structures that people use to make assessment, judgments or decisions involving opportunity evaluation, venture creation, and growth". Moreover, perceptions are an individual's mental understanding of the inputs received from the outside world. Perceptions, to put it another way, are mental representations of the external environment. Or we can say, mind's presentation of the external environment are called perceptions. The information provided by the sense organs is processed and interpreted by the mind. However, different persons may not process and analyse the same information in the same way, and the same data may be understood differently by the same person at various times. It might occur as a result of a shift in learning and conditioning. Entrepreneurial opportunity (Shane & Venkataraman, 2000), entrepreneurial alertness (Kirzner, 1979), perceived desirability (Shapero & Sokol, 1982), perceived behavioural control (Ajzen, 1991), entrepreneurial alertness (Kirzner, 1979), and perceived self-efficacy (Bandura, 1977) are all used to study the entrepreneurial process in the entrepreneurial cognition approach. These beliefs can assist in understanding why entrepreneurs launch new firms.

Perceived Opportunity

Compared to others, a person who sees the opportunity with more clarity increases the likelihood of becoming an entrepreneur. Shane & Venkataraman (2000), in this regard, (Kirzner, 1973, 1979, and Venkataraman (1997) defined origins of entrepreneurship as a study of opportunity, with opportunity recognition accounting

for most of the fundamental terms and aspects of entrepreneurism. Many studies have emphasised the importance of identifying opportunities for entrepreneurship. It should be noted that because women's socialisation processes differ from men's, their views of recognising entrepreneurial opportunities will also vary (DeTienne & Chandler, 2007). Some research, similar to those carried out by (Baughn, Chua,& Neupert, 2006;Langowitz, & Minniti, 2007;Baron, Markman, & Hirsa, 2001), believe that women's conventional roles are less desirable for them than men's in terms of fostering entrepreneurial activity. (Baughn et al. 2006; Baron et al. 2001; Langowitz, & Minniti, 2007). Other researchers have found that gender-based responsibilities for entrepreneurs lead to less entrepreneurial possibilities for women (Mousa & Wales, 2012; Wilson, Barker, & Brigandt, 2007; Schiller & Crewson, 1997and Yang, Li, &Wang, 2020). Similarly, research shows that more entrepreneurship opportunities are unavailable to women since the traditional environment favours their male counterparts (Nilsson 1997; Ahl 2002). As a result, the following proposal is made:
H2: Perceived opportunity enhance the likelihood of becoming a female agriculture entrepreneur.

Perceived Risk

Many experts agree that individuals are risk-averse; nonetheless, entrepreneurs' societal and economic roles are essential, and risk aversion is not an option (Shepherd & Douglas, 1999; Ekelund, Johansson, Järvelin, & Lichtermann, 2005; Kirby, 2004; Bhasin, 2012; Battistella, Biotto, & Toni, 2012).Every business strategy has the chance of failure, as well as additional risks. The capacity to take risks is the most crucial component of entrepreneurial purpose when turning a business concept into reality (Minniti & Nardone, 2007; Cegarra-Navarro, Sánchez-Vidal, & Cegarra-Leiva, 2011). One of the perceptual factors is the fear of failing linked to an individual's capacity to achieve goals and assess potential threats (Dinur 2011; Conroy, Willow, & Metzler, 2002). According to the literature, females have a higher fear of failure than males, explaining the gender difference in entrepreneurship (Wagner, 2007; Koellinger, Minniti, & Schade, 2007). As a result, we propose:
H3: The Perceived risk lowers the likelihood of becoming an agriculture entrepreneur who is female.

Perceived Capability

The relevance of female entrepreneurs' perceived talents is acknowledged in the literature. These self-perceived abilities are connected to a desire to start a business. (Duval-Couetil, Reed-Rhoads, & Haghighi, 2012;Boyd & Vozikis, 1994) and (Minniti & Nardone, 2007) play a significant influence in creating a firm. According

to previous research, men have a greater degree of capability than women (Wilson et al., 2007; Scherer, Brodzinski, & Wiebe, 1990; Curado, Henriques, & Bontis, 2011; Chen, Greene, & Crick, 1998) and this impression influences their intentions and behaviour in terms of entrepreneurship. It might be seen as a lack of female entrepreneurs compared to males, particularly in traditional industries (Kuip, & Verheul, 2003). Moreover, when exposed to male entrepreneur stereotypes, proactive female personalities struggle to recognise their entrepreneurial abilities, resulting in a drop in their entrepreneurial intentions (Huarng, & Yu, 2011; Gupta & Bhawe, 2007). Furthermore, according to other research, males have a more favourable opinion of their abilities than women. In contrast to their male colleagues, women have more entrepreneurial skills and trustworthiness (Menzies & Tatroff, 2006; Gatewood, Shaver, Powers, & Gartner, 2002). As a result, here's a hypothesis:
H4: Perceived capabilities enhance the likelihood of becoming a female agricultural entrepreneur.

Social Capital

Many authors (e.g., Aldrich, Zimmer, & Jones, 1986; Liao & Welsch, 2005; Afandi et al., 2017; Ramos-Rodrguez et al., 2012; Neira, Calvo, Fernández, & Portela, 2017; Davidsson & Honig, 2003) recognise the significance of factors of social capital determining entrepreneurial venture formation intentions and behaviour, as well as their eventual success or failure. According to two closely related ideas, people benefit from others in their social networks or groups (social capital theory and network theory) (Portes, 1998; Lin et al., 1981). Emerson (1972) defined social capital as "the facilitation of social interactions, the sharing of essential resources, and the valuable knowledge required for the start-up of a new firm" (Davidsson & Honig, 2003). We have added two aspects to the concept of social capital for this reason.

Networking

Human and social capital have also been used to investigate socio-demographic variables. (Davidsson & Honig, 2003) discovered that social capital is a reliable predictor of the chance of starting a business. Greene (2002) looked at self-employed women's social and human capital and found that social capital, such as knowing other entrepreneurs, substantially impacted entrepreneurial ambitions. In this area, formal and informal networks have been shown to affect women's EI (H. E. Aldrich & Martinez, 2001; Aldrich 1999) and forecast the relevance of role models for entrepreneurial decisions (Wagner & Sternberg, 2004; Kourilsky & Walstad, 1998). According to Langowitz, Sharpe, & Godwyn, (2006),women at various phases of

the entrepreneurial process place a high value on networks and role models. As a result, we propose
H5: Networking is positively linked with women's tendency to start new enterprises across all nations in our sample.

METHODOLOGY

This study examines the relevance of several factors of the entrepreneurial process based on the Global Entrepreneurship Monitor's data (GEM). The authors used a total of 121,218 interviews conducted with individuals (18–65 years of age) gathered in the project of GEM to evaluate the assumptions mentioned above. Reynolds, Bosma, Autio, Hunt, Bono, Servais, & Chin, (2005),for example, provide details on the processes used to collect and reconcile GEM data (2005). A questionnaire developed to evaluate the behaviour of young entrepreneurs was used in this study, which organisations conducted with experience in public opinion and market research. 33,711 people only were taken into account in the final sample due to missing data at the individual level.

Dependent Variable

An early-stage agribusiness entrepreneur is the study's dependent variable. We describe an agricultural entrepreneur as actively involved in starting a new venture in the agricultural sector, as defined by the GEM industry is termed as Agro-preneur. The ISIC (International Standard Industrial Classification) – 4th revision is used to determine the agriculture industry. The start-up entrepreneurs were evaluated based on binary variable value, i.e., if the respondent is involved in agricultural activity, including forestry and fishing, value 1 was awarded while otherwise a 0 value.

Independent Variables

All social, demographic, and cognitive, and responses of interviews are gathered in the following manner:

Factors of Demography

Year of Birth: Respondents were asked to indicate their birth year.

Income: "Lower 33 percentile," "Middle 33 percentile," & "Upper 33 percentile" was the response categories. The category for referencing was set at the upper 33 percentile.

The survey participants were requested to indicate their education of the highest level. Their replies were grouped into a five-category variable in each of the participating nations. A bachelor's degree from a university or a higher education institution is required, post-secondary, secondary degree, some secondary, and none are the five categories. The first group is chosen as the reference category in the logistic regression analysis.

Work Status: "not working, "student or retired, "Full: full or part-time" were the respondent categories. The final group was chosen as the reference group in the logistic regression analysis.

Factors of Cognition

Perception of an Opportunity: The response to the question "In the next six months, there will be strong chances to establish new enterprises in the region where you reside" a dichotomous variable was used to measure with a value of 1 if the respondent says yes, and 0 if they say no.

If the person answers yes in response to the question "Fear of failure would be a brake for you if you were to establish a business," this binary variable is 1; otherwise, it is 0. Capacity as perceived: If the responder says yes to the question "You have the requisite knowledge, abilities, and experience to start a new business," the dichotomous variable is 1; otherwise, it is 0.

The Factor of Social Capital

Knowing Other Entrepreneurs: If the responder says yes to the question "You know personally someone who has launched a new business in the last two years," the dichotomous variable is 1; otherwise, it is 0.

Logistic Regression

For predicting and explaining a binary categorical dependent variable, an excellent econometric approach would be logistic regression. Logistic regression is used to assess the influence of a collection of independent factors on a single non-metric (binary) dependent variable. The logistic regression variate resembles the multiple regressions variate (Hair, Celsi, Ortinau, & Bush, 2010).

X1+X2+X3+……+Xn = Y1

(non-metric and metric) (binary non-metric)

The most appropriate model for the relationship between dependent variables' characteristics and the predictors is developed using logistic regression. The logistic regression coefficients predict the logistic transformation of the probability of the presence of relationship features. We adopt the logistic regression technique in this empirical investigation since the early-stage entrepreneur (dependent variables) is a dichotomous dependent variable.. Most independent factors are categorical or dichotomous; we employ the logistic regression approach in this empirical study.

RESULTS AND DISCUSSION

Our research was separated into two sections: correlation and regression (logit model).

Table 1. Correlation

	1	2	3	4	5	6	7	8	9
1. Agricultural start up	1								
2. Age	-0.02	1							
3. Income	0.08	.10*	1						
4. Education	0.02	.15**	.12**	1					
5. Working status	.23**	-0.1	-0.00	.10*	1				
6. Perceived opportunity	.14**	-.11*	-0.01	-.21**	-0.07	1			
7. Perceived risk	-.10*	0.04	-0.01	0.08	0.02	-.13**	1		
8. Perceived capability	.18**	-0.1	0.04	0.03	-0.03	.23**	-.32**	1	
9. Entrepreneurial network	0.05	-0.1	.10*	-0.03	-.12**	.27**	-.10*	.20**	1
**. Correlation at 0.01 level of significance. *. Correlation at 0.05 level of significance.									

Logistic Regression

Risk perception only is not significant among all the perceptual variables. As a result, this finding does not support the proposed link between risk perception and early-stage entrepreneurial activity. As a result, hypothesis 3 is rejected. Since earlier research (Pindado & Sánchez, 2017; Arafat, et al. 2020a) studying this connection in the agricultural business has found that perception of risk acts in a different way of the agricultural sector, this finding suggests that perception of risk behaves differently in hotel and industry of restaurant and the agriculture sector, (Ramos Rodrguez et al., 2012), and (Arafat & Saleem, 2017a, 2017b; Neira et al., 2017) generic entrepreneurship, have discovered a similar connection.

As a result, becoming an agricultural entrepreneur has a favourable and significant relationship with opportunity perception. It backs up hypothesis 2. Individuals looking for chances are 89% more prone to be agricultural entrepreneurs than those blind to possibilities. According to this variable has an odds ratio of 1.893. This finding is consistent with studies looking at the same relationship in the context of agriculturial entrepreneurship (Arafat et al. 2020b; Pindado & Sánchez, 2017), as well as other studies looking at this relationship in the context of different types of entrepreneurs (Hunjo2015; Tsai, Chang, & Peng, 2016; Lián, Santos, & Fernández, 2011; Ramos-Rodríguez et al., 2012; Arenius & Minniti, 2005; Arafat & Saleem, 2017b; Ahmad, Xavier, & Bakar, 2014). Individuals' entrepreneurial inclination is influenced by their perceived capacity or confidence in their abilities and skill to start a new firm. This result backs up H4. This variable has a 2.530 odds ratio, showing that confident persons in their capacities are 1.53% more likely to start an agribusiness. Previous research (Arafat & Saleem, 2017b; Ramos-Rodrguez et al.,

Table 2 Logistic regression

	Sig.	Exp(B)
Demographic factors		
Age	0.06	1.02
Income	0.26	
-Middle income	0.16	0.65
-Upper income	0.12	0.60
Education level	0.52	
-primary	0.14	0.20
-Secondary	0.12	0.18
-Senior secondary	0.12	0.18
-Graduate or higher	0.21	0.25
Working status	0	
-Not working	0.34	0.33
-Retired or students	0.34	3.36
Cognition		
Perceived opportunity	0.01	1.9
Perceived risk	0.80	0.93
Perceived capability	0.002	2.50
Social capital		
Networking	0.75	1.079
Constant	0.4	4.6

2012; Tsai, Nie, Blancon, Stoumpos, Asadpour, Harutyunyan, & Mohite, 2016; Arafat, Ali, Dwivedi & Saleem, 2020c; Arenius & Minniti, 2005; Paray, Arafat, & Saleem, 2021) has shown similar results. In terms of social capital variables, the findings suggest that knowing other entrepreneurs has little impact on the likelihood of becoming an entrepreneur, contradicting H5.

This result appears to contradict the agriculture entrepreneurship research findings (Pindado & Sánchez, 2017) and entrepreneurship in many forms (Arenius & Minniti, 2005; Ramos-Rodrguez et al., 2012; Arafat & Saleem, 2017a; Neira et al. 2017). We can't discover any evidence that demographic variables have an impact on agricultural entrepreneurship.

IMPLICATION

Agriculture is the "foundation business" for other industries, accounting for one-third of the world gross domestic product in 2014 (GDP). According to a 2016 World Bank estimate, agriculture employs 65% of impoverished working adults, and by 2050, agriculture will feed 9.7 billion people. Furthermore, agriculture plays a critical role in emerging nations such as India, the world's second-largest agricultural goods producer the world's second-largest agricultural goods producer, accounting for 7.68% of world output. The current study focuses on forming new businesses in the agricultural sector, and it is possible to explain the entrepreneurial propensity of women to a more significant level. According to this study, the primary predictors of women's entrepreneurial inclination in agriculture are cognitive variables such as opportunity perception and perceived capabilities (self-efficacy). Policymakers in all countries should pay attention to these two cognitive variables and support women who want to start a business. Only work status has demonstrated a positive relationship with entrepreneurship among demographic variables, indicating that women with experience in agriculture are more likely to become entrepreneurs. Therefore, priority should be given to those women with expertise. Policymakers should enact strong measures to empower women to start businesses in the agricultural sector, which would help the country achieve its socioeconomic goals.

FUTURE DIRECTION AND LIMITATIONS

Global Entrepreneurship Monitor provided secondary data for the current study; nevertheless, the number of elements in this database relating to entrepreneurial ambitions and attitudes is limited. Because of the items contained in the GEM data, more precise statistical approaches, such as structural equation modelling, can't be used to illustrate the various connections between perceptions and intents. A single item was used to measure all of the constructions. Multi-item scales, on the other hand, give more accurate measurements. More research may be done using more precise statistical methodologies and a multi-item scale to propose more practical importance of demographic, cognitive, and social capital variables on women's propensity.

CONCLUSION

Female entrepreneurs are starting more businesses, but there is still a significant difference between their entrepreneurial efforts and men throughout the world. According to studies, researchers have given greater attention to female entrepreneurial characteristics' perceptual and social capital to design future policy for its promotion. Agriculture is the primary industry that contributes to decreasing poverty, increased earnings, and 80% of the world poor now have better food security. The authors' goal in this research article was to use a large sample of Global Entrepreneurship Monitor data of Adult Population Survey to examine the cognitive, social capital, and demographic characteristics of women entrepreneurship in the agriculture business experimentally. The findings that perceived capability and the concept of opportunity perception are the decisive factors in female entrepreneurship have been proven. Risk perception and social networks (knowing other entrepreneurs) were shown to be unimportant, in contrast to earlier research (Kim and Kang, 2014; Wagner, 2007; Freytag & Thurik, 2010; Dahl & Sorenson, 2009). Policymakers should create programmes to help people gain the ability to explain, recognise, and evaluate entrepreneurial opportunities among women entrepreneurs in the agricultural industry

REFERENCES

Acs, Z. J., Arenius, P., Hay, M., & Minniti, M. (2005). *Global Entrepreneurship Monitor. Executive Report 2004*. Global Entrepreneurship Monitor.

Acs, Z. J., Braunerhjelm, P., Audretsch, D. B., & Carlsson, B. (2009). The knowledge spillover theory of entrepreneurship. *Small Business Economics*, *32*(1), 15–30. doi:10.100711187-008-9157-3

Afandi, E., Kermani, M., & Mammadov, F. (2017). Social capital and entrepreneurial process. *The International Entrepreneurship and Management Journal*, *13*(3), 685–716. doi:10.100711365-016-0421-8

Ahl, H. J. (2002). *The making of the female entrepreneur: A discourse analysis of research texts on women's entrepreneurship* (Doctoral dissertation). Internationella Handelshögskolan.

Ahmad, S. Z., Xavier, S. R., & Bakar, A. R. A. (2014). Examining entrepreneurial intention through cognitive approach using Malaysia GEM data. *Journal of Organizational Change Management*, *27*(3), 449–464. doi:10.1108/JOCM-03-2013-0035

Ajzen, I. (1991). The theory of planned behavior. *Organizational Behavior and Human Decision Processes*, *50*(2), 179–211. doi:10.1016/0749-5978(91)90020-T

Aldrich, H., & Martinez, M. (2001). *Many are called, but few are chosen: An evolutionary*. Academic Press.

Aldrich, H. (1999). Promise, Failure, and Redemption: A Life Course Perspective on Teaching as a Career. *The Social Worlds of Higher Education: Handbook for Teaching in a New Century*, 533-42..

Aldrich, H., Zimmer, C., & Jones, T. (1986). Small business still speaks with the same voice: A replication of 'the voice of small business and the politics of survival'. *The Sociological Review*, *34*(2), 335–356. doi:10.1111/j.1467-954X.1986.tb02705.x

Ali, J. (2016). Performance of small and medium-sized food and agribusiness enterprises: evidence from Indian firms. *International Food and Agribusiness Management Review, 19*, 53-64.

Anwar, I., & Saleem, I. (2018). Effect of entrepreneurial education on entrepreneurial intention of Indian students. *International Journal of Research*, *5*(12), 2306–2316.

Anwar, I., Saleem, I., Islam, K. B., Thoudam, P., & Khan, R. (2020). Entrepreneurial intention among female university students: Examining the moderating role of entrepreneurial education. *Journal for International Business and Entrepreneurship Development*, *12*(4), 217–234. doi:10.1504/JIBED.2020.110254

Anwar, I., Thoudam, P., & Saleem, I. (2021). Role of entrepreneurial education in shaping entrepreneurial intention among university students: Testing the hypotheses using mediation and moderation approach. *Journal of Education for Business*, 1–13. doi:10.1080/08832323.2021.1883502

Arafat, M. Y., Ali, J., Dwivedi, A. K., & Saleem, I. (2020a). Social and Cognitive Aspects of Women Entrepreneurs: Evidence from India. *Vikalpa*, *45*(4), 223–239. doi:10.1177/02560909211015457

Arafat, M. Y., Khan, A. M., Saleem, I., Khan, N. A., & Khan, M. M. (2020b). Intellectual and cognitive aspects of women entrepreneurs in India. *International Journal of Knowledge Management Studies, 11*(3), 278–297. doi:10.1504/IJKMS.2020.109092

Arafat, M. Y., & Saleem, I. (2017a). *Determinants of entrepreneurial startups in India: A study using GEM data*. Academic Press.

Arafat, M. Y., & Saleem, I. (2017b). Examining start-up Intention of Indians through cognitive approach: A study using GEM data. *Journal of Global Entrepreneurship Research, 7*(1), 1–11. doi:10.118640497-017-0073-3

Arafat, M. Y., Saleem, I., & Dwivedi, A. K. (2020c). Understanding entrepreneurial intention among Indian youth aspiring for self-employment. *International Journal of Knowledge and Learning, 13*(3), 185–200. doi:10.1504/IJKL.2020.109882

Arafat, M. Y., Saleem, I., Dwivedi, A. K., & Khan, A. (2020d). Determinants of agricultural entrepreneurship: A GEM data based study. *The International Entrepreneurship and Management Journal, 16*(1), 345–370. doi:10.100711365-018-0536-1

Arenius, P., & Minniti, M. (2005). Perceptual variables and nascent entrepreneurship. *Small Business Economics, 24*(3), 233–247. doi:10.100711187-005-1984-x

Bandura, A. (1977). Self-efficacy: Toward a unifying theory of behavioral change. *Psychological Review, 84*(2), 191–215. doi:10.1037/0033-295X.84.2.191 PMID:847061

Barnes, A. P., Hansson, H., Manevska-Tasevska, G., Shrestha, S. S., & Thomson, S. G. (2015). The influence of diversification on long-term viability of the agricultural sector. *Land Use Policy, 49*, 404–412. doi:10.1016/j.landusepol.2015.08.023

Baron, R. A., Markman, G. D., & Hirsa, A. (2001). Perceptions of women and men as entrepreneurs: Evidence for differential effects of attributional augmenting. *The Journal of Applied Psychology, 86*(5), 923–929. doi:10.1037/0021-9010.86.5.923 PMID:11596808

Battistella, C., Biotto, G., & De Toni, A. F. (2012). From design driven innovation to meaning strategy. *Management Decision, 50*(4), 718–743. doi:10.1108/00251741211220390

Baughn, C. C., Chua, B. L., & Neupert, K. E. (2006). The normative context for women's participation in entrepreneruship: A multicountry study. *Entrepreneurship Theory and Practice, 30*(5), 687–708. doi:10.1111/j.1540-6520.2006.00142.x

Bhasin, S. (2012). An appropriate change strategy for lean success. *Management Decision, 50*(3), 439–458. doi:10.1108/00251741211216223

Blanchflower, D. G. (2004). *Self-employment: More may not be better*. Academic Press.

Blanchflower, D. G., & Oswald, A. J. (1998). *Entrepreneurship and the youth labour market problem: A report for the OECD*. Report to OECD.

Boyd, N. G., & Vozikis, G. S. (1994). The influence of self-efficacy on the development of entrepreneurial intentions and actions. *Entrepreneurship Theory and Practice, 18*(4), 63–77. doi:10.1177/104225879401800404

Brush, C. G. (1990). Women and enterprise creation: Barriers and opportunities. *Enterprising women: Local initiatives for job creation, 37*.

Brush, C. G. (1992). Research on women business owners: Past trends, a new perspective and future directions. *Entrepreneurship Theory and Practice, 16*(4), 5–30. doi:10.1177/104225879201600401

Carter, N. M., Williams, M., & Reynolds, P. D. (1997). Discontinuance among new firms in retail: The influence of initial resources, strategy, and gender. *Journal of Business Venturing, 12*(2), 125–145. doi:10.1016/S0883-9026(96)00033-X

Carter, S., & Rosa, P. (1998). Indigenous rural firms: Farm enterprises in the UK. *International Small Business Journal, 16*(4), 15–27. doi:10.1177/0266242698164001

Cegarra-Navarro, J. G., Sánchez-Vidal, M. E., & Cegarra-Leiva, D. (2011). Balancing exploration and exploitation of knowledge through an unlearning context: An empirical investigation in SMEs. *Management Decision*.

Conroy, D. E., Willow, J. P., & Metzler, J. N. (2002). Multidimensional fear of failure measurement: The performance failure appraisal inventory. *Journal of Applied Sport Psychology, 14*(2), 76–90. doi:10.1080/10413200252907752

Cowling, M., & Taylor, M. (2001). Entrepreneurial women and men: Two different species? *Small Business Economics, 16*(3), 167–175. doi:10.1023/A:1011195516912

Curado, C., Henriques, L., & Bontis, N. (2011). Intellectual capital disclosure payback. *Management Decision, 49*(7), 1080–1098. doi:10.1108/00251741111151154

Dahl, M. S., & Sorenson, O. (2009). The embedded entrepreneur. *European Management Review, 6*(3), 172–181. doi:10.1057/emr.2009.14

Davidsson, P., Delmar, F., & Wiklund, J. (2006). Entrepreneurship as growth; growth as entrepreneurship. *Entrepreneurship and the Growth of Firms, 1*(2), 21–38.

Davidsson, P., & Honig, B. (2003). The role of social and human capital among nascent entrepreneurs. *Journal of Business Venturing, 18*(3), 301–331. doi:10.1016/S0883-9026(02)00097-6

DeTienne, D. R., & Chandler, G. N. (2007). The role of gender in opportunity identification. *Entrepreneurship Theory and Practice, 31*(3), 365–386. doi:10.1111/j.1540-6520.2007.00178.x

Dinur, A. R. (2011). Common and un-common sense in managerial decision making under task uncertainty. *Management Decision.*

Dubé, L., Pingali, P., & Webb, P. (2012). Paths of convergence for agriculture, health, and wealth. *Proceedings of the National Academy of Sciences of the United States of America, 109*(31), 12294–12301. doi:10.1073/pnas.0912951109 PMID:22826252

Duval-Couetil, N., Reed-Rhoads, T., & Haghighi, S. (2012). Engineering students and entrepreneurship education: Involvement, attitudes and outcomes. *International Journal of Engineering Education, 28*(2), 425.

Ekelund, J., Johansson, E., Järvelin, M. R., & Lichtermann, D. (2005). Self-employment and risk aversion—Evidence from psychological test data. *Labour Economics, 12*(5), 649–659. doi:10.1016/j.labeco.2004.02.009

Emerson, R. M. (1972). Exchange theory, part II: Exchange relations and networks. *Sociological Theories in Progress, 2*, 58-87.

Evans, D. S., & Jovanovic, B. (1989). An estimated model of entrepreneurial choice under liquidity constraints. *Journal of Political Economy, 97*(4), 808–827. doi:10.1086/261629

Freytag, A., & Thurik, R. (2010). Introducing entrepreneurship and culture. In *Entrepreneurship and culture* (pp. 1–8). Springer. doi:10.1007/978-3-540-87910-7_1

Gatewood, E. J., Shaver, K. G., Powers, J. B., & Gartner, W. B. (2002). Entrepreneurial expectancy, task effort, and performance. *Entrepreneurship Theory and Practice, 27*(2), 187–206. doi:10.1111/1540-8520.00006

Grande, J. (2011). New venture creation in the farm sector–Critical resources and capabilities. *Journal of Rural Studies, 27*(2), 220–233. doi:10.1016/j.jrurstud.2011.02.003

Gupta, V. K., & Bhawe, N. M. (2007). The influence of proactive personality and stereotype threat on women's entrepreneurial intentions. *Journal of Leadership & Organizational Studies, 13*(4), 73–85. doi:10.1177/10717919070130040901

Hair, J. F., Celsi, M., Ortinau, D. J., & Bush, R. P. (2010). *Essentials of marketing research* (Vol. 2). McGraw-Hill/Irwin.

Hassan, A., Saleem, I., Anwar, I., & Hussain, S. A. (2020). Entrepreneurial intention of Indian university students: The role of opportunity recognition and entrepreneurship education. *Education + Training*, *62*(7/8), 843–861. doi:10.1108/ET-02-2020-0033

Huarng, K. H., & Yu, T. H. K. (2011). Entrepreneurship, process innovation and value creation by a non-profit SME. *Management Decision*, *49*(2), 284–296. doi:10.1108/00251741111109160

Hunjo, H. (2015). A Critical Discourse Evaluation of Decolonisation and Democratisation: Issues in Africa as Exemplified in Soyinka's Non-fictional Texts. *African Literature and the Future*, 49.

Kelley, D., Singer, S., & Herrington, M. (2016). Global entrepreneurship monitor. Global Report, Global Entrepreneurship Research Association, London Business School.

Khan, A. M., Arafat, M. Y., Raushan, M. A., Khan, M. S., Dwivedi, A. K., Khan, M. M., & Saleem, I. (2019b). Role of intellectual capital in augmenting the start-up intentions of Indians—An analysis using GEM data. *Journal of Global Entrepreneurship Research*, *9*(1), 1–14. doi:10.118640497-019-0146-6

Khan, A. M., Arafat, M. Y., Raushan, M. A., & Saleem, I. (2020). Examining the relevance of intellectual capital in improving the entrepreneurial propensity among Indians. *International Journal of Knowledge Management*, *16*(1), 123–140. doi:10.4018/IJKM.2020010106

Khan, A. M., Arafat, M. Y., Raushan, M. A., Saleem, I., Khan, N. A., & Khan, M. M. (2019a). Does intellectual capital affect the venture creation decision in India? *Journal of Innovation and Entrepreneurship*, *8*(1), 1–15. doi:10.118613731-019-0106-y

Khan, A. M., Arafat, M. Y., Raushan, M. A., Saleem, I., Khan, N. A., & Khan, M. M. (2019c). Does intellectual capital affect the venture creation decision in India? *Journal of Innovation and Entrepreneurship*, *8*(1), 1–15. doi:10.118613731-019-0106-y

Kirby, D. A. (2004). Entrepreneurship education: Can business schools meet the challenge? *Education + Training*, *46*(8/9), 510–519. doi:10.1108/00400910410569632

Kirzner, I. M. (1979). *Competition & Entrepreneurship*. University of Chicago Press.

Koellinger, P., Minniti, M., & Schade, C. (2007). "I think I can, I think I can": Overconfidence and entrepreneurial behavior. *Journal of Economic Psychology*, *28*(4), 502–527. doi:10.1016/j.joep.2006.11.002

Kourilsky, M. L., & Walstad, W. B. (1998). Entrepreneurship and female youth: Knowledge, attitudes, gender differences, and educational practices. *Journal of Business Venturing*, *13*(1), 77–88. doi:10.1016/S0883-9026(97)00032-3

Krueger, N. F. (2003). The cognitive psychology of entrepreneurship. In *Handbook of entrepreneurship research* (pp. 105–140). Springer. doi:10.1007/0-387-24519-7_6

Lado-Sestayo, R., & Vivel-Bua, M. (2018). Women and Entrepreneurship: An Analysis of the Determinants in the Tourism Sector. In Nascent Entrepreneurship and Successful New Venture Creation (pp. 203-214). IGI Global.

Langowitz, N., & Minniti, M. (2005). Gender differences and nascent entrepreneurship. *Documento de trabajo*.

Langowitz, N., & Minniti, M. (2007). The entrepreneurial propensity of women. *Entrepreneurship Theory and Practice*, *31*(3), 341–364. doi:10.1111/j.1540-6520.2007.00177.x

Langowitz, N. S., Minniti, M., & Arenius, P. (2005). *Global entrepreneurship monitor: 2004 report on women and entrepreneurship*. University of Illinois at Urbana-Champaign's Academy for Entrepreneurial Leadership Historical Research Reference in Entrepreneurship.

Langowitz, N. S., Minniti, M., & Arenius, P. (2005). *Global entrepreneurship monitor: 2004 report on women and entrepreneurship*. University of Illinois at Urbana-Champaign's Academy for Entrepreneurial Leadership Historical Research Reference in Entrepreneurship.

Lans, T., Blok, V., & Wesselink, R. (2014). Learning apart and together: Towards an integrated competence framework for sustainable entrepreneurship in higher education. *Journal of Cleaner Production*, *62*, 37–47. doi:10.1016/j.jclepro.2013.03.036

Lans, T., Seuneke, P., & Klerkx, L. (2017). Agricultural entrepreneurship. In *Encyclopedia of creativity, invention, innovation and entrepreneurship* (pp. 1–7). Springer. doi:10.1007/978-1-4614-6616-1_496-2

Liao, J., & Welsch, H. (2005). Roles of social capital in venture creation: Key dimensions and research implications. *Journal of Small Business Management*, *43*(4), 345–362. doi:10.1111/j.1540-627X.2005.00141.x

Liñán, F., Santos, F. J., & Fernández, J. (2011). The influence of perceptions on potential entrepreneurs. *The International Entrepreneurship and Management Journal, 7*(3), 373–390. doi:10.100711365-011-0199-7

Menzies, T. V., & Tatroff, H. (2006). The propensity of male vs. female students to take courses and degree concentrations in entrepreneurship1. *Journal of Small Business and Entrepreneurship, 19*(2), 203–223. doi:10.1080/08276331.2006.10593367

Minniti, M., & Nardone, C. (2007). Being in someone else's shoes: The role of gender in nascent entrepreneurship. *Small Business Economics, 28*(2), 223–238. doi:10.100711187-006-9017-y

Mitchell, R. K., Busenitz, L., Lant, T., McDougall, P. P., Morse, E. A., & Smith, J. B. (2002). Toward a theory of entrepreneurial cognition: Rethinking the people side of entrepreneurship research. *Entrepreneurship Theory and Practice, 27*(2), 93–104. doi:10.1111/1540-8520.00001

Mousa, F. T., & Wales, W. (2012). Founder effectiveness in leveraging entrepreneurial orientation. *Management Decision, 50*(2), 305–324. doi:10.1108/00251741211203588

Neira, I., Calvo, N., Fernández, L., & Portela, M. (2017). Entrepreneur: Do social capital and culture matter? *The International Entrepreneurship and Management Journal, 13*(2), 665–683. doi:10.100711365-016-0418-3

Nilsson, T. (1997). *Kundageranden i små tillverkande företag* (PhD diss.). Luleå University of Technology, Luleå.

Pindado, E., & Sánchez, M. (2017). Researching the entrepreneurial behaviour of new and existing ventures in European agriculture. *Small Business Economics, 49*(2), 421–444. doi:10.100711187-017-9837-y

Portes, A. (1998). Social capital: Its origins and applications in modern sociology. *Annual Review of Sociology, 24*(1), 1–24. doi:10.1146/annurev.soc.24.1.1

Ramos-Rodríguez, A. R., Medina-Garrido, J. A., & Ruiz-Navarro, J. (2012). Determinants of hotels and restaurants entrepreneurship: A study using GEM data. *International Journal of Hospitality Management, 31*(2), 579–587. doi:10.1016/j.ijhm.2011.08.003

Reynolds, P. D., Bygrave, W. D., Autio, E., & Arenius, P. (2004). *GEM 2003 global report*. Babson College.

Saleem, I., Arafat, M. Y., Balhareth, H. H., Hussain, A., & Dwivedi, A. k. (2021). Cognitive aspects of disadvantaged entrepreneurs: Evidence from India. *Community Development (Columbus, Ohio)*. Advance online publication. doi:10.1080/15575330.2021.197

Scherer, R. F., Brodzinski, J. D., & Wiebe, F. A. (1990). Entrepreneur career selection and gender: A socialization approach. *Journal of Small Business Management*, *28*(2), 37.

Schiller, B. R., & Crewson, P. E. (1997). Entrepreneurial origins: A longitudinal inquiry. *Economic Inquiry*, *35*(3), 523–531. doi:10.1111/j.1465-7295.1997.tb02029.x

Schmidhuber, J., & Tubiello, F. N. (2007). Global food security under climate change. *Proceedings of the National Academy of Sciences of the United States of America*, *104*(50), 19703–19708. doi:10.1073/pnas.0701976104 PMID:18077404

Senger, I., Borges, J. A. R., & Machado, J. A. D. (2017). Using the theory of planned behavior to understand the intention of small farmers in diversifying their agricultural production. *Journal of Rural Studies*, *49*, 32–40. doi:10.1016/j.jrurstud.2016.10.006

Shapero, A., & Sokol, L. (1982). *The social dimensions of entrepreneurship*. University of Illinois at Urbana-Champaign's Academy for Entrepreneurial Leadership Historical Research Reference in Entrepreneurship.

Shepherd, D. A., & Douglas, E. J. (1999). Attracting Equity Investors. *Sage (Atlanta, Ga.)*.

Smallbone, D., & Welter, F. (2001). The distinctiveness of entrepreneurship in transition economies. *Small Business Economics*, *16*(4), 249–262. doi:10.1023/A:1011159216578

Sofer, M., & Saada, M. T. (2017). Entrepreneurship of women in the rural space in Israel: Catalysts and obstacles to enterprise development. *Sociologia Ruralis*, *57*, 769–790. doi:10.1111oru.12125

Taylor, A. M. (1996). *International capital mobility in history: the saving-investment relationship*. Academic Press.

Tsai, H., Nie, W., Blancon, J. C., Stoumpos, C. C., Asadpour, R., Harutyunyan, B., Neukirch, A. J., Verduzco, R., Crochet, J. J., Tretiak, S., Pedesseau, L., Even, J., Alam, M. A., Gupta, G., Lou, J., Ajayan, P. M., Bedzyk, M. J., Kanatzidis, M. G., & Mohite, A. D. (2016). High-efficiency two-dimensional Ruddlesden–Popper perovskite solar cells. *Nature*, *536*(7616), 312–316. doi:10.1038/nature18306 PMID:27383783

Tsai, K. H., Chang, H. C., & Peng, C. Y. (2016). Refining the linkage between perceived capability and entrepreneurial intention: Roles of perceived opportunity, fear of failure, and gender. *The International Entrepreneurship and Management Journal, 12*(4), 1127–1145. doi:10.100711365-016-0383-x

Van der Kuip, I., & Verheul, I. (2003). *Early development of entrepreneurial qualities: the role of initial education* (No. N200311). EIM Business and Policy Research.

Venkataraman, S., & Shane, S. (2000). The promise of entrepreneurship as a field of research. *Academy of Management Review, 25*(1), 217–226.

Wagner, J. (2007). What a difference a Y makes-female and male nascent entrepreneurs in Germany. *Small Business Economics, 28*(1), 1–21. doi:10.100711187-005-0259-x

Wagner, J., & Sternberg, R. (2004). Start-up activities, individual characteristics, and the regional milieu: Lessons for entrepreneurship support policies from German micro data. *The Annals of Regional Science, 38*(2), 219–240. doi:10.100700168-004-0193-x

Wilson, R. A., Barker, M. J., & Brigandt, I. (2007). When traditional essentialism fails: Biological natural kinds. *Philosophical Topics, 35*(1/2), 189–215. doi:10.5840/philtopics2007351/29

Yang, M. M., Li, T., & Wang, Y. (2020). What explains the degree of internationalization of early-stage entrepreneurial firms? A multilevel study on the joint effects of entrepreneurial self-efficacy, opportunity-motivated entrepreneurship, and home-country institutions. *Journal of World Business, 55*(6), 101114. doi:10.1016/j.jwb.2020.101114

Yaseen, A., Saleem, M. A., Zahra, S., & Israr, M. (2018). Precursory effects on entrepreneurial behaviour in the agri-food industry. *Journal of Entrepreneurship in Emerging Economies.*

Zhao, H., Seibert, S. E., & Hills, G. E. (2005). The mediating role of self-efficacy in the development of entrepreneurial intentions. *The Journal of Applied Psychology, 90*(6), 1265–1272. doi:10.1037/0021-9010.90.6.1265 PMID:16316279

Chapter 11
Analyzing the Role of Micro-Entrepreneurship and Self-Help Groups (SHGs) in Women Empowerment and Development:
A Bottom-of-Pyramid Perspective

Veer P. Gangwar
Lovely Professional University, India

Shad Ahmad Khan
https://orcid.org/0000-0001-7593-3487
University of Buraimi, Oman

ABSTRACT

In India, societal development is a cynosure, and thus, it is not a truism but a reality. Women's entrepreneurship emerging through self-help groups (SHGs) contributes to the economic well-being, sustainability, and in poverty reduction. The growth of SHGs is evidence in itself. In the recent decade, micro enterprises and SHGs have come up as prominent solutions to the entrepreneurial crisis existing in the country, and their role in empowering women and their development needs to studied. This study uses a descriptive statistic, reliability, and correlation analysis through SPSS and structural equation modeling (SEM) as an analytical tool to explore linkages between empowerment effected by SHG and micro-entrepreneurship. The research study results show that the contributory role of women entrepreneurs to the society is considerably worth appreciation. The causal relationship has also surfaced demonstrating the connection between women empowerment and development brought about by SHG and micrŏ-entrepreneurship from a bottom-of-pyramid perspective.

DOI: 10.4018/978-1-6684-2349-3.ch011

INTRODUCTION

Societal Development is strongly linked with the income generating capacity of its constituent members. In a country like India, it is not a truism but a reality. Though India has been demonstrating fair economic growth in the last couple of decades yet its rank in the Human Development Index (HDI) has not increased considerably; instead it has been centering round 130. Despite all this, one of the most important challenges being faced by India is eradication of poverty. The poverty is increasing chasm between male and female.

In terms to fight against poverty, encouragement of entrepreneurship among women, particularly of rural areas emerged as an effective way. New business is considered as a crucial driver of economy (Schumpeter, 1936). Entrepreneurship is often considered as a mechanism for socio-economic development, poverty reduction, and generation of employment (Galindo and Méndez-Picazo, 2013).

Omoruyi, et.al. (2017) shows that entrepreneurship positively explains the variations in the growth and further it is reasonable to contend that entrepreneurship in developing economies is instrumental to unlocking economic growth, create employment and reduce poverty.Employment creation in rural areas by launching entrepreneurship will help people to earn their daily bread and also result in inclusive growth.

The "Base of Pyramid (BoP) 2.0" model developed by London and Hart (2011) briefs about various openings even to the last person of the society in the pyramid model, aiming at earning US2$ per day. This concept helps us to give ideas for involving people who are in need of opportunities for their overall development and socio-economic progress of our country. In 2012, Hall, Matos, Sheehan, & Silvestrequoted that "Generating entrepreneurial opportunities for people at the base of the pyramid is important for inclusive growth". In a country like India, generating employment facilities at BoP by undertaking various entrepreneurship/business activities will lead to the inclusive development of India and this will be more effective, if the employment creation is women-centric particularly in agriculture dominant areas.

Many researchers have proven that women played a key role in economic growth of the country and they will remain playing this major role in the future too. If there is equality between men and women, the economy can grow faster following the reduction in poverty.

Lortie,Castrogiovanni, and Cox (2016), reported that 'n' of firms launched and spearheaded by womenfolk for the last two decades have grabbed the attention of many practitioners and scholars. Such firms played an important role in proving that 'gender' has a greater impact on the social performance of any organization.

The Gross Domestic Product (GDP) growth rate measures the growth of the economy and economic output of a country. Many studies have reported that the GDP growth rate can be increased by as much as 34% in the countries, with increasing female-to-male employment levels. A report from the United Nations Industrial Development Organization (UNIDO) says—with women-to-men employment levels, the productivity may be increased by as much as 25%. The reason behind this is—when a woman starts earning, she spends her money to provide education to her children and to look after them in their needs, Doepke and Terlilt (2011) said adding that to stimulate economic growth, women create human capital as well as jobs.

As per the UNIDO report, "As many as 870 million women will enter the economic mainstream for the first time both as consumers, producers, employees, and entrepreneurs at the end of 2020". The economic influence of this fact can be witnessed significantly on the overall development process. Duflo (2011) emerged with a concept of "Bi-directional relationship between empowerment and economic development". The researcher found that both empowerment and economic development are parallel to each other and entrepreneurship can stimulate both women empowerment and economic development.

It is the fact that both male and female come across hurdles in establishing the enterprises, however, women, especially those with rural backgrounds face more obstacles, as they lack access to proper educational facilities, face financial problems and other issues. In such cases, Self-help groups (SHGs) formed across India with social and economic purposes are emerging successful in solving some of these problems.

SHG, an informal group of people facing similar problems and trying to meet their demands in one particular area or region are helping a large number of people, particularly rural women in a large manner. They have become an effective tool for eradication of poverty from society and an instrument for social and economic upliftment of every member of the group.

As SHGs are groups with a minimum of 10 to 20 members who strive hard to seek solutions to their common problems, start saving a small portion of their income in the name of SHG. Every member of the group invests and contributes an equal amount on a regular basis to develop a habit of small savings, with an objective to help the needy in emergency conditions, which ultimately facilitates access to credit.

On one hand, the members can fulfil their needs and demands with the money saved by them and on the other hand, they can witness the growing credit amount every week or month. This encourages them to establish their small scale enterprise to be self-reliant and economically stable.

The "Micro, Small & Medium Enterprises Development Act (2006)" defines micro-enterprise as a small business. If it is a manufacturing sector, wherein the investment should not exceed Rs 25 lakh (including both plant and machinery) and if it is a service sector, the investment should not exceed Rs10 lakh.

Empowerment of the female community is being measured with their ability to make decisions about their kids' education, their marriage, make their own decisions, access the properties and resources available in their surroundings.

Many researchers have conducted detailed studies on the empowerment aspects of self-help groups and most of their studies were focused on the role of the SHGs in empowering women. Priority was also given to study the effect of empowerment on economic development of the country. But it is a matter of concern that not many studies were carried out to show how the women empowered with the help of SHGs are shaping their career in small-scale business as micro-entrepreneurs.

The research carried by us conducts the studies in various ways. Our study mainly focuses on various ways and means being implemented by the SHGs to empower women. So that they become micro-entrepreneurs.

The study strands are interwoven with the theory and development of hypotheses, model of research, research methodology used, discussion of empirical results, limiting factors of the research, future scope and discussion of results followed by findings.

THEORETICAL FRAMEWORK AND DEVELOPMENT OF HYPOTHESES

The National Bank for Agriculture and Rural Development (NABARD), India started the Self-Help Group (SHG) as a pilot project in 1992, with a basic objective of giving surety-free credit to the poor.

The amount collected after the contribution of every member of the SHG, can be used for lending it to one of the members to accomplish their needs and to facilitate any member to take up any small business. As per the Act, there is a provision of adding up to 20 people in one SHG and these groups will be formed purely on the basis of mutual trust for each other.

In the initial stage of developing the corpus fund, each member of the group contributes a small amount of money every week and the cash credit, which will be six times of the savings, will be given to them after six to eight months. After the successful completion of the project, the government will decide to scale it up.

Apart from NABARD, SHGs can be formed under the "Swarnajayanti Gram SwarojgarYojana" (SGSY). Various State governments across the country had collectively agreed to a policy for starting SHGs as an initiative to eradicate poverty from India.

The view of establishing SHGs also stands to ensure feasible economic development which will be labour-centric, unbiased, and included with the development of social sectors, especially concentrated towards the overall progress of financially backwards.

The process of involving people with feeble financial backgrounds, especially women in the SHGs has been administered by implementing various programmes launched by various departments of both the central as well as the state governments. The programmes like—" SHG-Bank Linkage Programme" (BLP) launched by NABARD, and the "Social Intermediation Programme" and several programmes initiated by various Non-Governmental Organizations (NGOs) are striving hard for the active participation of women in SHGs.

More than one lakh SHGs have been formed in West Bengal with the collective efforts of government and NGOs. Implementation of 'SGSY-SHG-based programme' by the Rural Development Department has been attributed to the establishment of nearly 55,699 self-help groups.

Replacing the SGSY Scheme, the Rural Development Ministry implemented the 'National Rural Livelihood Mission' (NRLM) and enforced it from April 1, 2019. After this development, NRLM has been considered as a flagship programme of the Central government to root out poverty. Since then, the primary section of NRLM has been dealing with the formation of SHGs for women on the basis of mutual understanding. The study by Latha, M., & Kumar, G. C. (2012) discovered that agricultural women SHGs were doing well in terms of obtaining and repaying microcredit, which had aided in their socioeconomic empowerment and improved living conditions. Kalra et al (20013) find that Self-help groups that engage in development activities have the potential to empower their members by providing sustainable agricultural information, skills, motivation, and competences.

According Shah, Rao, and Shankar (2007),"In the financial system of India, an important role has been played by SHGs. In 2004, Nobel prize winner MohdYunus wrote "The importance of collateral-free loan on the poor has been proved by Grameen Bank experience". Based on the topic "The impact of Grameen Bank and two other microcredit organizations of Bangladesh on labor supply, schooling, household expenditures, and assets", conducted by Pitt and Khandker in 1998 concluded that where women are programme participants, the programme credit has a huge impact on the behaviour of poor families in a under developing nation like Bangladesh.

But there are also many research studies which challenge the role played by self-help groups in exterminating poverty and developing the concept of empowerment in the country. Many studies have also found that SHGs have yielded poorer results than the expected level. Micro-credits have negatively impacted the "decision-making power of women", regarding loans and financial transactions previously managed by their husbands(Banerjee, et. al., 2013;Garikipati, 2012).

Some studies carried out by experts in Andhra Pradesh have not found any remarkable growth through SHGs to boost empowerment activities. Garikipati (2012) conducted a study in 2001-2003 by interviewing a total of 291 married couple of two villages in Mahabubnagar district in Andhra Pradesh, found that handing over the credit sum of the SHG to women can only help them to diversify their families' livelihood and also help them to reduce the family's susceptibility to shocks, but completely fail to empower them.

After reviewing all the studies and research conducted by experts mentioned above, in the conclusion, we can say that SHGs played a vital role for the empowerment of women by making them economically stable, and on the other hand, they have failed to yield the result as expected.

However, the study conducted by us is on the basis of the positive role played by SHGs in the successful conversion of women, rural women in large, into micro-entrepreneurs, which further helps them to be successful in various sectors of life. The role of SHGs in empowering women and becoming a way for economic development of the country has been discussed in detail.

Mittal (1999) referred SHG as a people's scheme and its organization as a significant step toward empowering women. The study viewed SHG from the human resource development perspective. With regards to development, Gore (2001)consider it as a collective result of social, economic, and political aspects. In other words, development cannot be complete without the development of the human resource of a given community.

Empowerment can result in alternative development initiatives like reservation in politics, awareness on various aspects related to the political system in the country, property rights, and various other important interventions for empowering women. Further, the factors such as legal and regulatory framework; and social norms and culture also play a significant role in the empowerment process (Beteta,2006).

Swain and Wallentin (2012) investigated the factors which lead the SBLP members towards empowerment found "Economic factor is among the most significant factors for women empowerment. Apart from this, autonomy of the members in decision making and social attitudes are other factors that contribute significantly to empowering the SBLP female clients".

Hashemi, Schuler, & Riley, (1996)analysed the changes in the empowerment of women by conducting a quantitative survey and ethnographic study. Investigating the sample data of nearly 1,300 womenfolk to find the Grameen Bank impact and "Bangladesh Rural Advancement Committee", they created an indicator with the eight criteria—such as mobility, making small purchases, economic security, large purchases, relative freedom from family domination, involvement in major household decisions, legal & political awareness and participation in political campaigns & public protests to gauge empowerment. Setting up indicators to measure women empowerment has proven as an inefficient and non-credible way as it permits the usage of arbitrary weights.

Pitt, Khandker, & Cartwright, (2006) conducted a comprehensive study using the "Item Response Theory", in which, "the element of analysis is the whole pattern of a set of binary indicators that proxy for women's autonomy, decision-making power, and participation in household and societal decision-making". Their study also found that Credit giving programmes had a greater role in inspiring and uplifting the morale of the women in making decisions as well as a greater access to monetary resources. This gives a much needed freedom and building networks.

Awareness about politics, day-to-day happenings, political development, participation in political activities, getting memberships in any parties, and getting hold of power through any position can be considered as ways to achieve political empowerment. While women are being associated with SHGs, they will be ensured a firm base for dialogue and get co-operation from government departments, co-operatives, finance organisations, and Panchayat Raj Institutions to organize programmes in joint association with other institutions. In this whole process, awareness is being created among women to make them understand the political factors and involve them in the political affairs too.

Swain and Wallentin, (2009) says that "Improved networking, better communication, and greater mobility" are determinants for women's empowerment. Participation in SHGs as active members encourages women to take part in different activities at community level, enhancing their capabilities facilitating them to be more active in the democratic system. Bardhan and Klasen (1999) too expressed similar views that female participation and increased share of responsibility outside their homes will empower them and help them in taking part in several economic activities.

The generated hypotheses that are to be checked are based on theoretical underpinnings. Relevant hypothesis like there is a positive linkage among various dimensions of empowerment – social, economic, political, external communication, and network with SHGs and lastly all the dimensions tend to meet at a point i.e. empowerment of women. There is an interconnection between SHGs and women empowerment and in that process micro-entrepreneurs play pivotal role.

Sustainable forms of livelihoods are created by entrepreneurial women as part of SHGs. (Parasuraman, 2017). The Swarnjayanti Gram SwarozgarYojana (SGSY) was implemented in 1990, as an integrated scheme to generate employment facilities for the poor people with rural backgrounds. The programme helps in involving the target group into the Self-Help Groups (SHGs) and the government provides the guidance and assistance to them. The main objective of the scheme is producing a considerable number of small scale business exponents for viable generation of income. Priority is accorded for training, marketing, credit, as well as infrastructure (SGSY website).

Programmes such as SHG-BLP, which were conducted recently are being expanded at present with an aim of reaching 100 million women, says a report by NABARD in 2016. The primary objective of SHG-BLP is to promote social empowerment with the help of financial skill development.

Self-Help Groups by engaging in to micro-enterprises like beauty parlours, dairies, candle manufacturing, carpet making, farm-based activities, animal husbandry which includes goat rearing, floriculture, handicrafts making-Zari, Shola, Kantha, and others, sericulture, manufacture of "dhoop", and mat production, have been helpful in giving rise to a source of income for rural women and an effective way towards eradication of poverty. SHGs have been capable of producing income and result in social development.

Micro-enterprises enhance the level of confidence in women and facilitate them to take active part in theaffairs of family and give opportunities to take a stand to raise their voice against social evils. At the end, women with strong mindset and financial stability emerge successfully by setting up their own enterprises. Women micro-entrepreneurs are the end result of SHGs at the core. Women entrepreneurs are being considered as engines of growth (UNIDO). Thus the another hypothesis that is to be tested is empowered women of SHGs get turned into micro-entrepreneurs. SHGs help making empowered women micro-entrepreneurs. There is considerable SHGs related literature on entrepreneurial women and their contribution towards women entrepreneurs and economic development of society. But only several studies were found to throw light on the role played by self-help groups in the creation of small scale entrepreneurs.

The proposed research model is built on structural model using second-order factor which is women empowerment whose hidden indicators are economic empowerment, social empowerment, political empowerment, and external communication and networks. Further, the second-order underlying factor is responsible for the start of micro-entrepreneurship and serves as a mediator. The concise suggested framework is as follows: -Economic, Social, Political and Communication and Network→Women Empowerment → Micro entrepreneurship.

METHODOLOGY

As part of methodology for the study range of questions along with reasonable sample size is required to estimate hidden factor like empowerment of women. Appropriate data collection and related constraints have posed a quite challenge for the study. A survey questionnaire with suitable coding is used to collect data based experimental design. The study has interviewed 450 members of SHG of Uttrakhand and Uttar Pradesh with varied Household characteristics and characterized by financial inclusion. All the SHGs that were studied had operations in banks and had been before 2018 and after liberalized regime. An attempt has been made to assess any mindset change toward entrepreneurship among the women associated with the groups including changed political regime.

As for measurements and scale, the data was collected with self-administered questionnaire using five-point scale Likert scale based on Swain and Floro (2012) that is followed by pilot survey in Uttrakhand and Uttar Pradesh. Cronbach's α (Cronbach, 1951) is put to use for checking responses' reliability and internal consistency indicating robust figures (Economic (.72) Social (.67), Political (.62) and Communication and Network (.78) and Women Empowerment (.69).

RESULTS AND DISCUSSION

After joining the SHG, there is an incremental contribution property-related matters, family planning related matters in around 30 percent cases and increase is also noticed in case of girl child school joining to the tune 50 percent, reveals the questionnaire survey. The CFI value should be 0.90 or higher (Hu and Bentler, 1999), and 0.901 is the value for tested model and 0.041 is the RMSEA value, while 0.05 or less is the recommended score.(Browne and Cudeck, 1992).The indicators of model fit (CMIN/ DF 2.671 CFI 0.991 RMSEA) indicate that the model shows a good approximate fit implying estimates reliability. different dimensions of empowerment indicating statistically significant results. There is significant link with Women empowerment dimensions that are in tune with the existing theory and relevant studies.

Social dimension, network, and communication are the significant contributor to women's empowerment for all three models. The covariance coefficients of the economic as well non-economic parameters can be included with the SEM. The structural model has been found to be significant enough with all its variables leading to empowerment. The co-variance matrix has also revealed the linkage between the empowerment of women and the factors.

The second phase has demonstrated that economic empowerment can be experienced by women micro- entrepreneurs. Sometimes, it has reported that the financial aid given to the women members of self-help groups to take up their own business was appropriated by men in their families mainly for two reasons-- either for selling or for their own use. The study revealed less impact on the use of women's time as well as her access to financial aspects. Thus this can be solved by empowering women over financial matters with tailor made loans. That in most of the cases, men hold power on the loans taken by women, leading women to depend upon them (husbands) for repayment of the debt and domestic discord is explained by Gotez and Gupta (1996). In terms of stopping such practices by the male, who are being considered as the first persons in the family, there is a need for creating awareness among the family members of SHG members. The connection between small business and empowerment has emerged as a ray of hope for sustainable and longtime development. With the creation of an economic approach just like small-scale business, it would be possible for "ladder transformation', which further briefs about the concept of development.

PRACTICAL INFERENCES

As it has been clearly mentioned in the introduction part of this research study, the main objective of the research is to substantiate the merging of various aspects of empowering womenfolk with the constant association with self-help groups. The thick connection between women empowerment and the above mentioned dimensions gives us a clear picture to the strategy makers, to implement this concept in terms to encourage mass formation. Along with that, a supportive environment should be created by the government, enabling the SHGs to flourish.

The second priority has been given to explain the connection between micro-entrepreneurship and women empowerment. The positive impact of women empowerment will be noticed on both economic and social aspects.

The survey conducted to analyze all these factors revealed that there is a lack of interest to start own business is likely because of the problems to be faced in marketing, supply, and distribution.

LIMITATIONS AND FUTURE RESEARCH SCOPE

Though sincere efforts were made to put forth many findings, there are certain restrictions for this study. To start with, there is only data of two State of the country, which restricts the general comparison and analysis of the findings. For the collection of primary data, questionnaire was the only method, which was opted. Hence, regarding the impact of common method bias, interpretation of results is a must. This practical/experimental investigation is cross-sectional, hence, long lasting interchange of empowerment concepts as well as small-scale/micro entrepreneurship was not allowed. It is to clarify that the present research is not longitudinal by nature.

For future policy design, studies should be extended to different states across the country. This will help in checking replication of similar results. This will give a holistic idea towards policy designing in the coming days. Qualitative research equipped with detailed and in-depth studies will lead to the generation of an extensive idea on women empowerment with regard to entrepreneurship. However, it is acknowledged that inferences drawn are from two states of northern India with representative of broader perspective inside agriculture dominant India.

CONCLUSION

The final findings of the research study concentrate on various empowerment aspects connected with self-help groups (SHGs). Along with emphasizing the entrepreneurship literature, the study briefs how empowerment of women can be a way for the overall development of small-scale business, which has never been considered as an important aspect. The literature further focuses on incorporation of convergence aspects of various dimensions of the empowerment being emerged by the SHGs in agriculture dotted regions of both the states. Microcredit generated by self-help groups (SHGs) in agriculture is a widely used tool for poverty alleviation and empowerment of women. Public policy of setting up self-help groups has a greater role in gender development and this was proved by capability broadening and direct influence of SHGs on women empowerment in agri-regions of the states.

Absolute empowerment is a far cry and a distant dream. The Government should make efforts to have a holistic environment where women get abundant support to start their own business and experience a holistic growth in all aspects. This thought is ably supported by the previous findings

REFERENCES

Banerjee, A., Duflo, E., Glennerster, R., & Kinnan, C. (2015). The miracle of microfinance?Evidence from a randomized evaluation. *American Economic Journal. Applied Economics*, *7*(1), 22–53. doi:10.1257/app.20130533

Bardhan, K., & Klasen, S. (1999). UNDP's gender-related indices: A critical review. *World Development*, *27*(6), 985–1010. doi:10.1016/S0305-750X(99)00035-2

Cueva Beteta, H. (2006). What is missing in measures of women's empowerment? *Journal of Human Development*, *7*(2), 221–241. doi:10.1080/14649880600768553

Bardhan, K., & Klasen, S. (1999). UNDP's gender-related indices: A critical review. *World Development*, *27*(6), 985–1010. doi:10.1016/S0305-750X(99)00035-2

Browne, M. W., & Cudeck, R. (1992). Alternative ways of assessing model fit. *Sociological Methods & Research*, *21*(2), 230–258. doi:10.1177/0049124192021002005

Cronbach, L. J. (1951). Coefficient alpha and the internal structure of tests. *Psychometrika, 16*(3), 297-334.

Doepke, M., & Tertilt, M. (2011). *Does female empowerment promote economic development? IZA Discussion Papers 5637. Institute for the Study of Labor*. IZA.

Duflo, E. (2011). *Women empowerment and economic development.* NBER Working Paper No. 17702.

Galindo, M. Á., & Méndez-Picazo, M. T. (2013). Innovation, entrepreneurship and economic growth. *Management Decision, 51*(3), 501–514. doi:10.1108/00251741311309625

Garikipati, S. (2012). Microcredit and women's empowerment: Through the lens of time-use data from rural India. *Development and Change, 43*(3), 719–750. doi:10.1111/j.1467-7660.2012.01780.x

Gore, M. S. (1994). *Indian education: Structure and process*. Rawat Publications.

Goetz, A. M., & Gupta, R. S. (1996). Who takes the credit? Gender, power, and control over loan use in rural credit programs in Bangladesh. *World Development, 24*(1), 45–63. doi:10.1016/0305-750X(95)00124-U

Government of India. (2019). *Swarnajayanti Gram SwarozgarYojna*. Retrieved from https://www.india.gov.in/swarnjayanti-gram-swarozgar-yojana

Hall, J., Matos, S., Sheehan, L., & Silvestre, B. (2012). Entrepreneurship and innovation at the base of the pyramid: A recipe for inclusive growth or social exclusion? *Journal of Management Studies, 49*(4), 785–812. doi:10.1111/j.1467-6486.2012.01044.x

Hashemi, S. M., Schuler, S. R., & Riley, A. P. (1996). Rural credit programs and women's empowerment in Bangladesh. *World Development, 24*(4), 635–653. doi:10.1016/0305-750X(95)00159-A

Hu, L. T., & Bentler, P. M. (1999). Cutoff criteria for fit indexes in covariance structure analysis: Conventional criteria versus new alternatives. *Structural Equation Modeling, 6*(1), 1–55. doi:10.1080/10705519909540118

Kalra, R. K., Anil, B., Tonts, M., & Siddique, K. H. M. (2013). Self-help groups in Indian agriculture: A case study of farmer groups in Punjab, Northern India. *Agroecology and Sustainable Food Systems, 37*(5), 509–530. doi:10.1080/10440046.2012.719853

Latha, M., & Chandrakumar, G. (2012). A study on agricultural women Self Help Groups (SHGs) members' micro credit analysis in Trichy district, Tamil Nadu. Academic Press.

London, T., & Hart, S. L. (2010). *Next generation business strategies for the base of the pyramid: New approaches for building mutual value*. FT Press.

Lortie, J., Castrogiovanni, G. J., & Cox, K. C. (2017). Gender, social salience, and social performance: How women pursue and perform in social ventures. *Entrepreneurship and Regional Development*, *29*(1-2), 155–173. doi:10.1080/08985626.2016.1255433

Ministry of Rural Development - Govt. of India. (2019). *DeendayalAntyodayaYojana-National Rural Livelihoods Mission (DAY-NRLM)*. Retrieved from https://aajeevika.gov.in/en/content/social inclusion development

Mittal, R. K. (1999). *Entrepreneurial Management: Marketing Strategy and Product Portfolio*. Fine Art Press.

NABARD. (2019). *Report on Status of Micro Finance in India 2019-20*. Available at: www.nabard.org/ auth/writereaddata/tender /1307174808Status%20of%20Microfinance /1307174808Status%20of%20Microfinance%20in%20India%202016-17.pdf

Omoruyi, E. M. M., Olamide, K. S., Gomolemo, G., & Donath, O. A. (2017). Entrepreneurship and economic growth: Does entrepreneurship bolster economic expansion in Africa. *Journal of Socialomics*, *6*(4), 1–11.

Chatterjee, S., Gupta, S. D., & Upadhyay, P. (2018). Empowering women and stimulating development at bottom of pyramid through micro-entrepreneurship. *Management Decision*.

Pitt, M. M., Khandker, S. R., & Cartwright, J. (2006). Empowering women with micro finance: Evidence from Bangladesh. *Economic Development and Cultural Change*, *54*(4), 791–831.

Schumpeter, J. (1934). *The theory of economic development*. Harvard University Press.

Shah, M., Rao, R., & Shankar, P. V. (2007). Rural credit in 20th century India: Overview of history and perspectives. *Economic and Political Weekly*, •••, 1351–1364.

Bali Swain, R., & Wallentin, F. Y. (2012). Factors empowering women in Indian self-help group programs. *International Review of Applied Economics*, *26*(4), 425–444.

Swain, R. B., & Wallentin, F. Y. (2009). Does microfinance empower women? Evidence from self-help groups in India. *International Review of Applied Economics*, *23*(5), 541–556.

ReportU. N. I. D. O. (2017). *Women and Youth*. Available at: https://www.unido.org/fileadmin/user_media_upgrade/What_we_do/Topics/Women_and_Youth/Brochure_low_resolution.pdf

Yunus, M. (2004). Grameen Bank, microcredit and millennium development goals. *Economic and Political Weekly*, •••, 4077–4080.

Chapter 12
Rural Innovation Using Technology in Progressive Agriculture:
Empowering Women Towards Sustainable Rural Development

Swati Bhatt
SRM Institute of Science and Technology, India

Sharat Sharma
SRM Institute of Science and Technology, India

Pallavi Jain
SRM Institute of Science and Technology, India

ABSTRACT

Sustainable technology is an individual and group effort through the experts and professionals in rural as well as urban people. Rural innovation involves efficient execution of ideas, information, imagination, and innovative initiative in fulfilling social requirements, and new ideas are converted into something useful for the rural development. It starts with the proper utilization of resources, rural innovation, and modernization of agricultural activities. The study attempts to identify the use of sustainable technology with rural innovation, progressive agriculture, and women empowerment for rural development. The required data collected led to a discussion that the practices of modern farming with the active involvements of rural population and promoting individual performance and fine-tuning of green initiative among rural livelihoods. The results of this study include the difficulties in progressive effort on rural innovation vis-a-vis planning and executing of advanced farming, protecting traditional arts, and crafting through maximum involvement of rural women.

DOI: 10.4018/978-1-6684-2349-3.ch012

INTRODUCTION

Rural development is the major issue to build a balanced and sustainable society. The rationale of the present study considers the development and execution of sustainable technology in rural segments (Amba-Rao et. al. 2008). The present study also gives priority of rural innovation especially in the field of modern agriculture and women empowerment.

Sustainable technology ensures the proper utilization of renewable energy, sustainable living, solid waste management, sewage treatment, energy conservation etc. for efficient industrial activities (C. Hedberg et. al. 2003). The urban infrastructure, electricity supply, expert human resources, modern incubation centers facilitates and encourage sustainable technology, but the rural livelihoods stills struggling for the basic requirements to fulfill and using obsolescence methods for various functions related to different sectors (Carroll, A. et. al. 2005) (Ishaya, S. et. al. 2008).

Rural Innovationrepresents the innovation in the areas of continuous improvement likehandicrafts, organic farming, crop rotation, organic fertilizers, poultry farming, vocational training and micro finance, etc (Suplab K. Podderet. al. 2018). The globalization concept and the entry of multi-national corporations in Indian markets, the rural livelihoods under the threats of manufacturing traditional production process and forced to improve the product and services quality time to time. With the limited sources of funds and infrastructure facility, the rural innovation restricted to few areas only (John F. Morton et. al. 2007).

Modern agriculture represents the area of agribusiness, intensive farming, organic farming, smart farming and sustainable agriculture which are suitable for the cultivation and ensure the maximum productions (Crowther, D. et. al. 2008). Modern agriculture interrelated to rural innovation and sustainable technology that facilitates development of agricultural land, selection of suitable crops and the cultivation based on the soil characteristics and production capacity, protection from adverse climatic conditions and development and utilization of hybrid seed (Z Chvatalová et. al. 2011).

SUSTAINABLE TECHNOLOGY – THE CONCEPT

Sustainable technology is the process of avoiding the precariousenthusiasm and protect from the negative impact of repeated social changes. Sustainable technology related to renewable energy like electricity generation, air and water heating/cooling, transportation, and rural energy services, reduction of carbon footprint, energy conservation and proper waste management (SuplabKantiPodderet. al. 2019). The eco-friendly environment is highly influenced by sustainable technology and its executions.

Factors Influencing Sustainable Technology

Various factors that influence Sustainable Technology are as follows:

1. *Long run survival:*Sustainable technology is used for long run survival of business operations. Rural livelihoods can change their living standards and create examples for the generation next (Suplab K. Podder et. al. 2018).
2. *Perfect Competition*: Competitions are increasing day to day in all sectors. The sustainable technology can ensure the better position of any business operations (Hart S. L.et. al. 2003).
3. Continuous Improvement: Continuous improvement is the main principle of rural innovation through sustainable technology. The rural livelihoods need to update their operation methods for modern agriculture, smart farming, agri-business, foresting, handicraft products manufacturing etc. for attracting the potential customers.
4. Energy Efficiency: There are few sources of energy but using for multiple purposes. The sustainable technology gives priority on optimum utilization of energy for maximum outcomes.
5. Work Life Balance: Modern technology gives more stress for the operators because of short term project-based work. But the technology should be stress free, comfortable and ensure work life balance.
6. Economic Condition: Economic condition is another influencing factor towards sustainable technology. The rural livelihoods concentrate on financial benefits of sustainable technology ant is implementations.

Sustainable Technology Leads to Rural Development

Sustainable Technology has great impact on rural development and builds a sustainable society. The technology creates awareness among the rural livelihoods about the progressive technology and keeps away from the harmful technology (Crowder, D. et. al. 2009).

The implementation of progressive technology related to renewable energy like electricity generation, air and water heating/cooling, transportation, and rural energy services, reduction of carbon footprint, energy conservation and proper waste management are the sustainable technology initiatives. With the support of sustainable technology, the rural livelihoods gets the idea of proper utilization of natural resources and increases their income sources that leads to development of economic condition.

RURAL INNOVATION – THE CONCEPT

Rural innovation is theusage of information, imagination and initiative in driving greater on different values from resources and includes all processes by which new ideas are generated and converted into something useful for the rural development. Rural development starts with the proper utilization of resources, rural innovation, and modernization of agricultural activities with the active involvement of women empowerment (Dyllick, T. et. al. 2002).

Importance of Rural Innovation

Importance of rural innovation can be summarized as follows:

1. Improvement of rural livelihoods: Rural innovation for the rural livelihoods, with the proactive involvement of them. The rural innovation improves their way thinking towards sustainability.
2. Resources Utilization: The involvement and proper utilization of various resources that lead to rural innovation. Using biogas, wind mills and solar panels for generating electricity, using dig trenches and build earthen dams for water conservation, cultivating more Ayurveda medicine plants are the well-known resources that can be utilized properly. The unskilled people need to be trained for organic farming, crop rotation, modern handicrafts etc (S K Podder et. al. 2019).
3. Energy Efficiency: Rural innovation helps to use the technology and methods that ensures the energy efficiency. Using energy efficiency engine, reducing repeated work, work blueprint and assign work load ensure energy efficiency towards successful rural innovation.
4. Water conservation: Water is the life of all types of rural functions like, agriculture, foresting, poultry farming, arts and crafts etc. Rural innovation gives special provisions for water conservation especially in rural areas.
5. Systematic Education: Innovation is the outcomes of systematic education and the applications of knowledge towards sustainable society.

Sources of Rural Innovation

Various sources of innovation can be summarized as follows:

1. Scientific Education:Scientific education is the important source of rural innovation. The information, imagination and initiative are derived from scientific education. The applications of knowledge reflect the innovative ideas and finally can be used for rural development.
2. Curiosity: The curiosity of the researchers leads to innovation towards sustainable development.
3. Demand and Supply: When the demand increases comparable to availability or supply, the innovative practices are initiated. The demand of organic fruits, vegetable and other food items, the scientists and researchers take initiative of innovative techniques of modern smart farming that facilitates increased productivity.
4. Competition: The rural innovation is highly influenced by global competition. It is the challenge for the rural livelihoods to face the global competition but the systematic vocational training builds the confidence among them (P. Hardi et. al. 1997).
5. Economic development: Rural development is inter-connected with the economic conditions of rural livelihoods. The improvement of living standards, employability, increasing income sources etc. are the indications of economic development.

MODERN AGRICULTURE IN BRIEF

Modern agriculture represents the area of agribusiness, intensive farming, organic farming, smart farming and sustainable agriculture which are suitable for the cultivation and ensure the maximum productions. Modern agriculture interrelated to rural innovation and sustainable technology that facilitates development of agricultural land, selection of suitable crops and the cultivation based on the soil characteristics and production capacity, protection from adverse climatic conditions and development and utilization of hybrid seed (Kenneth L. Deavers et. al. 2014).

Techniques of Executing Modern Agriculture

Various techniques of executing Modern Agriculture are as follows:

1. Organic Farming: Organic Farming is the modern agriculture technique, in which the chemical fertilizers are not used for entire cultivation process. These techniques of cultivation ensure intent attention to promote the health and meet the behavioural needs of human being.

2. Crop Rotation: Crop rotation is the technique of executing modern agriculture, in which different crops are growing in succession on a specific land to avoid damaging the production capacity of soil and helps to control weeds, pests, and diseases (Mitchell, T. D. et. al. 2005).
3. Organic Fertilizer:Modern agriculture depends on the organic fertilizer which is produced from waste vegetable, animal manure, human manure, unused plants and leaps.
4. Soil Test and Development:Soil Test and Development are the techniques of modern agriculture to understand the nature and characteristics of the composition of soil and capacity to grow specified crops. If the land is not cultivable, the necessary actions need to be taken care to develop and convert the soil into something useful.
5. (e) Development of hybrid seed: Modern agriculture is promised to grow more crops for fulfilling the prospect social demand. The hybrid seed which is used to grow by cross-pollinated plants. Hybrid seeds are the outcomes of advanced technique of modern agriculture that can be cultivated in diverse weather conditions.

Modern Agriculture and Rural Livelihoods

Rural Livelihoods are suffering from unexpected climate changes and migrating other suitable places to surviving. They sale their lands or leave it unused and unproductive conditions which is dangerous for the social balancing.

Modern agriculture is the revolution of rural livelihoods especially for the farmers. Different techniques like crop rotation, smart farming, organic farming, developing of soil and hybrid seed and using manure for cultivation ensures less expenses in cultivation and growing more crops. The modern agriculture also ensure the storage of crops for longer period of time to sale the crops in high rate in future (Mahua B.et. al. 2019).

WOMEN EMPOWERMENT – THE CONCEPT

Women's empowerment is the concept of educating, recognizing and strengthens the women in the area of knowledge, skills, idea, research and development, finance and in the contribution towards economic development. Women empowerment creates strong backbone of a nation and multiple the production and services.

In India, maximum portion of women either educated or uneducated remain unemployed and the male dominated society neglects them. Women empowerment is the only way of making awareness and strengths of those women are capable to work especially in rural segments. The women can take part of creativity and innovation that can be executed in rural development and modern agriculture (Hart, S. L. et. al. 1997).

Activities Related to Women Empowerment

Various activities related to Women Empowerment are:

1. Scientific Education: Scientific education is the process of understanding and analyzing the requirements of human being and society at large regarding any product of service. The research and development activities build the learning environment for the women, so that they can initiative for solving problems by using scientific technique. The educated women can take part of these activities.
2. Systematic Training: The educated as well as uneducated women can learn through the systematic on the job training and execute for the real life operations. After completion of family work, this segment of women can get training for proper utilization of time and capacity for themselves and the society (Daly, H. et. al. 1999).
3. Group Development: Group development is the important initiative for women empowerment. Individual woman can do little but in group, they can do so many activities. The rural women can take part of handicrafts product manufacturing, agricultural activities, poultry farming, vegetable growing and foresting with the developing a group.
4. Income Generation Activities: The most important area of women empowerment is to generate income sources. The self-help group becomes stronger when it generates income sources and secures the individual and family life.
5. Micro Credit and Implementation: The micro credit facility and its proper utilization build confidence among the women. When the women take micro credit facilities and proper guidelines for utilizing that money in productive purpose, they create income opportunities for them and for the other women.
6. Establishment of linkages: The social media is very strong in the sense of generating business and professional links. The rural women also can create and access the urban women empowerment link that facilitates the building confidence, fight for the human rights and strengthen their future generation.

Women Empowerment – Contribution towards Rural Development

Women Empowerment and contribution towards rural development are:

1. Building self-help groups: Self-help group is the revolutionary initiatives for women empowerment in Rural India. A group of women creates self-help group and provide a social support system for members, which is especially helpful for people with conditions that tend to isolate them (Demeke, M. et. al. 2010).
2. Micro Credit and Business Initiatives: When the women take micro credit facilities and proper guidelines for utilizing that money in productive purpose, they create income opportunities for them and for the other women.
3. Initiative for modern agriculture: Rural women can take part of modern agriculture especially in organic farming, vegetable growing etc.
4. Development of Arts and Crafts: The traditional arts and crafts are developed through women empowerment. The systematic training program creates awareness among the rural women to develop modern arts and crafts and manufacture in a group and finally promote in the national and international markets (Elliot, S. R. et. al. 2005).

METHODS OF DATA COLLECTION

The required data were collected both from primary and secondary sources. Primary data was collected from 40 respondents selected at random from five villages in Karnataka namely Alambur, Kodihalli, Rayasandra, Palahalli and Agalagurki which were the representative of most undeveloped villages in Karnataka. Questionnaire and semi-structured interview schedules were employed to collect the primary data. The secondary data sources included web literature, journals, periodicals and newspaper reports to get a picture about the prevailing context. The data collected was recorded and organized for further analysis.

RESULTS AND DISCUSSION

The summary of results includes the Sustainable Technology through Rural Innovation, Modern Agriculture and Women Empowerment towards Rural Development.

Figure 1. Shows the relationship among women empowerment, rural innovation, modern agriculture and sustainable technology towards rural development

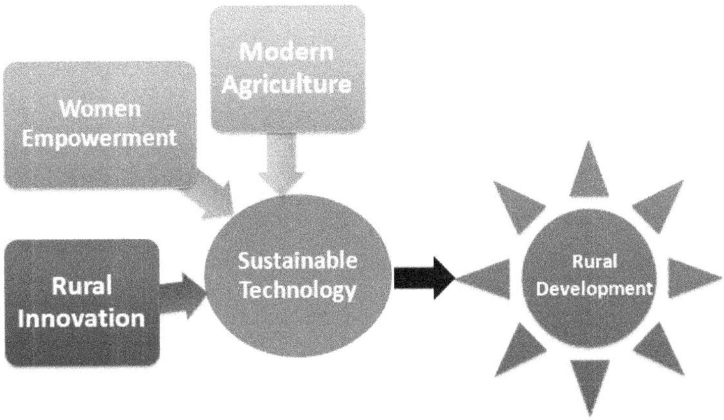

Table 1 shows the various factors that Influence Sustainable Technology towards rural development. Energy Efficiency considers first priority of 12% followed by 10% long run survival, 7% perfect competition, 5% continuous improvement, 4% economic condition and 2% work life balance.

Table 1. Factors that influence sustainable technology

SL. No.	Factors that Influence Sustainable Technology	No. of Respondents	Percentage (%)
1	Energy Efficiency	12	12
2	Long run survival	10	10
3	Perfect Competition	7	7
4	Continuous Improvement	5	5
5	Economic Condition	4	4
6	Work Life Balance	2	2

(N=40)
Source: compiled by the researcher

Figure 2. Graphical representation of factors that influence sustainable technology

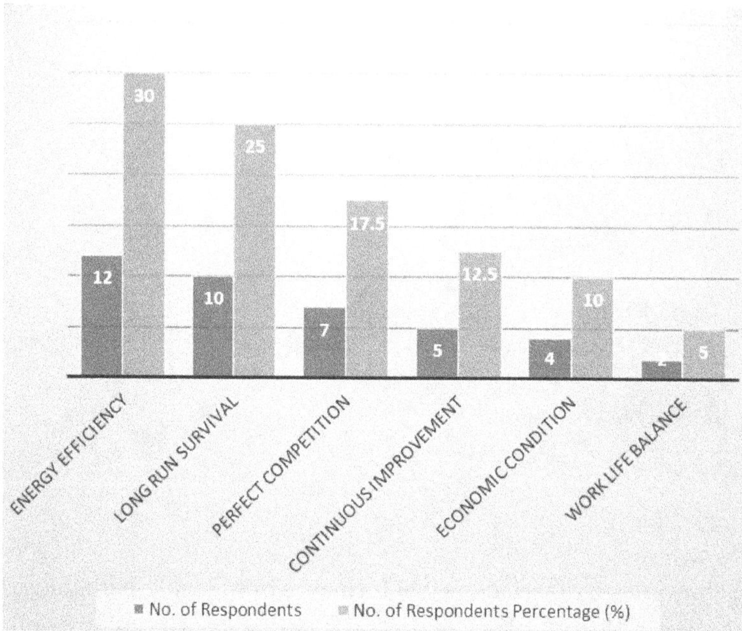

Table 2 lists the 4 R's of environmental concerns as well as sustainable technology initiatives as applied to the rural innovation, modern agriculture and women empowerment. Each of these dimension had the maximum score of 10. These four dimensions and the total score are presented both for individual practices as well as group level practices. It is interesting to observe that there is very high degree of agreement at both individual as well as group level in terms of Green HRM.

This is evident from the similarity in the descending order both at individual as well as at group level sustainable technology practices. This phenomenon may be explained by the fact that sustainable technology practices as they presently exist are at the cognitive level and the affect component need to consolidate in these practices.

Table 3 shows the summary of t-test results comparing the extent of individual and group level Sustainable Technology initiatives. It was found that individual level practices were significantly higher than group level Sustainable Technology initiatives.

These results indicate that individual level concerns about the sustainable technology did not get nurtured into group level efforts due to lack of support from the local government or organizational systems.

Table 2. Mean levels of sustainable technology initiatives towards rural development

Sl. No.	Dimensions	Mean Sustainable Technology initiatives Score	
		Individual Practices	Collective or Group Practices
1	Degree of "Reduce" Component	6.30	5.09
2	Degree of "Recycle" Component	5.87	4.89
3	Degree of "Reuse" Component	5.68	4.62
4	Degree of "Refuse" Component	5.12	4.58
5	Total Sustainable Technology initiatives	23.87	19.65

Figure 3. Graphical representation of mean levels of sustainable technology initiatives towards rural development

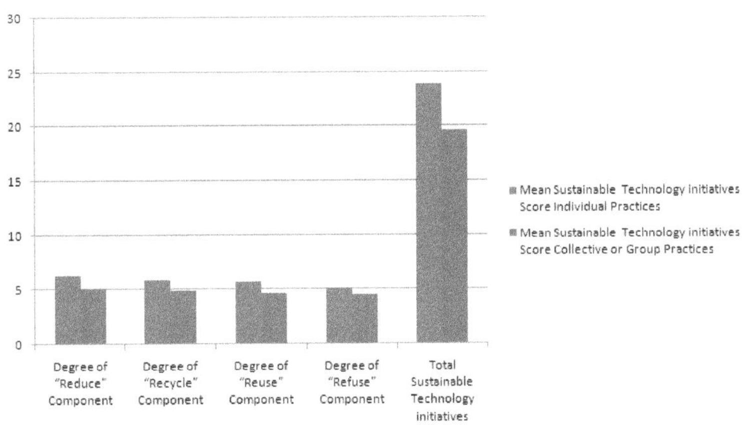

Table 3. Summary of t-test results of individual practices and collective practices

Category	Mean	SD	df	t-value
Individual Practices	23.87	4.03	39	4.3803**
Collective or Group Practices	19.65	4.57	39	

(N=40)
**Significant at 0.01 level of significance

Figure 4: Graphical representation of Individual practices and collective practices

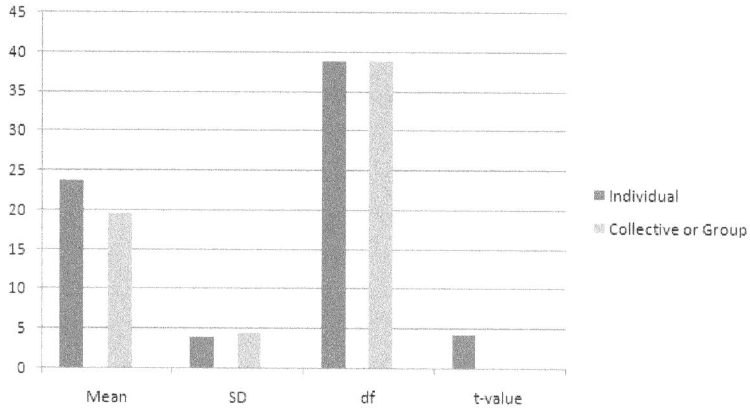

CONCLUSION

Sustainable technology ensures the proper utilization of renewable energy, sustainable living, solid waste management, sewage treatment, energy conservation etc. for improving rural livelihoods. But, the lack of interest of individuals and group of peopletowards sustainable technology initiatives remain unchanged of rural livelihoods. Rural Innovation represents the innovation in the areas of continuous improvement like handicrafts, organic farming, crop rotation, organic fertilizers, poultry farming, vocational training and micro finance, etc. which results for long term benefits. But, the rural livelihoods suffering problems in the present situation and they expect the immediate results of rural innovation which are invisible. Modern agriculture represents the area of agribusiness, intensive farming, organic farming, smart farming and sustainable agriculture which are suitable for the cultivation and ensure the maximum productions. The lack of infrastructure and modern instruments are the major problems of executing modern agriculture, because the farmers are not showing interest on changing their cultivation process towards modern agriculture. Micro Finance, vocational training, increase income level, and economic progress are the important factors that influence women empowerment in rural segments.

REFERENCES

Amba-Rao, S. C. (2008). Multinational Institution Social Responsibility, Ethics, Interactions and Third World Governments: An Agenda for the 1990s. *Journal of Business Ethics*, *12*(3), 553–572.

Carroll, A. (2005). Impact of Institution Social Responsibility and Higher Education. *Journal of Business and Society*, *38*(3), 268–295. doi:10.1177/000765039903800303

Chvatalová, Kocmanová, & Dočekalová. (2011). Institution Sustainability Reporting and Measuring Institution Performance. *International Symposium on Environmental Software Systems*, 245-254.

Crowder, D. (2009). The roles of Institution Reporting, Stakeholders and the Internet: Mapping the New Institution Landscape. *Journal of Urban Studies*, *37*(10), 1837–1848. doi:10.1080/00420980020080451

Crowther, D. (2008). *The Social and Environmental Accounting and its applicability*. The Financial Times.

Daly, H. (1999). The Allocation, Distribution and Scale towards an Economics that is Efficient, Just and Sustainable. *Journal of Ecological Economics*, *6*(3), 185–193. doi:10.1016/0921-8009(92)90024-M

Deavers. (1997). Outsourcing: A corporate competitiveness strategy, not a search for low wages. *Journal of Labour Research, 18*(4), 504.

Demeke, M. & Meschi, M. (2010). Technology adoption in the presence of constraints: the case of fertilizer demand in Ethiopia. *Review of Development Economics, 7*, 58–70.

Dyllick, T., & Hockerts, K. (2002). Beyond the Business Case for Institution Sustainability. *Journal of Business Strategy and the Environment*, *11*(2), 130–141. doi:10.1002/bse.323

Elliot, S. R. (2005). Sustainability: An Economic Perspective, Resources Conservations and Recycling. *Journal of Economics*, *44*(1), 263–277.

Hardi, P., & Zdan, T. (1997). Assessing sustainable development: principles in practice. Principles in Practice.

Hart, S. L. (1997). Beyond Greening: Strategies for a Sustainable World. *Harvard Business Review*, *75*(1), 66–76.

Hart, S. L., & Milstein, M. B. (2003). Creating Sustainable Value. *The Academy of Management Executive*, *17*(2), 56–67. doi:10.5465/ame.2003.10025194

Hedberg & Malmborg. (2003). The Global Education and Institution sustainability reporting in Swedish companies. *Institution Social Responsibility and Environmental Management, 24*(2), 153-164.

Ishaya, S., & Abaje, I. (2008). Indigenous people's perception of climate change and adaptation strategies in Jema's local government area of Kaduna State, Nigeria. *Journal of Geography and Regional Planning*, *1*, 138–143.

Mahua, Podder, & Samanta. (2019). Factors that influence Sustainable Education with respect to Innovation and Statistical Science. *International Journal of Recent Technology and Engineering, 7*(5S2).

Mitchell, T. D., & Jones, P. D. (2005). An improved method of constructing a database of monthly climate observations and associated high-resolution grids. *Journal of Climatology*, *25*(6), 693–712. doi:10.1002/joc.1181

Morton, J. F. (2007). The impact of climate change on smallholder and subsistence agriculture. *Proceedings of the National Academy of Sciences of the United States of America*, *104*(50), 680–685. doi:10.1073/pnas.0701855104 PMID:18077400

Podder & Samanta. (2019). Impact of Climate Changes on Rural Livelihoods and Re-Orienting the Situation through Human Resources towards a Sustainable Society. In *International Conference on Sustainable Computing in Science, Technology & Management (SUSCOM-2019)*. Elsevier.

Podder, S. K. (2019). Impact of Business Analytics for Smart Education System and Management Functions. In *Proc. of International Conference on Advancements in Computing & Management (ICACM-2019)*. Springer

Suplab, K. & Podder, A. B. K. (2018). Comparison of effectiveness of employee engagement through permanent employees or outsourced employees. *International Journal of Academic Research and Development, 3*(2), 1406-1408.

Suplab, K., & Podder, A. B. K. (2018). Contribution of HR sub-functions outsourcing in the improvement of quality and innovation of Education. *Primax International Journal of Commerce And Management Research, 3*(1), 11-14.

Chapter 13
Social Capital and Entrepreneurship

Ermal Bino
University of Buraimi, Oman

Ferdinand Epoc
University of Buraimi, Oman

Ilya Bystrov
University of Buraimi, Oman

ABSTRACT

This chapter presents a social capital point of view of entrepreneurship and how the assets and value embedded in entrepreneur social relations could support the success of the enterprise, especially small ones. Social connections are unique in nature, are personal and stable in the long run. These features make them very unique and difficult to imitate. Therefore, if used properly, based on the position of the entrepreneur in the social hierarchy, it can generate or at least behave as a very unique and inimitable source of competitive advantage. Capitalizing on such resources could be of help to entrepreneurs especially in times of high competitive rivalry.

INTRODUCTION

It is a constant struggle for the firms, particularly the small ones, to access resources vital for firms' development (Carter & Van Auken, 2006; Jansen *et al.*, 2011). Lacking such access on resources includes financial ones that negatively impact the firm's development, technological ones that affect the firm's competitive advantage and

DOI: 10.4018/978-1-6684-2349-3.ch013

managerial capabilities (management skills, planning, record keeping, professional advice, educational level, learning) that impacts entrepreneurs' ability to improve firm's performance in general (Apostolos, 2016).

Social capital has the potential to mitigate the negative impact of such limitation the firm faces because it has the potential to function as a resource upon which the entrepreneur can build a competitive advantage (Aldrich& Kim, 2007). Social capital refers to the value embedded in the connections between individuals. This is true for individuals, social groups, or other social entities, where the firm or organizations are one example.

Capitalizing on social capital, which means making use of resources embedded in the connections, is vital for new entrepreneurs of small enterprises. It is often the case that the owner also manages the firm, particularly in the initial phase after establishing the firm, therefore making the resources embedded in the entrepreneur network an integral and important part of the firm's resources. Capitalizing on such connections might appear in the form of being part of professional associations that facilitate access to trade platforms, accessing technological resources (Teckchandani, 2014), better relationships with suppliers that may impact firm's flexibility in pricing (Sherry &Stubberud, 2013) and personal, informal links and connections that often time serve as a reservoir for innovation (Komulainen, Mainela, & Tahtinen, 2006).

It is of interest to mention that concepts similar to social capital as we know it today have been presented by Ibn Khaldun, the Tunisian Arab philosopher, born in 1332, in his monumental work "The Muqaddimah", The Introduction to History (Ibn K & Rosenthal, 1967). Ibn Khaldun, in his book, presents the idea that civilization holds in itself the seeds of its destruction. The same idea is presented later by Karl Marx. Ibn Khaldun introduces the term "asabiya," loosely translated as "group feeling," to describe the bond of connection among people within a community. The concept behind"asabiya" is very similar to the concept of social capital. This social bond is present at different levels of intensity in a given community or social entity, be it a tribe, a nation, an empire, a civilization or an organization. Ibn Khaldun argues that each society, as a social entity (the organization is similar in nature), holds the seeds of its demise within itself. He continues explaining that communities with strong community bonds (high levels of social capital) appear at the peripheries of the great empires, and they use their social bond in their favour. Eventually, these communities with high levels of social bond (high levels of social capital) will establish themselves into empires or civilizations, living an abundant and comfortable life. A luxurious, productive and comfortable life will weaken and undermine the social bond (social capital) among the community or social organization, leading them to become less competitive, less disciplined and more dependent on others.

Suppose we allow ourselves to draw some parallels between civilizations and organizations, since both are social organizations. In that case, we can find striking similarities of very successful firms that were once the leading players in their industries, declining and exiting the industry they once were the epitome of success. Nokia and Kodak are just two significant examples of how their strategic capabilities transformed into rigidities and eventually became weaknesses and liabilities, leading the company to decline and exit the market. Apple's and android's successes in disrupting the smartphone industry pose questions on what drove their inventions and success. Apple had been a very troubled company for an extended period of time. It hasn't always been successful as it is now. On the other hand, Android is not the end-product innovation of a tech empire, but, as Ibn Khaldun said, it is the innovation of "peripheral tribes" characterized by strong social bonds.

The disruptive innovations of the companies (peripheral tribes) in the examples reflect Ibn Khaldun's cyclical theory, where non very successful companies stay edgy and can recognize the changes and needs in the business landscape. The same business environmental changes and needs were ignored by, then, successful companies.

SOCIAL CAPITAL VIEW OF ENTREPRENEURSHIP

Social capital offers to the community members (the firm is a type of social organization) to access assets available in the network (Nahapiet&Ghoshal, 1998). It is a type of capital of [1]a nature that can provide a kind of competitive advantage for the individuals and groups that have access to it to pursue their goals (Burt, 2000). In other words, better social connections better opportunities to achieve desired results.

Social capital theory differentiates between three main types of network connections:"bonding, bridging" and "linking" social capital. Bonding social capital refers to individuals in a group with close relations to each other. It is of a horizontal nature between individuals with high levels of trust, reciprocity, group feeling, acohesiveness that facilitates the exchange of resources among them. It enables the pursuit of common goals and is accessible only to group members. Bridging social capital refers to horizontal ties of individuals pertaining to different social groups of varying backgrounds yet common interests. This type of social capital is weaker than bonding social capital and is called structural holes (Burt, 2000).Despite being known as weak ties, they are more diverse and offer accessto new connections and, therefore, access to potential assets outside the firm. These assets might be of relevant importance since connections beyond the firm help understand the differential advantage of other operators within the same industry or market. Linking social capital is also considered to be weak, regulated by the formal institutionalized power structure. It is a rare type of social capital since the formal hierarchical structures make it difficult to surpass and establish a strong connection. Being a rare type of capital, linking social capital can serve as a unique resource for the individual and, therefore, a valuable, difficult to imitate resource for the firm. The embedded value and resources within the network of connections or relationships are linked to the hierarchy and strength of ties. The individual will be able to access the assets within the network subject to ones position in the structure and the willingness of members (strong ties) to allow you to benefit from assets. The individual or social group needs to access the resources that reside in the networks.

An entrepreneur's network is informal and personal in nature and manifests a high degree of uniqueness since relations are difficult to imitate. Individual social capital, networks and connections are highly informal and stable in the long run (Arregl, Hitt, Sirmon, &Very, 2007). The unique nature of an entrepreneur's social capital is difficult to imitate. It is very personal and an outcome of mutual trust, culture and cooperation. This type of network is not something other operators can access. They are not public formal institutions and associations open to all.

The opposite is the nature of formal networks. They are public in nature and easily accessible by all operators in a certain market or industry. Suppliers, distributors, banks, media, professional bodies are easily accessible by all competitors. The degree of relationship between these formal organizations, institutions and individuals has a degree of diversity. However, it does not make them unique and inimitable.

Social capital displays qualities of a unique strategic resource, especially in periods of high competitive rivalry when unique, inimitable resources are crucial for the firm to succeed. Entrepreneur's private social capital might generate knowledge, ideas, innovative or improved products or services. Knowledge as an asset is embedded in connections, networks and communities. Social capital becomes the tool via which the entrepreneur accesses knowledge and innovation because high levels of social capital and network density positively acknowledge sharing that in turn impact innovation, consequently improving firm's performance (Kim, N. & Shim, C., 2018).

An entrepreneur's social capital is not a solution to all issues related to entrepreneurship. Depending on competitive rivalry, they need to invest in the right connections upon which to capitalize. When intensity rivalry is higher, personal social capital would be a strategic source for building competitive advantage. That seems to lie in the fact that personal social capital tends to be difficult to imitate and manifests qualities of distinctive capabilities that are important and needed, especially when competitive rivalry in a certain industry is high.When the competitive rivalry is not high, institutional networks and links might be a valuable source to access for improved firm performance.

LITERATURE REVIEW

Initially, scholars and practitioners perceived social capital as an element of an enterprise's intangible assets, andthe latter gained considerable power in the innovation economy. Paired with human capital, this part of intangible assets augments the advancement of the knowledge economy and defines the level of its development across the world. In its turn, utilization of intangible assets nowadays requires multiple pre-requisites. Among them isthe adequate social and institutional infrastructure that includes quality human capital, personal and business security and other factors that generate creativity and innovation in many aspects, including entrepreneurship. According to Fukuyama (1999), the key economic role of social capital in a free-market liberal democracy is to reduce transactional costs associated with contracts, regulations, bureaucracy, hierarchies and similar.

If we look at the side of entrepreneurship, the key element of entrepreneurial activity is innovation (Schumpeter, 1934). The scholar viewed the categories of "development", "innovation", and "entrepreneur" as closely related: development is achieved through innovation, which entrepreneurs lead. In this way, growth is hardly possible without the participation of an entrepreneur in some way.It is worth mentioning, he agreed with Karl Marx on the self-destructive nature of capitalism, naming the decadence of the entrepreneurial function, destruction of the institutional framework of capitalist society and disintegration of bourgeoisie family as the fundamental forces of this process.

Interestingly, through these forces, we see the relationship with the concept of social capital – social capital reduces transactional costs, which, in turn, allows an entrepreneur to create and implement innovations easier, generate profit for themselves, and promote the development of the economy as a result whole.

The category of social capital is actively utilized in contemporary studies of societies – operations of companies and businesses, interrelations of network structures, solidarity inside a society and multiple other ways. Creating this unique form of capital requires three major elements: social networks, standard norms, and trust. The special characteristic of social capital is incarnated not in subjects and objects but rather in relationships between them. It can't be purchased, neither can it change ownership. A significant role in developing this concept belongs to P. Bourdieu, R. Putnam and F. Fukuyama; most contemporary studies are either based or inspired by these three pillars. Bourdieu (1986), viewed social capital as resources built on family relations and relations within groups. Social capital is incarnated in social networks, social norms of behavior, and cooperation to obtain mutual benefits. He also views it as a way of inequality generation. Fukuyama (2001), perceived social capital as an instantiated informal norm that promotes cooperation between individuals. In the economic sphere, it reduces transaction costs. The political sphere encourages the kind of associational life necessary for the success of limited government and modern democracy. Robert Putnam (1995, 2001) viewed social capital as a uniting factor in American society, but focuses on negative trends that he traces back to 1960s. According to him, the decline in social capital leads to rising rates in unhappiness and political apathy. Their approaches differ in many aspects; however, their positions are common in one aspect – state institutions are excluded from social capital analysis, while civil society institutions are an indispensable part of the discourse.

The most recent studies focus on specific aspects of relationship between entrepreneurship and social capital. For example, research of Setini, *et al.*, (2020) examines the role of social capital in developing the creative industry of women entrepreneurs in Bali, Indonesia, by looking the role of information sharing and innovations. The study performed on 200 female respondents suggests that social capital positively influences the business performance of women entrepreneurs. Another survey of Harjanti&Noerchoidah (2017) suggests that social capital significantly affects innovation capability, while high level of knowledge collecting and knowledge donating can lead to high level of innovation capability.Escandon-Barbosa, Urbano-Pulido, & Hurtado-Ayala, (2019).examined how social capital as a characteristic of the institutional environment affects the relationship between formal and informal institutions and entrepreneurial activities, differentially, in developing and developed economies and concluded that in developing countries, this influence is greater in the relationship between property rights, access to credit, subjective insecurity, and entrepreneurial activity.

FINDINGS

A useful instrument in the analysis of social capital and entrepreneurship is the European Social Survey. By the moment this publication was prepared, the last dataset available was for the 2018 and covered most European countries. A glance at some data available for analysis allows one to formulate multiple hypotheses in this area and perform further in-depth studies. For example, it will enable to get a snapshot of the male and femaleself-employed or is a part of the labor market as employees (see Table 1).

As we can see in the table, there is a different prevalence of male population in the Self-Employed category, compared to the female population of European countries. The reason of this can be discussed and require further study. An exciting picture could be revealed also, if this data is studied with respect to country. An important characteristic of social capital is related to values of entrepreneurs. In Table 2, the value of generation of new ideas and being creative with respect to employment status is demonstrated.

Table 1. Share of male and female population in different types of employment.

Gender	Male	Female	Total	N=
Employment relation				
Employee	45.2	54.8	100	37034
Self-employed	62.6	37.4	100	4784
Working for own family business	46.4	53.6	100	789
Total	47.2	52.8	100	42607

Table 2. Value of creativity and importance of new idea generation related to type of employment

Important to think new ideas and being creative	Very much like me	Like me	Somewhat like me	A little like me	Not like me	Not like me at all	Total	N=
Employment relation								
Employee	16	32.5	26.5	14	8.7	2.3	100	36206
Self-employed	27.7	34.9	21.5	9.9	4.5	1.5	100	4698
Working for own family business	18.9	31.9	26.8	14.3	6.9	1.2	100	771
Total	17.4	32.8	26	13.5	8.2	2.2	100	41675

The table suggests that creativity and the generation of new ideas can be seen in European entrepreneurial group identity more often than in the identity of employees.

REFERENCES

Aldrich, H. E., & Kim, P. H. (2007). Small worlds, infinite possibilities? How social networks affect entrepreneurial team formation and search. *Strategic Entrepreneurship Journal*, *1*(1/2), 147–165. doi:10.1002ej.8

Arregle, J. L., Hitt, M. A., Sirmon, D. G., & Very, P. (2007). The development of organizational social capital: Attributes of family firm. *Journal of Management Studies, 44*(1), 73–95. doi:10.1111/j.1467-6486.2007.00665.x

Bourdieu, P. (1986). The forms of capital. In J. G. Richardson (Ed.), *Handbook of Theory and Research for the Sociology of Education* (pp. 241–258). Greenwood Press.

Burt, R. S. (2000). The network structure of social capital. *Research in Organizational Behavior, 22*, 345–423. doi:10.1016/S0191-3085(00)22009-1

Carter, R., & Van Auken, H. (2006). Small firm bankruptcy. *Journal of Small Business Management, 44*(4), 493–512. doi:10.1111/j.1540-627X.2006.00187.x

Escandon-Barbosa, D., Urbano-Pulido, D., & Hurtado-Ayala, A. (2019). Exploring the Relationship between Formal and Informal Institutions, Social Capital, and Entrepreneurial Activity in Developing and Developed Countries. *Sustainability, 11*(2), 550. doi:10.3390u11020550

Fukuyama, F. (1999). *Social capital and civil society*. Paper prepared for delivery at the International Monetary Fund Conference on Second Generation Reforms.

Fukuyama, F. (2001). Social capital, civil society and development. *Third World Quarterly, 22*(1), 7–20. doi:10.1080/713701144

Greve, A., & Salaff, J. W. (2003). Social networks and entrepreneurship. *Entrepreneurship Theory and Practice, 28*(1), 1–22. doi:10.1111/1540-8520.00029

Harjanti, D., & Noerchoidah, N. (2017). The Effect of Social Capital and Knowledge Sharing on Innovation Capability. *Jurnal Manajemen dan Kewirausahaan, 19*, 72–78.

Hernández-Carrión, C., Camarero-Izquierdo, C., & Gutiérrez-Cillán, J. (2016). Entrepreneurs' Social Capital and the Economic Performance of Small Businesses: The Moderating Role of Competitive Intensity and Entrepreneurs' Experience. *Strategic Entrepreneurship Journal, 11*(1), 61–89. doi:10.1002ej.1228

Ibn, K., & Rosenthal, F. (1967). *The Muqaddimah: An introduction to history*. Princeton University Press.

Kim, N., & Shim, C. (2018). Social capital, knowledge sharing and innovation of small- and medium-sized enterprises in a tourism cluster. *International Journal of Contemporary Hospitality Management, 30*(6), 2417–2437. doi:10.1108/IJCHM-07-2016-0392

Nahapiet, J., & Ghoshal, S. (1998). Social capital, intellectual capital, and the organizational advantage. *Academy of Management Review*, *23*(2), 242–266. doi:10.5465/amr.1998.533225

Pirolo, L., & Presutti, M. (2010). The impact of social capital on the start-ups' performance growth. *Journal of Small Business Management*, *48*(2), 197–227. doi:10.1111/j.1540-627X.2010.00292.x

Putnam, R. D. (1995). Bowling alone: America's declining social capital. *Journal of Democracy*, *6*(1), 65–78. doi:10.1353/jod.1995.0002

Putnam, R. D. (2001). Social Capital: Measurement and Consequences. *Isuma: Canadian Journal of Policy Research*, *2*, 41–51.

Putnam, R. D. (2001). *Bowling Alone: The Collapse and Revival of American Community*. Touchstone.

Schumpeter Joseph, A. (1934). *The Theory of Economic Development, Cambridge*. Harvard University Press.

Setini, M., Yasa, N. N. K., Gede Supartha, I. W., Ketut Giantari, I. G. A., & Rajiani, I. (2020). The Passway of Women Entrepreneurship: Starting from Social Capital with Open Innovation, through to Knowledge Sharing and Innovative Performance. *Journal of Open Innovation*, *6*(2), 25. doi:10.3390/joitmc6020025

Sherry, R., & Stubberud, H. A. (2013). Partnerships for innovation among European small businesses. *Journal of International Business Research*, *12*(2), 97–108.

Teckchandani, A. (2014). Do membership associations affect entrepreneurship? The effect of type, composition, and engagement. *Nonprofit and Voluntary Sector Quarterly*, *43*(2_suppl), 84S–104S. doi:10.1177/0899764013502580

Zaridis. (2016). The Relation Between Entrepreneur's Characteristics and Competitive Advantage. A Roadmap to SME's Success. *International Journal of Emerging Research in Management & Technology*.

Chapter 14
Role and Prospects of Women Entrepreneurs in the Agriculture Sector of India

Faran Ahmad Qadri
King Abdulaziz University, Saudi Arabia

Saima Shadab
Aligarh Muslim University, Aligarh, India

Arifa Khan
Aligarh Muslim University, Aligarh, India

ABSTRACT

This chapter mainly focuses on the role and prospects of women entrepreneurs (or women agri-entrepreneurs) in India's agriculture sector. India has witnessed unprecedented growth in the total number of entrepreneurs and innovations over the past many years. Despite this appreciable growth, the role of women entrepreneurs remains devitalised and underutilised, which requires proper attention by the government and other stakeholders of the country. However, over the past many years, the government has taken various crucial initiatives to promote the role of women entrepreneurs, especially in the agriculture sector. As a result, there has been a remarkable transformation in the share and contribution of women entrepreneurs engaged in the agriculture sector. Therefore, the chapter examines the trend and pattern of women agri-entrepreneurs in India and highlights their challenges.

DOI: 10.4018/978-1-6684-2349-3.ch014

INTRODUCTION

It is an undeniable fact that the participation of entrepreneurs plays a vital role in the Indian economy. Vesper (1990) defined the broad field of entrepreneurship as creating new ventures by individuals or small groups. Further, Marshal expanded entrepreneurship as a) Structuring of an organisation through division of labour, b) Formulating plans and policies to be executed by a subordinate, c) Innovations, and d) Bearing of risks. Similarly, Hagen (1960) described entrepreneurship as creating opportunities for investment and production, establishing an organisation capable of introducing production and non-production process, collection of raw materials, accumulation of capital, searching for new sources of raw materials, innovating new production techniques and new products, and to choose the efficient manager in order to run the day-to-day affairs of the organisation.

Entrepreneurship is paramount to our country's economy as it is a major driving force behind economic transformation and diversification. Over the past many years, the Indian economy has witnessed appreciable growth in many young entrepreneurs present. In this context, entrepreneurs engaged in the agriculture sector (or agri-entrepreneurs) are providing a significant economic contribution to their families and society in India. Alongside this, there has also been a significant increase in the role of women in the country's level of entrepreneurship and innovations in the agriculture sector (also known as women agri-entrepreneurs). Women entrepreneurs are instrumental for sustainable and stable economic growth. There has been a rise in the number of women entrepreneurs in India, especially since the 1990s. In the case ofIndian society, entrepreneurship, specifically, women agri-entrepreneurs, face severe challenges in the economy. Entrepreneurship needed an entrepreneur to be a superior capability who pursues her goals despite several obstacles, opposition, setbacks, and failure. The emergence of women agri-entrepreneurs and their contribution to the national economy is quite visible in India. Women entrepreneurs need to be applauded for their increased utilisation of modern technology, finding a niche in the export market, increased investments,creating sizable employment for others and setting the trend for other women entrepreneurs in the organised sector (John and Kispotta, 2016).

In order to encourage the participation of women in agri-entrepreneurship, there are a significant number of initiatives and schemes implemented by the Government over the past many years that have been highlighted in one of the subsequent sections of this study. However, to date, several factors serve as a significant impediment in this process, and women-led entrepreneurship requires continuous focus and efforts in the right direction. Various challenges such as gender disparities, social issues like societal and family pressure, crimes, lack of education, and improper implementation of laws that empower women are prime reasons behind the slow progress of women entrepreneurs in India, specifically in the rural sector. Specifically, the attention of the Government is much needed on improving the scenario of women entrepreneurs in the rural sector of India. Given the negative impact of the COVID-19 pandemic, one of the most viable solutions to revive the economy is encouraging women entrepreneurship as it embeds enormous potential to increase economic growth and development (Shah, 2020).

Due to secondary data limitations, this study also takes into account a review of a large number of existing studiesto analyse the significant factors that serve as significant impediments behind the low performance of women entrepreneurs with a specific focus on those who are based in the agriculture sector (also known as women agri-entrepreneurs).The importance of agri-entrepreneurship for India's economic growth has received great attention from policymakers and the Government. In the context of India's agriculture sector, women entrepreneurs are engaged in Agricultural Establishments that are most commonly engaged in the production of agriculture products (excluding production of crops or plantation by farmers), services related to agriculture, and other such activities. Several studies have been conducted to identify the challenges and opportunities that women agri-entrepreneurs face in India. For instance, Sivanesan (2014) examined variousproblems women entrepreneurs face in rural areas and analysed various motivating and de-motivating factors of women entrepreneurship in rural and urban areas in the Kanyakumari district of Tamil Nadu concerning the Garrett ranking method. The analysis suggested policies for eliminating and reducing constraints of women entrepreneurship in both rural and urban areas. **Sasikala (2015)** examined the socio-economic status of the women entrepreneurs in the Salem district of Tamil Nadu. The study analysedthat the prime problem of women entrepreneurs, as shown by the total and mean values of Garrett ranking technique, was the prevalence of frequent arguments with labours, which is followed by hurdles with administrative staff and feeling of restlessness quite often followed by the lack of support from husband/family members and the inability of women entrepreneurs to compete with men entrepreneurs.

Similarly, **Aggarwal and Jain (2014)** suggested that women entrepreneurs face numerous challenges like family obligations, male domination, lack of education, finance problems, limited mobility, self-confidence, limited managerial skills, etc.

Whereas the Indian Government has taken many policy measures yet, there is a long way to go. The paper highlighted the role of women entrepreneurs inthe 21st century and difficulties faced by Indian women entrepreneurs, how to overcome them and to quantify the policies of the Government of India for their empowerment. **Tamilarasi (2013)** examined the hurdles of women entrepreneurs in the Salem district of Tamil Nadu. The analysis was limited to 97 service enterprises and was chosen by random sampling. The study has applied the Garrett ranking technique to find out the problem faced by women entrepreneurs. They were asked to rank by both start-up and growth up phase of their enterprise. The study results showthat women entrepreneurs face many challenges and constraints at the start-up phase due to a lack of finance and poor business skills. At the growth phase, the major problem affecting women entrepreneurs is lack of finance, tax harassment and finding qualified labour. Regarding gender-related issues, mobility and time management between family and business have initially severely affected women, entrepreneurs.

In contrast, time distribution between family and business and the non-acceptance of women's authority have been the major problems for entrepreneurs at the growth phase. Thus, policymakers must consider while framing policies to ensure easy access to finance for women entrepreneurs. Business skills could be developed through experience only, and hence they could be educated to face the challenges with courage and confidence. Time distribution between family and business affairs has been the most significant constraint for women entrepreneurs at the start-up and growth phases.Sarma and Swamy (2012) studied exploring the introspection of women entrepreneurs with the help of the Garrett ranking approach. For the study of introspection, three components were selected; conceive the idea of doing business, the perception of women entrepreneurs on their entrepreneurial capability, and how they are perceived and face the business challenges and risks. The finding revealed from the study that women who expertise their hobby and further it may be converted into the business. "Ability to deal with the situation" is the most important talent they have for their success. Communication skills help women entrepreneurs for their succeed in the business. The 'entry of large business players' is the greatest problem of women entrepreneurs. Women entrepreneurs should perceive the risks as stepping stones for success and improve confidence in their entrepreneurial ability.

RECENT TREND OF WOMEN ENTREPRENEURS IN THE AGRICULTURE SECTOR

At present, female entrepreneurs in India secure a 17 per cent share in the country's total GDP. As per the Global Entrepreneurship Monitoring Report, 2020, India is one of the countries with the lowest rates of female entrepreneurs in early-stage entrepreneurial activity (Bosma et al., 2021). According to the report, ensuring inclusiveness in entrepreneurship in the country is essential. This is because, lack of inclusiveness in entrepreneurship impedes the level of income, a new range of products or services that could have been produced, and jobs that could have been created in the economy. Therefore, ensuring inclusiveness in entrepreneurship is of paramount importance to an economy such that resources can be utilised to their total capacity (Bosma, N et al., 2021). As per the latest data released by Global Entrepreneurship Monitor, India's performance in women entrepreneurs has been abysmally low as the Female/Male Total Early-Stage Entrepreneurial Activity (TEA) stood as low as 0.30 per cent in the year 2020. According to States' Startup Ranking, 2019, around 15.7 million MSMEs and agribusiness in India are owned by women entrepreneurs (Shukla, Bharti, and Dwivedi, 2021).

A comparative analysis of data recorded in the Fifth and Sixth Economic Census confirms the appreciable growth in the number of overall businesses in the agricultural sector in India. The total number of agricultural establishments (other than crop production and plantation) was 6.08 million in 2005, i.e., 14.54 per cent of the total establishments. From this, the total number of agricultural establishments increased to 13.13 million in the year 2014. This implies that over the period 2005-2014, the number of agricultural establishments has increased by 115.95 per cent.

Apart from this, of the total 21.81 million "own account establishments", around 5.13 million (i.e., 19.05 per cent) establishments were agricultural in 2005.

As per the Sixth Economic Census, of the total establishments in the country, 13.13 million establishments exist within the agriculture sector, which is around 22.45 per cent of the total establishments across the country. Further, out of the total 13.13 million agricultural establishments, around 92 per cent are concentrated in the rural sector. This highlights the significance of India's rural areas in the context of entrepreneurship in the Agriculture Sector. The importance of rural areas in terms of agri-entrepreneurship is not just limited to the total number of agricultural businesses. However, rural areas also offer an appreciable share of employment opportunities to both female and male workers.

The importance of women entrepreneurs in the Indian economy was given special attention in the Sixth Economic Census that consisted of a particular chapter covering data on this area. Following important points were highlighted in the chapter that shows the increase in the role of women agri-entrepreneurs in the economy:

- As per the latest data shared in the Sixth Economic Census, the percentage share of women out of the total number of entrepreneurs in India is 13.76 per cent. Further, the data reveals that the share of female workers employed in agricultural businesses in the rural areas stood at around 92.6 per cent as against 36.03 percentage share of female workers in agricultural establishments in the urban areas. This indicates the high concentration of women agri-entrepreneurs within the rural sector in India.
- Table 1 shows the share of establishments owned by women in terms of the top five major economic activities. As evident from the table, the agriculture sector secures the largest share of establishments owned by women entrepreneurs in terms of economic activities. The share of establishments in the agriculture sector is around 34.3 per cent of the total establishments in 2014. This indicates the importance of women entrepreneurs in the

agriculture sector and their contribution to its economic growth. As per the latest data revealed in the sixth economic census, presently, over 2.76 million establishments in the agricultural sector are owned by women. Furthermore, the manufacturing sector secured the second-largest share of establishments owned by women entrepreneurs in the top five economic activities (29.8 per cent of the total share). The third, fourth, and fifth largest share has been secured by the trade, other services and food & accommodation services sector as evident from table 1.

Table 1. Top five economic activities as per establishments owned by women entrepreneurs

Economic Activity	Agriculture	Manufacturing	Trade	Other Services	Accommodation and Food Services
Share of establishments in the sector (in percentage)	34.3	29.8	18.23	5.38	2.77

Source: Sixth Economic Census

- A similar trend and pattern were observed for the top five economic activities regarding the total number employed in establishments owned by women agri-entrepreneurs. The Agriculture sector holds the largest share in terms of the total number of people employed by women-owned establishments, followed by the Manufacturing sector, Trade, Education and Other Services sector. This again indicates the importance of the agriculture sector and women entrepreneurs engaged within this sector for the economy.
- Within the agriculture sector, the largest number of women-owned establishments are found in Livestock related agriculture activities that have around 92.20 per cent of the total women establishments. The share of establishments in agricultural activities such as forestry and logging, fisheries and aquaculture, and agriculture excluding crop production is relatively insignificant, as evident in table 2.

Table 2. Percentage share of total establishments in different agricultural economic activities

Type of Agricultural Economic Activity	Livestock	Forestry and Logging	Fisheries and Aquaculture	Agriculture excluding crop production
Percentage share of total establishments owned by women entrepreneurs	92.20	4.51	1.4	1.89

Source: Sixth Economic Census

- A similar pattern and trend were observed for the number of people employed in different agricultural establishments. The largest number of people employed (around 90.8 per cent) were in the Livestock related women-owned agricultural establishments, followed by forestry & fishing.
- The top five states in terms of the share of the total number of establishments owned by women entrepreneurs are secured by Tamil Nadu, having around 13.51 per cent of such establishments, followed by Kerala, Andhra Pradesh, West Bengal. Maharashtra has the lowest share, i.e. only 8.25 per cent of the total number of women-owned establishments.
- Another noteworthy point is that as per the sixth economic census, in terms of source of financing, 79 per cent of the women-owned establishments were self-financed. Only 3.4 per cent and 1.1 per cent of the total finances were sourced from 'Government Assistance and Borrowing from Financial Institutions'. This indicates the challenge of the lack of appropriate financial facilities for the women entrepreneurs engaged in both rural and urban sectors.

RECENT SCHEMES AND INITIATIVES FOR DEVELOPMENT OF WOMEN ENTREPRENEURSHIP IN AGRICULTURE SECTOR

Standup India Scheme: The primary focus under this scheme is on facilitating financial facility to greenfield enterprises in terms of bank loans of the amount Rs. 10 lakh to up to Rs. 1 crore to at least one woman, one Scheduled Caste (SC) or Scheduled Tribe (ST) entrepreneurs per bank branch. This scheme aims to reduce the challenges faced by entrepreneurs who are specifically Women, SC or ST.[1] The enterprise, however, can only be engaged in agro-allied manufacturing, services or trading activities. The scheme was launched in 2016 and aimed to provide entrepreneurial support to only those Women, SC and ST entrepreneurs engaged in the manufacturing, services, and trading activities. Recently, the Government of India took a laudable step of announcing in the Union Budget 2021 the inclusion of Agriculture related Activities (dairy, fishery, agriclinic, agribusinesses, agro-processing, Livestock, pisciculture, etc.) in the Stand-Up India Scheme as well in order to widen the outreach of the scheme.[2] Further, the margin money requirements for loans under this scheme have been reduced considerably for the Activities related to Agriculture. The scheme has also been extended up to 2025. Until July 2021, the Government has sanctioned over 1.17 Lakh applications with over 26.4 crore sanctioned amount for these applications. The main stakeholders of this scheme are two Connect Centres, i.e., SIDBI& NABARD, Lead District Managers, DLCC, Bank Branches, and lastly, Borrowers who are designated different roles and responsibilities for proper implementation of the scheme.

Udyogini Scheme: The term Udyogini means Women, entrepreneurs. The World Bank first put forward this concept in 1992 to help Indian women entrepreneurs face a barrier-free business environment. Udyogini Scheme is an initiative implemented by the Women Development Corporation under the Government of India to encourage and empower women entrepreneurs in rural and underdeveloped areas. It mainly extends credit facilities such as interest-free loans to women engaged in entrepreneurial activities in the agriculture sector. Earlier, women entrepreneurs in the agriculture sector could apply for up to Rs 1 Lakh of loan. However, this limit has been increased to Rs. 3 Lakh presently. It also provides the opportunity of training programs for such women entrepreneurs. Several agri-businesses such as Bakeries, Agarbatti Manufacturing, Cane and Bamboo Manufacturing, Flour mills, Flower shops, Beauty Parlour, Handicrafts, Silk Thread Manufacturing, etc., have been supported under this scheme.

Dena Shakti Scheme: This is another scheme for the empowerment of Women Entrepreneurs in India. It is a scheme that extends financial support to women entrepreneurs through Dena Bank to help them fulfil their working capital requirements. The scheme provides financial facilities to women agri-entrepreneurs and entrepreneurs engaged in retail, micro-credit, and manufacturing sectors. Loans of up to Rs 20 Lakhs are offered to women agri-entrepreneurs at a concessional rate, i.e., at a rate of interest that is 0.25 per cent lower than the base rates. The loan facility is provided for a period of one to three years.

Cent Kalyani Scheme: This scheme is an initiative offered by the Central Bank of India to women entrepreneurs engaged in agriculture. The scheme mainly offers women agri-entrepreneurs who seek to start a new business project or expand an already existing start-up. It offers financial support for fulfilling the working capital and capital expenditure requirements of women involved in agri-business. It offers collateral-free loans of up to Rs.100 Lakhs.

Training Programs: To boost women entrepreneurship in the agriculture and rural sector, National Bank for Agricultural and Rural Development (commonly known as NABARD) has been fully committed to organising and conducting various training programs and sessions. These sessions are conducted specially for women to help them become self-dependent and successful entrepreneurs. For instance, recently, the Government of Goa, along with NABARD, extended financial support for establishing Rural Marts or Supermarts to help women entrepreneurs improve their business activities (NewsDesk, 2021).

Similarly, NABARD has also been conducting training programs on Micro Food Processing of Pickles and Jam in Arunachal Pradesh for the empowerment of women entrepreneurs in the rural areas by providing them proper skill and knowledge of the use of modern technology, machinery, and packaging techniques that are compatible to be employed in rural areas.

Challenges Faced by Women Agri-Entrepreneurs

The structure of the traditional society was largely gender-biased, which severely affected women's productivityby the distribution of their time,various responsibilities, and their lack of access to essential inputs, including knowledge, i.e., lack of education, social barriers (Garg and Agarwal, 2017). Similarly, another study examined that women own less than 20 per cent of the world's land, a survey of 34 developing nations by the United Nations Food and Agriculture Organization puts that percentage as low as 10 (Villa, 2017), such that they are not able to get benefits from several schemes for that to require land entitlement for providing financial assistance to farmers of India. Thus, women agri-entrepreneurs face financial problems while buying the critical inputs of agriculture. Along with this, other barriers that restrict women from flounder are personal obstacles like self-confidence and fear of failure, less market awareness,lack of skill, lack of entrepreneurial aptitude, and other operational barriers. Moreover, due to the unequal opportunities, women's working day is longer than that of men, and many women have reached the limits of continuity in stretching their day. Also, despite their hard labour, the gap with men is steadily widening. Women are engaged in labour in the agriculture sector,despite significantly less remunerative or unpaid work, and much of rural women work is not captured and highlighted in the official statistics. Secondary data is extremely limited in this area.

CONCLUSION

Agriculture, the major economic base of the Indian economy, employs almost 50 per cent of the workforce, which consists of 63.1 per cent of women. Among the pool of farmers, 70 per cent are women. These women farmers act as principal food producers, devote maximum time to agriculture but remain unreported or underreported in statistics. To do justice to their heavy work, their socio-economic needs require to be properly channelised.

India is a developing country and has a mixed economy, male-dominated society and women are considered to be socially and economically dependent on males. However, rural women are the major contributors in agriculture and its allied fields. Women entrepreneurs have basic indigenous knowledge, skill, potential and resources to manage agriculture activities. However, simultaneously women entrepreneurs faced lots of problems like lack of education, social barriers, legal formalities, high cost of production, male-dominated society, limited managerial ability, lack of self-confidence, harassment and not the fulfilment of rules and regulation etc. Several factors like positive and negative reinforcement influencing women entrepreneurs.

REFERENCES

Aggarwal, V. K., & Jain, S. (2014). Role of women entrepreneurs in 21st century India: Challenges and empowerment. *International Journal of Advanced Research, 2*(9), 834–840.

Bosma, N. (2021). *Global Entrepreneurship Monitor Report, 2020/21*. Author.

Garg, S., & Agarwal, P. (2017). Problems and prospects of woman entrepreneurship–a review of literature. *IOSR Journal of Business and Management, 19*(1), 55–60. doi:10.9790/487X-1901065560

Hagen, E. E. (1960). The entrepreneur as rebel against traditional society. *Human Organization, 19*(4), 185–187. doi:10.17730/humo.19.4.tl57064355861441

John, R., & Kispotta, W. (2016). Constraints Faced by the Women Entrepreneurs in Agribusiness in District of Allahabad, U.P. *IOSR Journal of Business and Management, 18*(1), 13-17.

NewsDesk. (2021, July 18). Goa is Committed to Encouraging Women Entrepreneurs: CM. *Goa Chronicle*. Retrieved from https://goachronicle.com/goa-is-committed-to-encouraging-women-entrepreneurs-cm/

Sarma, Y. V. S., & Swamy, A. K. (2012). Introspection of women entrepreneurs: A garrette's approach. *Sumedha Journal of Management, 1*(4), 4–12.

Sasikala, T. (2015). *Socio economic status of women entrepreneurs in Salem district of Tamilnadu*. Academic Press.

Shah, V. (2020, November 3). Boosting entrepreneurship among women – A golden opportunity to bring the economy back on track. *The Times of India*. Retrieved from https://timesofindia.indiatimes.com/blogs/voices/boosting-entrepreneurship-among-women-a-golden-opportunity-to-bring-the-economy-back-on-track/

Newman, A., North-Samardzic, A., Bedarkar, M., & Brahmankar, Y. (2021). *Entrepreneurship in India*. Routledge.

Sivanesan, R. (2014). Impact of Brand image and advertisement on consumer buying behavior–Comparative study on rural and urban consumers. *International Journal of Research in Management & Business Studies*, *1*(2), 73–80.

Tamilarasi, R. (2013). Constraints of Women Entrepreneurs in Salem District. *Asia Pacific Journal of Marketing & Management Review*, *2*(8).

Vesper, K. H. (1990). *New venture strategies*. University of Illinois at Urbana-Champaign's Academy for entrepreneurial leadership historical research reference in entrepreneurship.

Villa, M. A. B., Jaramillo, I. D. T., & Alzate, B. A. (2017). EmprendimientoCorporativo e Innovación: Una Revisión y FuturasLíneas de Investigación. *Espacios*, *38*, 1–18.

ENDNOTES

1 https://www.standupmitra.in/Home/SchemeGuidelines

2 https://www.standupmitra.in/

Chapter 15
Identifying the Barriers and Drivers to Agriculture Entrepreneurship in India

Sablu Khan
Aligarh Muslim University, Aligarh, India

Hesham Magd
Modern College of Business and Science, Muscat, Oman

ABSTRACT

This study mainly focuses on the identification of barriers and drivers to agriculture entrepreneurship. Entrepreneurship not only includes the creation of new ventures or start-ups but includes innovation and growth. Agri-entrepreneurship can address the current issues of lower farm mechanisation and low productivity of the livestock sector. It can be helpful for farmers in general and agri-entrepreneur in particular. Understanding barriers and drivers to agri-entrepreneurship have become a necessity in current times. Agriculture entrepreneurship has the ability to raise the living standards and helps in the creation of wealth not only for the entrepreneurs but also for other farmers. Agriculture entrepreneurship is beneficial in the global food supply in least-developed and developing countries. The purpose of this chapter is to familiarise the readers with the meaning, prospects, barriers, and key drivers of agri-entrepreneurship to engage in agri-entrepreneurship.

DOI: 10.4018/978-1-6684-2349-3.ch015

INTRODUCTION

Agriculture and its allied sectors is a crucial sector because it has significant role in poverty reduction, employment generation, availability of food supply and income. This sector faces many challenges, including access to credit, insurance coverage for harvest and post-harvest losses, irrigation facilities, etc. are need to be addressed. Entrepreneurship not only includes the creation of new ventures or start-ups but includes innovation and growth also (Gartner, 1990). Agri-entrepreneurship needs to address the issue of lower farm mechanisation, low productivity of the livestock sector so that it can be helpful for farmers in general and agri-entrepreneur in particular. Though the food processing sector and livestock sector are rising at the rate of five per cent and 8 percent, respectively, more actions are needed to check wastages and create better marketing of the products. Agriculture incorporates agricultural activities, including the growing up of crops, horticulture, animal husbandry and dairying. Entrepreneurship is a kind of change consisting of 3 parts, i.e., recognizing, assessing and exploiting opportunity, a new setup that can be managed by facilitating producing and selling new goods or services and through value creation.

The share of the primary sector in India's total Gross Domestic Product (GDP) has declined due to the relatively high growth in the manufacturing and service sectors. This is a transitional result of the development process of any country because of the faster growth of non-agricultural sectors. In Table 1, it is indicated that the share of GDP of the agriculture sector in the GDP of the Indian economy is continuously decreasing with the fact that the share of other sectors of the economy is constantly increasing as India is moving from an agrarian economy to service economy.

Table 1. Share of agriculture as per the total GDP of India

Year	Agriculture & Allied Sectors (Crops, Livestock, Forestry, Fishing)	
	Share in Total GDP	Annual Growth Rate (%)
2014-15	18.2	-0.2
2015-16	17.7	0.6
2016-17	17.9	6.3
2017-18	17.2	5.0
2018-19	16.1	2.9
2019-20	16.5	2.8
Source: Economic Survey 2020-21		

Agri-Entrepreneurship can play a vital role in providing employment and agri-business opportunity to farmers. Agriculture mechanisation has increased the productivity of agricultural fields by making optimum use of inputs including fertilisers, pesticides, machinery for different activities besides reducing the cost of farming. With the reduction in land holding, water resources and labour force, the main focus is on the mechanisation of production, post-harvesting operations. Availability of farm power increases the farm yield concerning agricultural operations. In India, farm power availability is likely to increase from approximately 2 kW per hectare (2016-17) to 4.0 kW per hectare by the end of 2030. Also, the Indian tractor industry is the leading industry globally, having about 33 per cent of the entire world's output (Economic Survey, 2019-2020). However, agriculture accounts for 16 percent of India's GDP but provides livelihood to more than half of India's population. Many sectors of agriculture and allied activities, e.g., food processing industries, dairy industry and livestock sectors are growing incessant; still, the GDP growth in agriculture has to portray an up and downtrend due to El-Nino led drought years. Livestock is one of the key sectors of the agricultural-based economy of India. India is an agriculture-dominated country, and livestock is the leading subsidiary sector comprising twenty-eight per cent of agricultural and allied sector GDP. The livestock sector, including dairying, poultry farming, and leather-based industries, can play a significant role in national GDP. The dairying industry is consistently growing ata 15 per cent compound annual growth rate from 2016-20. Milk production has reached around 175 million tonnes in the year 2018. With the rising Internet and communication technology and the awareness among general customers, the inclination towards better alternatives has been rising among the new-age consumers.

Still, small and marginal farmers, which consist of more than 86 per cent of India's population, contributes about forty per cent of cereals production and more than 50 per cent of horticultural products and vegetable production. India exports rice and wheat to other countries of the world to fulfil the people's dietary needs.

AGRI-ENTREPRENEURSHIP AND ITS SIGNIFICANCE

Agri-ownership is primarily linked with the production and marketing of different agricultural products and agro-processed food items. Agri-entrepreneurs also includes generating inputs to agriculture and help the farmers for quality of seeds and rental machinery for better cultivation of crops or farming of different livestock. Agricultural entrepreneurs involved in dairy, poultry, forestry and horticulture. Agrarian entrepreneurship is a process in which agricultural products are produced and managed for marketing. Entrepreneurship is a way of life, a thought process to bring any sustainable change; the effort must be broader. The entrepreneurial orientation is assessed based on his risk-taking ability, quick decision-making ability, and better planning and organising abilities to raise the agri-outputs. The entrepreneurial behaviour of farmers has gained importance since farmers have to change their knowledge, skills and attitudes towards their selected enterprise.

Agriculture entrepreneurship is considered an individual's attitude and strategies for modernising farming and value-creation (Gartner, 1990; Condor, 2020).Many farmers respond better to the new ventures linked with a high yielding variety of seeds, chemical fertilisers and pesticides, and new agricultural equipment than the others, thus named agricultural enterprisers (Bhattacharya, 1986). Shirur, Shivalingegowda, Chandregowda, Manjunath, & Rana (2019) in their study, found that entrepreneurial behaviour of consists of innovativeness, result orientation, profit motivation, technical proficiency, risk-taking ability, information-seeking behaviour, scientific direction, leadership ability, management orientation and marketing strategy. Singh (1969) revealed that self farming by the farmers is entrepreneurship as they are independently growing crops and bears the risk. Agriculture entrepreneurship is pertinent for various reasons (Milestad, Ahnström, & Björklund, 2011). It has the ability to raise the living standards and helps in the creation of wealth not only for the entrepreneurs but also for related businesses. Agri-business is a combination of both agriculture and business activities (Bairwa, Lakra, Kushwaha, Meena, & Kumar, 2014). Hadimani (1985) found that agriculture entrepreneurship is growing crops and other farming activities with the help of the latest technology and earning the profit irrespective of its landholding size. Heredero (1979) outlined that an agricultural entrepreneur as a person who manages changes that directly or indirectly lead to higher agricultural output.

Agri-entrepreneurship includes new venturing (Hulsink & Dons, 2008; Grande, Madsen, & Borch, 2011) and non-agricultural activities (Richards & Bulkley, 2007). Agri-entrepreneurship can help the farmers reduce this high markup, ultimately helping them fetch better remuneration for their produce. Further, agri-entrepreneurship minimises the role of middlemen for trading and marketing agricultural produce, thereby empowering the Indian farmers. It helps to raise farmers' income, thereby reducing poverty (Nagler & Naude, 2014). There is more focus on agri-entrepreneurship in the developed world, especially in west European, North American countries (Fitz-Koch & Nordqvist, 2017; Hassink, Grin, & Hulsink, 2013, 2016). However, Agricultural entrepreneurship as a concept is still evolving in the Indian agrarian society. Agriculture entrepreneurship is very helpful in the global food supply in the least developed and developing countries (Wilkinson, 2009).

Table 2. Current status of agri startups in India

Category	Total No.
Agri-Tech	2888
Dairy Farming	293
Food Processing	732
Organic Agriculture	1344
Others	964

Source: Startup India (2021)

This year, 3.8% of the total start-ups were registered in agriculture, and Maharashtra (18%) has the highest number, followed by Karnataka, Delhi and Uttar Pradesh (IndSamachar, 2020). The highest category of the start-up is Agri-Tech, followed by Organic Agriculture. Agri-tech start-ups help in the improvement of the farm-wise yield, information-driven assessments to improve the quality of soil and data collection concerning crops health in agriculture fields through remote sensing and precision farming methodology. Equipment renting services are the game changer for small and marginalised farmers who otherwise cannot afford agricultural purposes and can benefit the agri-entrepreneurs in renting services start-ups.

BARRIERS TO AGRICULTURAL ENTREPRENEURSHIP

Due to the erratic pattern of climate and monsoon and hindrances in the market's trading system, agriculture is less profitable, which hampers the farmers' will in engaging in agri-business activities. Most farmers in India are small and marginalised; they don't have adequate capital, resources, and agricultural equipment for carrying out agri-business activities. This low asset base minimises their abilities involve in agri-entrepreneur for sustaining the profit and employment generation.

The socio-economic condition of agricultural farmers is poor because most farmers have small and fragmented land holdings and are engaged in subsistence farming. The rise in population at a faster pace makes land shrinking a major problem for farmers.

The natural calamities and their direct impact on Indian agriculture are impediments to the progress of agri-entrepreneurship developments. Recently, the primary effect of the COVID-19 pandemic on the agricultural economy hampered the Indian food sector, dairying sector and meat processing industry.

Collaboration among the marginalised and small farmers is a key factor in determining the success of agri-entrepreneurs (Mars & Schau, 2017). The cooperative spirit among the small and marginalised farmers is at bay. Their group formation hampers in one way and another in the absence of their formal training in soft skills, which ultimately affects the entrepreneurial orientation of farmers.

Infrastructure problems such as lack of metalled roads between villages and cities, lack of storage facilities to store the vegetables, fruits, poultry products and other dairy products at village level and high cost of storage place are the major impediments in the rural areas for the agri-entrepreneurs. Water and electricity facilities on their farm also lack in different parts of the country.

Rising Agri-input costs and stagnant prices of production affect the profitability of Indian agriculture. The presences of middlemen and the plurality of middlemen are the most important problems because middlemen purchase the produced materials at reducing the rate and by payment of cash.

Lack of finance and unavailability of different types of resources for starting the agri-starts up hinders the agri-entrepreneurship in India. Further, lack of knowledge about cropping patterns, crops diseases, and medicine hampers the yield of crops. Skills training for Indian farmers are very poor, and most of the farmers traditionally do agriculture. Lack of knowledge is the leading cause of excessive nitrogenous fertilisers in agriculture, which further deteriorates the soil quality.

Lower socio-economic status either due to possession of marginal landholdings, high dependency ratio hinders the agriculture entrepreneur from investing in the new ventures or starting the agribusiness in India.

Table 3. Summary of key barriers to agri-entrepreneurship by previous researchers

Key Barrier	Authors
Financial constraints	Petrova (2012); Paulson & Townsend (2004); Alessa, Zaabi, & Diab (2018).
Poor Infrastructure support	Verma, Sahoo, & Rakshit (2017); Esiobu & Ibe, (2015).
Lack of Awareness of Government Concessions	Moraru, Ungureanu, Bodescu & Donosă (2016); Verma, Sahoo, & Rakshit (2017); Simmons, Durkin, McGowan, & Armstrong (2007)
Non-availability of updated machinery	Dixit, Rawat, Sharma, Mann, & Kumar (2016); Shehrawat (2006).
Lack of confidence	Depositario, Aquino, & Feliciano (2011)
Low educational status	Onchiri, Maurice, & Patrick (2015)
Linkage to high-level market	Dung, Bonney, Adhikari, & Miles (2020).
Source: Prepared by Researcher	

More than 40% workforce is involved in agriculture, with the prevailing case of disguised employment. According to the Periodic Labour Force Survey (2017), only 14% workforce in 15-59 years has received formal training. The majority of them learned the work through self-experience, family or on-the-job training. Based on the previous work, the following key barriers to agri-entrepreneurship were identified by the earlier researchers in table 3.

DRIVERS TO AGRICULTURAL ENTREPRENEURSHIP

Much of the global share of food staples such as rice and wheat come from India, and almost half of the population in India depends on agriculture for their livelihood. The dairy sector plays an important role in improving their economy and had much more prospects in the future. Indian agriculture is known for the high markup (i.e. the difference between the price that farmers pursue their produce and what consumers pay) for agriculture produce. It stands around 65% for India.

Farmers can avail the agricultural loan for short term farm needs under the Kisaan Credit Card Scheme at a subsidised rate. They can utilise the same for better farm inputs during the cropping season and meet the needs of the post-harvest season.

Getting the correct information about agriculture provided by D.D. Kisan Channel, Choupals, and information about various government schemes related to loans by banks can help the entire agri-entrepreneurs e.g. Digital advisory application by ITC's Agribusiness team like "e-Chaupal" to procure products from the farmers disseminates information to farmers about the prevailing prices of food and helps them fetch competitive prices.

COVID-19 pandemic led to unemployment in the informal sector. The agriculture sector received many migrant workers; agri-entrepreneurs can make use of surplus labour and reduce the cost of agri-business activities can help the agri-entrepreneur in reaping the benefit.

Reserve Bank of India has updated Priority Sector Loans norms in 2015, categorising a minimum quota of 10% for agriculture farmers and 8% for small and marginal farmers. The government of India has launched two schemes, i.e. Mahila Samriddhi Yojana and Standup India, to assist the entrepreneurs in providing concessional loans.

The recent developments in market facilitation for agri-products like e-NAM for connecting Agricultural Produce Market Committee (APMC) mandis for barrier-free trade in agriculture. Also, India's export of agri-commodities will touch the mark of $60 billion by 2022. Based on the previous work, the following key drivers to agri-entrepreneurship were identified by the earlier researchers (Table 4).

Table 4. Summary of key drivers to agri-entrepreneurship by previous researchers

Key Drivers	Authors
Profit Margin	Vijayan and Shivkumar (2020); Yang and An (2002)
Government Support for agri-entrepreneurship	Michael and Pearce (2009).
Self Development	Wang, Chang, Yao, and Liang (2016)
Job Security	Deepthi (2016).
Market Reforms	Manjula (2021); Kelegama and Hirimuthugodage (2019); Kumar (2017).
Source: Prepared by Researcher	

WAY FORWARD

Farmers' skill training for setting up agri-startup has become necessary to incorporate modern farm technology and information and communication technology for farm practices. Agri-entrepreneurs should also focus on inter-organisation networking to link horizontal and vertical networks to develop agricultural activity.

As most of the Indian workforce is engaged in agricultural activities, there is an urgent need to shift all the disguised unemployed workers from agriculture to manufacturing/ service sectors after proper skill training so that the non-agriculture sector can absorb the labour from a growing population. New training centres should be formed to promote agri-entrepreneurship in India, focusing on growing the farmers' income by rationalising the input cost and better remuneration. In the developed countries, the workforce involved in agriculture is very low.

Small and marginalised farmers should be equipped with the latest knowledge of high yielding seeds, fertilisers and credit facilities with reasonable subsidies and concessional interest rates and their requirements. Collaboration should be developed between the veterinary colleges and universities and agriculture entrepreneurs to provide knowledge and training for the development and maintenance of dairy and poultry farms and educate them about modern and scientific techniques. Infrastructural facilities should be made available to agri-entrepreneurs regarding cold storage, transport, updated information and post-harvest processing in nearby towns and villages. Cooperative farming should be enhanced to raise a progressive and productive agricultural system. Collateral free loans should be provided to agriculture entrepreneurs to start their agribusiness, and they should assist them in facilitating credit accessibility from the financial institutions.

REFERENCES

Alessa, A. A., Zaabi, E. A., & Diab, A. M. (2018). Impact of Environmental Factors on Financing Agriculture Entrepreneurs. *International Journal of Entrepreneurship*, *22*(4), 1–13. https://www.abacademies.org/articles/Impact-of-environmental-factors-on-financing-agriculture-entrepreneurs-1939-4675-22-4-228.pdf

Bairwa, S. L., Lakra, K., Kushwaha, S., Meena, L. K., & Kumar, P. (2014). Agripreneurship development as a tool to the upliftment of agriculture. *International Journal of Scientific and Research Publications*, *4*(3), 1–4. https://www.academia.edu/download/33331824/ijsrp-p2770_agripreneurship.pdf

Bhattacharya, S. N. (1986). *Rural economic growth in India in theory and practice*. Metropolitan Book Company. Retrieved from https://www.abebooks.com/9788120000797/Rural-economic-growth-India-theory-812000079X/plp

Condor, R. (2020). Entrepreneurship in agriculture: A literature review. *International Journal of Entrepreneurship and Small Business*, *40*(4), 516. doi:10.1504/IJESB.2020.109013

Deepthi, V. (2016). *A critical study on entrepreneurial behaviour of agripreneurs in Andhra Pradesh* (Doctoral dissertation). Acharya NG Ranga Agricultural University.

Depositario, D. P. T., Aquino, N. A., & Feliciano, K. C. (2011). Entrepreneurial skill development needs of potential agri-based technopreneurs. Journal of ISSAAS.

Dixit, A. K., Rawat, I., Sharma, S., Mann, S., & Kumar, R. (2016). Entrepreneurship through EDP and Licensing of Post-Harvest Technology: An Impact Assessment. *Indian Journal of Economics and Development*, *12*(4), 679–686. doi:10.5958/2322-0430.2016.00191.8

Dung, T. Q., Bonney, L. B., Adhikari, R. P., & Miles, M. P. (2020). Entrepreneurial orientation, knowledge acquisition and collaborative performance in agri-food value-chains in emerging markets. *Supply Chain Management*, *25*(5), 521–533. doi:10.1108/SCM-09-2019-0327

Economic Survey. (2020). *State of the Economy*. https://www.indiabudget.gov.in/economicsurvey

Esiobu, N. S., & Ibe, G. O. G. (2015). Analysis of entrepreneurship development in agriculture among arable crop farmers in Imo State, Nigeria. *International Journal of African and Asian Studies*, *7*(3), 92–99.

Fitz-Koch, S., & Nordqvist, M. (2017). The Reciprocal Relationship of Innovation Capabilities and Socioemotional Wealth in a Family Firm. *Journal of Small Business Management*, *55*(4), 547–570. doi:10.1111/jsbm.12343

Gartner, W. B. (1990). What are we talking about when we talk about entrepreneurship? *Journal of Business Venturing*, *5*(1), 15–28. doi:10.1016/0883-9026(90)90023-M

Grande, J., Madsen, E. L., & Borch, O. J. (2011). The relationship between resources, entrepreneurial orientation and performance in farm-based ventures. *Entrepreneurship and Regional Development*, *23*(3-4), 89–111. doi:10.1080/08985620903183710

Hadimani, R. N. (1985). *Dynamics of industrial entrepreneurship*. APH Publishing.

Hassink, J., Grin, J., & Hulsink, W. (2012). Multifunctional Agriculture Meets Health Care: Applying the Multi-Level Transition Sciences Perspective to Care Farming in the Netherlands. *Sociologia Ruralis*, *53*(2), 223–245. doi:10.1111/j.1467-9523.2012.00579.x

Hassink, J., Hulsink, W., & Grin, J. (2016). Entrepreneurship in agriculture and healthcare: Different entry strategies of care farmers. *Journal of Rural Studies*, *43*, 27–39. doi:10.1016/j.jrurstud.2015.11.013

Heredero, J. M. (1979). Chapter. In T. V. Rao & T. K. Moulik (Eds.), Agricultural Entrepreneurship: Identification and selection of small scale entrepreneur (pp. 151–159). IIM.

Hulsink, W., & Dons, H. (2008). *Venturing and Clustering in Agri-Food and High-Technology Hot Spots*. Wageningen UR Frontis Series. doi:10.1007/978-1-4020-8768-4_1

Kelegama, S., & Hirimuthugodage, D. (2019). Food Security Policy Framework in Sri Lanka: Market reforms, diversification and safety nets. In Regional Cooperation for Sustainable Food Security in South Asia (pp. 227-247). Routledge India.

Kumar, P. (2017). Agro-Industry, Agricultural Marketing, Entrepreneurship, Agri-Business, Trade and Innovations. *Indian Journal of Agricultural Economics*, *72*(1), 76–80.

Manjula, M. (2021). The Smallholder in the Agriculture Market Reforms in India. *Economic and Political Weekly*, *56*(15), 23.

Mars, M. M., & Schau, H. J. (2016). Institutional entrepreneurship and the negotiation and blending of multiple logics in the Southern Arizona local food system. *Agriculture and Human Values*, *34*(2), 407–422. doi:10.100710460-016-9722-3

Michael, S. C., & Pearce, J. A. II. (2009). The need for innovation as a rationale for government involvement in entrepreneurship. *Entrepreneurship and Regional Development, 21*(3), 285–302. doi:10.1080/08985620802279999

Milestad, R., Ahnström, J., & Björklund, J. (2010). Essential multiple functions of farms in rural communities and landscapes. *Renewable Agriculture and Food Systems, 26*(2), 137–148. doi:10.1017/S1742170510000529

Moraru, R. A., Ungureanu, G., Bodescu, D., & Donosă, D. (2016). Motivations and challenges for entrepreneurs in agritourism. *Agronomy Series of Scientific Research/ Lucrari Stiintifice Seria Agronomie, 59*(1).

Nagler, P., & Naudé, W. (2014). *Non-Farm Enterprises in Rural Africa: New Empirical Evidence*. Policy Research Working Papers. doi:10.1596/1813-9450-7066

Onchiri, M. A., Maurice, S., & Patrick, M. (2015). Social economic factors and access to finance: The case of women managed agri-based small and medium enterprises Nyamira County, Kenya. *International Journal of Marketing and Technology, 5*(9), 56–74.

Paulson, A. L., & Townsend, R. (2004). Entrepreneurship and financial constraints in Thailand. *Journal of Corporate Finance, 10*(2), 229–262. doi:10.1016/S0929-1199(03)00056-7

Periodic Labour Force Survey. (2017). *Annual Report, PLFS, 2017-18*. www.mospi.nic.in/publication/annual-report-plfs-2017-18

Petrova, K. (2012). Part-time entrepreneurship and financial constraints: Evidence from the Panel Study of Entrepreneurial Dynamics. *Small Business Economics, 39*(2), 473–493. doi:10.100711187-010-9310-7

Richards, S. T., & Bulkley, S. (2007). Agricultural Entrepreneurs: The First and the Forgotten? SSRN *Electronic Journal*. doi:10.2139/ssrn.1020697

Shehrawat, P. S. (2006). Agro Processing Industries-A Challengeing Entrepreneurship For Rural Development. *Journal of Asia Entrepreneurship and Sustainability, 2*(3), 111.

Shirur, M., Shivalingegowda, N. S., Chandregowda, M. J., Manjunath, V., & Rana, R. K. (2019). Critical dimensions of entrepreneurship and entrepreneurial behaviour among mushroom growers: Investigation through Principal Component Analysis. *Indian Journal of Agricultural Research*. doi:10.18805/IJARe.A-5261

Simmons, G. J., Durkin, M. G., McGowan, P., & Armstrong, G. A. (2007). Determinants of internet adoption by SME agri-food companies. *Journal of Small Business and Enterprise Development*, *14*(4), 620–640. doi:10.1108/14626000710832730

Singh, N. P. (1969). n/ACH among Successful-Unsuccessful and Traditional-Progressive Agricultural Entrepreneurs of Delhi. *The Journal of Social Psychology*, *79*(2), 271–272. doi:10.1080/00224545.1969.9922421

Verma, R. K., Sahoo, A. K., & Rakshit, S. (2017). Opportunities in agri-preneurship in India: Need challenges and future prospects. *Rashtriya Krishi*, *13*(1), 71–72.

Vijayan, B., & Shivkumar, P. S. (2020). Pro-Agripreneurial Factors for the Formation of Agri-startups in India. *Asian Journal of Agricultural Extension. Economia e Sociologia*, 1–8.

Wang, J. H., Chang, C. C., Yao, S. N., & Liang, C. (2016). The contribution of self-efficacy to the relationship between personality traits and entrepreneurial intention. *Higher Education*, *72*(2), 209–224. doi:10.100710734-015-9946-y

Wilkinson, J. (2009). The Globalization of Agribusiness and Developing World Food Systems. *Monthly Review (New York, N.Y.)*, *61*(4), 29. doi:10.14452/MR-061-04-2009-08_4

Yang, D. T., & An, M. Y. (2002). Human capital, entrepreneurship, and farm household earnings. *Journal of Development Economics*, *68*(1), 65–88. doi:10.1016/S0304-3878(02)00006-8

Chapter 16
An Insight Into Research and Investigations of Gray Mold Focused on *Botrytis cinerea*

Sunil Jayant Kulkarni
Gharda Institute of Technology, India

ABSTRACT

Modern day agriculture focuses on the crop yield, cost, and space. Most of the investigations in the agricultural sector are aimed at rendering environmental friendliness to the application method and synthesizing new varieties of crops that are resistant to the attacks of microorganisms. Also, new fertilizers and pesticides are being developed. Plants such as grapes, raspberries, and strawberries are considered cash crops. Many useful compounds are synthesized from them. Pharmaceutical intermediates and products are synthesized from plants which have medicinal properties. Climatic conditions and soil properties are manipulated for confined and safe cultivation of these plants. It results in reduction in temperature and change in humidity, which in turn causes growth of undesirable species and diseases in the plants. The most common among these species is the fungi Botrytis cinerea. Many crops are affected adversely due to Botrytis cinerea (B. cinerea). In this chapter, a review on studies and investigations on gray mold for its effects on plants and control is carried with focus on Botrytis cinerea.

INTRODUCTION

Biotechnology has played key role in an attempt of the investigators to find out solution to the food problem of the growing population across the world. The

DOI: 10.4018/978-1-6684-2349-3.ch016

advantages of biotechnology such as synthesis of useful products from whole organisms or parts of organisms with sustainable pathways, make it most important technology in agricultural research. modern day agriculture focuses on the crop yield, cost and space. recent development in technology allows real time measurement of humidity, temperature, wind and other meteorological parameters(ayaz et al., 2019; lakhiar et al., 2018; sierra et al., 2018; xia & sun, 2002; zhang et al., 2018). Application of computational fluid dynamics helps to reduce dispersion of spray and helps to provide optimum dosage of pesticides and fungicides accurately(azeta et al., 2019; ginige & sivagnanasundaram, 2019; nandhini et al., 2019; saiz-rubio & rovira-más, 2020; zhu et al., 2018). This helps to reduce environmental footprints of pesticides and fungicides. rendering environmental friendliness to the application method and synthesizing new varieties of crops have been objectives of most of the investigations in agricultural sector. plants such as grapes, raspberries, strawberries are considered as cash crops. many useful compounds are synthesized from them. Pharmaceutical intermediates, products are synthesized from plants which have medicinal properties. Climatic conditions and soil properties are manipulated for confined and safe cultivation of these plants. it results in reduction in temperature and change in humidity, which in turn causes growth of undesirable species and diseases in the plants. The most common among these species is the fungi botrytis cinerea. Many crops are affected adversely due to botrytis cinerea (b. cinerea). Botrytis cinerea is one of the necrotrophic plant pathogens which can affect more than 500 plant species(li et al., 2018). Worldwide the losses because of this species are estimated to be around $10 billion to $100(anco & michael, 2011). Botrytis blight is one important disease caused due to botrytis cinerea in onion, flower and fruit crops (anco & michael, 2011; chilvers & toit, 2006; hansen, 2016; schraufnagel & hudelson, 2004). It has adverse effect on wine grapes and many other plant species. Botrytis cinerea is a phytopathogenic fungus which causes infection in crops. Cash crop like strawberry gets adversely affected due to this fungus(petrasch et al., 2019). Gray mold is one of the major reasons for rejection of crops in the market. climate resistance of the microorganisms is becoming a severe challenge to the investigators (weber & entrop, 2011). Many investigators consider botrytis cinerea as model organism for investigation on the microbial species that spoil crops and plants. Higher degrading potential and huge economical footprints of botrytis cinerea on crop yield make it ideal representative of such spoiling agents(de simone et al., 2020). Also, it has ability to develop resistance to fungicides. The worrying factor for investigators is the multiple fungicide resistance in fungicides with different plant hosts(sabrina et al., 2017).

CONTROL

Antibiosis is the predominant phenomenon in the mechanism of inhibitory effects of bacteria strains on the development of Botrytis fungi. In this phenomenon, two or more organisms come in contact and cause detrimental effect to at least one organism. Biological control of pathogenic fungi is an important aspect of the investigations. One such investigation was carried out by Bryk et al. (Bryk et al., 2004). They used isolated strains of Pseudomonas spp. derived from fruits and leaves of apple. Also, they isolated Botrytis cinerea Pers from apples. When mixed in the proportion 4:5, the inhibition of the germination of spores in mixture was found to be 90 percent.

Inappropriate and wrong use of fungicides causes huge loss. It is always envisaged to know effect of necrotrophic pathogens on the various fungicides. Kim et.al. used B. cinerea from strawberry fruit after isolation(Kim et al., 2016). They employed six fungicides namely fludioxonil, iprodione, pyrimethanil, tebuconazole, fenpyrazamine, and boscalid. Their results indicated that Fludioxonil had maximum inhibition effect. In detail investigation on the molecular mechanism and management for fungicide resistance in strawberry for in Botrytis cinerea was carried out by Grabke (Grabke, 2014). He surveyed the occurrence and prevalence of fungicide resistance by dicarboximide iprodione and hydroxyamide fenhexamid. Effect of the fungicide on the crop may alter based on the location as the soil quality differs(Grabke, 2014; Reglinski et al., 2010). Specific proteins are being explored for their application for plant immunization against gray mold (Mouekouba et al., 2013).

Chitosan is one of the important antimicrobial substance. It has the potential to control the activities of the gray molds. According to the research carried out by Reglinski et.al., Botrytis cinerea can inhibit growth of Chitosan liquid culture (Reglinski et al., 2010). In these investigations, they observed that a malt extract with Chitosan completely halted growth of B.cinerea. *Conidia* are asexually produced spores that are borne externally to the cells that produce them. These were affected by Chitosan. Chitosan can be viable alternatives to synthetic fungicides.

Botrytis Cinerea also adversely affects Tomato plants. Fungicides used for control of gray mold have two additional effects. One is its residual presence or perception regarding the same and rapid rise in resistance in pathogens. Biocontrol agents such as Clonostachys rosea (C.rosea) are very effective. Development and sporulation potential of B.cinerea is affected negatively due to biocontrol agents. Nutrient competition causes this resistance to B cinerea. Mouekouba et.al. have carried investigation on Botrytis Cinerea in Tomato Leaves (Mouekouba et al., 2013). This investigation supports earlier claims regarding use of C. rosea as an effective bio-fungicide for management of gray mold. Superoxide dismutase (SOD) plays important role in this activity. The level of nitric oxide and hydrogen peroxide is an important factor in the response of plants against pathogenic agents. According to these investigations, this activity(the response of plants against pathogenic agents) is at its peak after 24 hours.

Chickpea (CicerarietinumL.) is another important food crop cultivated globally. It is also affected adversely by gray mold. The abiotic and biotic stresses affect the crop badly. The most important diseases caused to Chickpea is botrytis gray mold (BGM), caused by Botrytis cinerea. Hosen et.al. have carried out an investigation on physiological variability and in vitro antifungal activity against Botrytis cinerea(Hosen et al., 2010). According to them, the colony diameter was highest for all isolates at 20°C. The maximum growth was observed by them at pH value of 4.5.

According to Rooi and Holz, the reduction in inoculum at the correct infection court and convenient developmental stage can facilitate the control of B. cinerea infection by cultural, chemical and biological means(Rooi & Holz, 2003). Their investigation was focused on studying the efficacy of fungicides at different morphological stages and various parts of the plant. This study was carried out in different seasons as seasonal variation like temperature and humidity plays vital role. Studies indicate that asymptomatic mature leaves carry high amounts of B. cinerea. Various crops, vegetables and fruits are nowadays made available throughout the year. In case of off-season cultivation and cultivation in the areas where outdoor cultivation is difficult, protective cultivation gains importance. The temperature and the humidity are kept under control. The major disadvantage of the protected cultivation is the conditions favorable to molds. Plant age and the site can affect the susceptibility (Adnan et al., 2019). In their investigations, Rooi and Holz observed that, the isolate from tomato plants was not aggressive in stems (Rooi & Holz, 2003). Many important factors to inoculate commercial crops and the fungicidal resistance are key factors affecting profitability(Adnan et al., 2019; Bi et al., 2021; Borges et al., 2014)

Lowering the leaf wetness period can decrease the risk of infections. The effect of B. cinerea infection depends on temperature and humidity. The use of small polythene tunnels was explored by Evenhuis and Wanten for control the fruit rot incidence(Evenhuis & Wanten, 2006). The incidences of Botrytis fruit rots were reduced by 47 to 37 percent when small polythene tunnels were used. They predicted that the lack of precipitation under polythene cover was one of the reasons for this along with the leaf wetness period. Choice of the proper cultivator becomes very important factor as the cultivators differs in susceptibility. Use of plastic tunnels in organically grown strawberries reduces the infection risk of B. cinerea on flowers.

Identifying virulence genes involved in infection has become one important development in the research (Barbosa et al., 2006; Collado et al., 2007; Cotoras et al., 2009; Pegg et al., 1987). It can be used to develop plants that are resistant to the diseases . An investigation conducted by Aguayo et.al. indicated that during filamentous growth, Bchex is very highly expressed gene (Aguayo et al., 2011). They also carried out Southern blot analysis of genomic DNA. They found that lesion development can be reduced by inoculating the tomato leaves with Bchex mutant.

It is difficult to maintain the non-toxicity of the fungicides except for the target organism. An investigation based on these principals was carried out by Adongo et al.(Adongo, 2012). A vast array of secondary metabolites is generated by Basidiomycete's fungi. They found that out of 23 active strains, 15 were reproducible. This study confirmed the presence of naturally occurring antifungal substances. These substances were developed from certain strains of basidiomycete fungi. They concluded that the active crude extracts contain the antibotrytis compounds which can offer an effective replacement for the current synthetic fungicides.

Inadequacies in present control practices can be overcome by using the microorganism that opposes the gray mold (Pratella & Mari, 1993; Keinath et al., 1991; Sutton & Peng, 1993; Sutton, 1994; Sutton et al., 1997; Bika et al., 2021). According to Sutton et.al., Gliocladium roseum is one of those antagonistic microorganisms (Sutton et al., 1997). It is found in tropical and desert areas and in a thin-walled hollow cavity. This exists in an animal or plant which contains liquid secretion in form of a sac, vesicle, or bladder. These cavities are termed as cysts. G. roseum has diverse association with various parts of plants. This is pathogenic to many fruits and vegetables. His studies indicated that B. cinerea showed very high effectiveness in biocontrol. Also, it was effective for many commercial fruits.

Integration of new practices with existing management strategies can help in providing effective results in ornamental production systems(Bika et al., 2021). These include various chemical methods and their modes of action. Recent studies indicate that detail studies on the microorganisms and their effect on the fungicides can help in developing the fungicides which can increase crop yield and are immune to the resistance by the molds(Favre—Godal et al., 2021; Dawadi et al., 2021; Traversari et al., 2021).

ANALYSIS

Advanced investigations on Botrytis cinerea for its electron microscopic study was carried out in 1963(Hawker & Hendy, 1963). Prompt diagnosis of the diseases in plants is envisaged and various investigations are making the diagnosis fast and sophisticated (Dawadi et al., 2021; Traversari et al., 2021; Hawker & Hendy, 1963; Pourreza et al., 2015; Qin et al., 2011; Sankaran et al., 2010; Maxwell & Johnson, 2000). It is always desired to have real time technologies for diagnosis. Larenas et al. studied emission obtained from five isolates of B. cinerea in two fixed positions (Larenas et al., 2015). Their research indicated that B. cinerea fluoresces mainly in the red-infrared area. Cauliflower is considered as the healthiest foods on the Earth. It is rich in phytochemicals contents having health enhancing effects. It also has ability to inhibit cancer, heart disease and brain disease. It is one of the most important cool season crops (Sankaran et al., 2010; Maxwell & Johnson, 2000; Larenas et al., 2015). In the modern day's optimization of the crop fertilization and irrigation is important aspect of the investigations. It is evident that the temporal and spatial rooting pattern of a crop needs to be taken care of (Bragg et al., 1983; Barraclough, 1984). Well calibrated crop growth models can readily predict the routing patterns. Kage et.al. investigated modeling of root growth measurements for Brassica oleracea L. botrytis plant (Kage & Stützel, 1999a). They found that for tap root and shoot, dry weights were related to each other. Logarithms of their rates were equal. They developed an expo-linear function of temperature sum for the development of rooting depth. Such studies help in deciding time required for pesticide application. This is helps for more precise targeting in pesticide spraying(Kage & Stützel, 1999a; Kage & Stützel, 1999b; González-Fernández et al., 2015; Kage et al., 2000).

Proteomics, along with other techniques has become important tool for obtaining crucial information about pathogenic properties and virulence factors. Proteomics is detailed and comprehensive study of proteins(González-Fernández et al., 2015). These studies are done particularly for their structures and functions. Various investigations are reported for intracellular proteomics, subproteomic, and secretomics. Fernandez and Novo have reported one such investigation (Fernandez & Novo, 2010). According to them, genetic varieties and survival potential makes Botrytis cinerea, one of the most dangerous pathogenic species. Fernandez-Acero et al. have carried out similar investigation (Fernandez-Acero et al., 2009). In their investigations, they developed a proteomic database. This database was very broad with large number of identified proteins from B. cinerea. Their study described number of identified proteins. This was first of its kind proteomic study of the phytopathogenic fungus B. cinerea. This paved way for molecular parthenogenesis. Proteomic Analysis is one of the most important tools nowadays for Identifying Pathogenicity Factors(Padliya & Cooper, 2006; Fernandez-Acero, Jorge, Calvo et al, 2007; Fernandez-Acero, Carbu´, Garrido et al, 2007; Dhingra et al., 2005; Myung et al., 2007; Tournu et al., 2005).

Bar-Nun has reported the effect of changes in mycelial structure of Botrytis cinerea (Bar-Nun et al., 2007). These changes in the structures were induced by removal of the glucan matrix. This glucan matrix plays important role in the extra-cellular enzyme activity. It also contributes to the virulence of the fungus. The gluconese produced Botrytis mycelia was more sensitive to the fungicides, as with many other investigations.

Many Investigations were reported on genetic characterization of Botrytis cinerea populations (Bar-Nun et al., 2007; González-Fernández et al., 2020; Williamson et al., 2007; Giraud et al., 1999; Diolez et al., 1995; Levis et al., 1997). Wahab carried out characterization of Botrytis cinerea isolates from plants (Wahab, 2015). His study was based on the isolates from three host plants namely grape, strawberry and lettuce. According to him, transposable elements (TEs) are the fungal genomic variability sources. These isolates were very responsive to the hydroxyanilide fungicide and neither having any host preference nor any bearing on the aggressiveness. However, B. cinerea isolates at morphological and cellular level shows variation.

Finkers et al. have carried out investigation on the susceptibility of five seedlings to the Botrytis cinereal (Finkers et al., 2008). Botrytis cinerea can also have adverse effect due to infection on onion seed heads and cause brown stain (Wukasch & Hofstra, 1977). The Botrytis Interactions for tomato and other commercial crops are investigated by many investigators(Wahab, 2015; Finkers et al., 2008; Wukasch & Hofstra, 1977; Baker & Wilcox, 1961). This fungus has wide presence throughout the world. It has the potential to infect leek, garlic and chive. (Finkers et al., 2007; Van Heusden et al., 1999). Post-harvest gray mold on apple can be effectively controlled by lycorine application(Zhao et al., 2020). Methyl thujate also has been investigated for its inhibitory effects on mycelial growth of, botrytis cinerea(Ji et al., 2018). Various investigations have resulted in innovative ways to control the destructive effects of botrytis cinerea. Lycorine, which is commonly used for human drug design was effective in the control of botrytis cinerea(Zhao et al., 2020). Grape seed proanthocyanidins have been used for In vitro antifungal activity of a new bioproduct derived from it on Botrytis Cinerea(Nechita et al., 2018). The byproduct is obtained from polymeric proanthocyanidins extracted from grape seeds of 'Fetească neagră' variety. Investigation by Nechita et al. indicated that such bioproducts can enhance the effect natural fungicides for biological control against Botrytis cinerea fungus(Nechita et al., 2018). Research at various geographical locations on Botrytis Cinerea on different plants underlines the importance of controlling gray mold effects on these crops(Lopes et al., 2017). Researchers are widening their investigation to study the effects various strains on post-harvest fungal phytopathogens(Alforja et al., 2021). Also, the studies and investigations focused on increasing crop yield and curb on the pathogen effects are yielding new technics for control of gray mold().

CONCLUSION

Climatic conditions and soil properties are manipulated for confined and safe cultivation of cash crops. It results in reduction in temperature and change in humidity, which in turn causes growth of undesirable species and diseases in the plants. Many Crops are affected adversely due to Botrytis cinerea (B. cinerea). In present control practices, hindrance to the application of the fungicides are factors such as, their residual presence or perception regarding the same and rapid rise in resistance in pathogens. Researchers are widening their investigation to study the effects various strains on post-harvest fungal phytopathogens. Also, the studies and investigations focused on increasing crop yield and curb on the pathogen effects are yielding new technics for control of gray mold. Research at various geographical locations on Botrytis Cinerea on different plants underlines the importance of controlling gray mold effects on crops.

FUNDING

This research received no specific grant from any funding agency in the public, commercial, or not-for-profit sectors.

REFERENCES

Adnan, M., Hamada, M. S., Hahn, M., Li, G.-Q., & Luo, C.-X. (2019). Fungicide resistance of *Botrytis cinerea* from strawberry to procymidone and zoxamide in Hubei, China. *Phytopathol Res*, *1*(1), 17. doi:10.118642483-019-0024-8

Anco, D. J., & Michael, A. E. (2011). Fruit Rots of Blueberry: Alternaria, Anthracnose, and Botrytis, *Agri. and Nat. Res. HYG-3213-11*, 1-2. https://ohioline.osu.edu/factsheet/plpath-fru-44

Ayaz, M., Ammad-Uddin, M., Zubair Sharif, Z., Ali Mansour, A., & Aggoune, E.M. (2019). Internet-Of-Things (IoT)-Based Smart Agriculture: Toward Making the Fields Talk. *Special Section on New Technologies for Smart Farming 4.0*, *7*, 129551–129583. doi:.2019.2932609 doi:10.1109/ACCESS

Azeta, J., Bolu, C. A., Alele, F., Daranijo, E. O., Onyeubani, P., & Abioye, A. A. (2019). Application of Mechatronics in Agriculture: A review. *International Conference on Engineering for Sustainable World Journal of Physics: Conference Series*, *1378*, 1-11. 10.1088/1742-6596/1378/3/032006

Bi, K., Scalschi, L., Jaiswal, N., Mengiste, T., Fried, R., Sanz, A. B., Arroyo, J., Zhu, W., Masrati, G., & Sharon, A. (2021). The *Botrytis cinerea* Crh1 transglycosylase is a cytoplasmic effector triggering plant cell death and defense response. *Nature Communications*, *12*(1), 2166. doi:10.103841467-021-22436-1 PMID:33846308

Bryk, H., Barbars, D., & Piotr, S. (2004). Inhibitory Effect of Pseudomonas spp. on the Development of Botrytis cinerea and Penicillium expansum. *Plant Protection Science*, *40*(4), 128–134. doi:10.17221/467-PPS

Chilvers, M. I., & Toit, L. J. (2006). Detection and Identification of Botrytis Species Associated with Neck Rot, Scape Blight, and Umbel Blight of Onion, Plant Management Network. *Plant Health Progress*, 1–20.

De Simone, N., Pace, B., Grieco, F., Chimienti, G., Tyibilika, V., Santoro, V., Capozzi, V., Colelli, G., Spano, G., & Russo, P. (2020). Botrytis cinerea and Table Grapes: A Review of the Main Physical, Chemical, and Bio-Based Control Treatments in Post-Harvest. *Foods*, *9*(9), 1138. doi:10.3390/foods9091138 PMID:32824971

Ginige, A., & Sivagnanasundaram, J. (2019). Enhancing Agricultural Sustainability through Crowdsensing: A Smart Computing Approach. *Journal of Advanced Agricultural Technologies*, *6*(3), 161–165. doi:10.18178/joaat.6.3.161-165

Grabke, A. (2014). *Fungicide Resistance in Botrytis cinerea from Strawberry - Molecular Mechanisms and Management. In Agronomy and Crop Sciences Commons, Biology Commons, and the Plant Path.* Comm.

Hansen, M. A. (2016). *Botrytis Blight of Peony. Communications and Marketing, College of Agriculture and Life Sciences, Virginia Tech.*

Hosen, M., Ahmed, A. M. R., & Islam, M. (2010). Physiological Variability and in Vitro Antifungal Activity against Botrytis Cinerea Causing Botrytis Gray Mold of Chickpea (Cicer arietinum L.). *Spanish Journal of Agricultural Research*, *8*(3), 750–756. doi:10.5424jar/2010083-1274

Kim, J., Shin, J., Gumilang, A., Chung, K., Choi, K., & Kyoung Su Kim, K. (2016). Effectiveness of Different Classes of Fungicides on Botrytis cinerea Causing Gray Mold on Fruit and Vegetables. *Plant Path. Jour.*, *32*(6), 570–574. doi:10.5423/PPJ.NT.05.2016.0114 PMID:27904464

Lakhiar, I. A., Jianmin, G., Syed, T. N., Chandio, F. A., Buttar, N. A., & Qureshi, W. A. (2018). Monitoring and Control Systems in Agriculture Using Intelligent Sensor Techniques: A Review of the Aeroponic System. *Journal of Sensors*, 1–18. doi:10.1155/2018/8672769

Li, H., Chen, Y., Zhang, Z., Li, B., Qin, G., & Tian, S. (2018). Pathogenic mechanisms and control strategies of *Botrytis cinerea* causing post-harvest decay in fruits and vegetables. *Food Quality and Safety*, *2*(3), 111–119. doi:10.1093/fqsafe/fyy016

Mouekouba, L., Zhang, Z., & Erinle Kehinde Olajide, E. (2013). Biological Control of Botrytis Cinerea in Tomato Leaves. *2013 International Conf. on Agri. and Biotech. IPCBEE*, *60*, 64-68. 10.7763/IPCBEE

Nandhini, S., Bhrathi, S., Goud, D. D., & Krishna, K. P. (2019). Smart Agriculture IOT with Cloud Computing, Fog Computing and Edge Computing. *International Journal of Engineering and Advanced Technology*, *9*(2), 3578–3583. doi:10.35940/ijeat.B2600.129219

Petrasch, S., Knapp, S. J., van Kan, J. A. L., & Blanco-Ulate, B. (2019). Grey mould of strawberry, a devastating disease caused by the ubiquitous necrotrophic fungal pathogen Botrytis cinerea. *Molecular Plant Pathology*, *20*(6), 877–892. doi:10.1111/mpp.12794 PMID:30945788

Reglinski, T., Elmer, P., Taylor, H., Wood, P., & Hoyte, S. (2010). Inhibition of Botrytis cinerea Growth and Suppression of Botrytis Bunch Rot in Grapes using Chitosan. *Plant Pathology*, *59*(5), 882–890. doi:10.1111/j.1365-3059.2010.02312.x

Rooi, C. V., & Holz, G. (2003). Fungicide Efficacy Against Botrytis cinerea at Different Positions on Grape Shoots. *Afr. J. Enol. Vitic*, *24*(1), 11–15.

Sabrina, R., Weber Roland W. S., Rieger Daniel, R., Peter, D., & Hahn Matthias, H. (2017). Spread of Botrytis cinerea Strains with Multiple Fungicide Resistance in German Horticulture. *Frontiers in Microbiology*, *7*(2075). doi:10.3389/fmicb.2016.02075

Saiz-Rubio, V., & Rovira-Más, F. (2020). From Smart Farming towards Agriculture 5.0: A Review on Crop Data Management. *Agronomy (Basel)*, *10*(207), 1–21. doi:10.3390/agronomy10020207

Schraufnagel, J., & Hudelson, B. (2004). Gray Mold (Botrytis Blight). *Wisconsin Garden Facts*, 1. wihort.uwex.edu

Sierra, J. E., Medina, B., & Vesga, J. C. (2018). Management system in intelligent agriculture based on Internet of Things. *Revista Espacios*, *39*(08), 20.

Weber, R., & Entrop, A. (2011). Multiple Fungicide Resistance in Botrytis: A Growing Problem in German Soft-Fruit Production. In Fungicides - Beneficial and Harmful Aspects. Academic Press.

Xia, B., & Sun, D. (2002). Applications of computational fluid dynamics (CFD) in the food industry: A review. *Computers and Electronics in Agriculture*, *34*(1-3), 5–24. doi:10.1016/S0168-1699(01)00177-6

Zhang, G., Choi, C., Bartzanas, T., Lee, I.-B, & Kacira, M. (2018). Computational Fluid Dynamics (CFD) research and application in Agricultural and Biological Engineering. *Computers and Electronics in Agriculture*, *149*, 1-2. .2018.04.007 doi:10.1016/j.compag

Zhu, N. Y., Liu, X., Liu, Z. Q., Hu, K., Wang, Y. K., Tan, J. L., Huang, M., Zhu, Q., Ji, X., Jiang, Y., & Guo, Y. (2018). Deep learning for smart agriculture: Concepts, tools, applications, and opportunities. *International Journal of Agricultural and Biological Engineering, 11*(4), 32–44. doi:10.25165/j.ijabe.20181104.4475

<unknown>29Borges, A. V., Saraiva, R. M., & Luiz, A. M. (2014). Key factors to inoculate Botrytis cinerea in Tomato plants. *Summa Phytopathol., Botucatu.* 40(3), 221-225. 10. 1590/0100 -5405 /1929</unknown>

Evenhuis, A., & Wanten, P. J. (2006) Effect of polythene Tunnels and Cultivars on Grey Mould Caused by Botrytis cinerea in Organically Grown Strawberries. *Agri. Consp. Scient.* 71(4), 111-114. https://hrcak.srce .hr/ 7873

Barbosa, A. C., do Carmo, A. E., Graf, L., Tomaz, R., de Souza, C. F., Mendes, J., Randi, M. A. F., Buchi, D., & Schadeck, R. J. G. (2006). Morphology and Lipid Body and Vacuole Dynamics during Secondary Conidia Formation in Colletotrichum Acutatum: Laser scanning confocal analysis. *Canadian Journal of Microbiology, 52*(2), 117–124. doi:10.1139/w05-104 PMID:16541147

Collado, I. G., Sa'nchez, A. J. M., & Hanson, J. R. (2007). Fungal Terpene Metabolites: Biosynthetic Relationships and the Control of the Phytopathogenic Fungus Botrytis cinerea. [Google Scholar]. *Natural Product Reports, 24*(4), 674–686. doi:10.1039/b603085h PMID:17653354

Cotoras, M., Garcı'a, C., & Mendoza, L. (2009). Botrytis Cinerea Isolates Collected from Grapes Present Different Requirements for Conidia Germination. [Google Scholar]. *Mycologia, 101*(3), 287–295. doi:10.3852/08-012 PMID:19537202

Lucas, J., & Knights, I. (1987). Spores on Leaves: Endogenous and Exogenous Control of Development. In: Pegg GF, Ayres PG(Eds) Fungal Infection of Plants. Symposium of the British Mycological Society. Cambridge University Press, Cambridge, 45–59. Google Scholar

Aguayo, R., Riquelme, J., Valenzuela, P. D. T., Hahn, M., & Silva Moreno, E. (2011). Bchex Virulence Gene of Botrytis Cinerea: Characterization and Functional Analysis. [Google Scholar]. *Journal of General Plant Pathology, 77*(4), 230–238. doi:10.100710327-011-0311-4

Adongo, J. O. Alice, Njue A.W., Omolo J.O.: Cheplogoi P.K., Dan O. Otaye D.O.(2012). In Vitro Inhibition of Botrytis Cinerea - Causative Agent for Grey Mold by Crude Extracts of Basidiomycetes Fungi. Science Journal of Biotech., Vol. 2012, Article ID sjbt-175, 3 Pages. doi: . Google Scholar doi:10.7237jbt/175

Pratella, G. C., & Mari, M. (1993). Effectiveness of Trichoderma, Gliocladium, and Paecilomyces in Postharvest Fruit Protection. *Postharvest Biol.* [Google Scholar]. *Tech.*, *3*, 49–56.

Keinath, A. P., Fravel, D. R., & Papavizas, G. C. (1991). Potential of Gliocladium Roseum for Biocontrol of Verticillium Dahliae. [Google Scholar]. *Hytopathology*, *81*(6), 644–648. doi:10.1094/Phyto-81-644

Sutton, J. C., & Peng, G. (1993). Manipulation and Vectoring of Biocontrol Organisms to Manage Foliage and Fruit Diseases in Cropping Systems. [Google Scholar]. *Annual Review of Phytopathology*, *31*(1), 473–493. doi:10.1146/annurev.py.31.090193.002353

Sutton, J. C. (1994). Biological Control of Strawberry Diseases. [Google Scholar]. *Adv. Strawberry Res.*, *13*, 1–12.

Sutton, J. C., Li, D., Peng, G., Yu, H., Zhang, P., & Valdebenito-Sanhueza, R. (1997). Gliocladium Roseum-A Versitile Adversary of Butrysis Ceneria in Crops. [Google Scholar]. *Plant Disease*, *81*(4), 316–328. doi:10.1094/PDIS.1997.81.4.316 PMID:30861808

Bika, R., Baysal-Gurel, F., & Jennings, C. (2021). *Botrytis cinerea* management in ornamental production: A continuous battle. *Canadian Journal of Plant Pathology*, *43*(3), 345–365. doi:10.1080/07060661.2020.1807409

Favre-Godal, Q., Schwob, P., Lecoultre, N., Hofstetter, V., Gourguillon, L., Riffault-Valois, L., Lordel-Madeleine, S., Gindro, K., & Choisy, P.Favre—Godal. (2021). Plant-microbe features of *Dendrobium fimbriatum* (Orchidaceae) fungal community. *Symbiosis*. Advance online publication. doi:10.100713199-021-00786-0

Dawadi, S., Baysal-Gurel, F., Addesso, K. M., Liyanapathiranage, P., & Simmons, T. (2021). Fire Ant Venom Alkaloids: Possible Control Measure for Soilborne and Foliar Plant Pathogens. *Pathogens (Basel, Switzerland)*, *10*(6), 659. doi:10.3390/pathogens10060659 PMID:34071926

Traversari, S., Cacini, S., Galieni, A., Nesi, B., Nicastro, N., & Pane, C. (2021). Precision Agriculture Digital Technologies for Sustainable Fungal Disease Management of Ornamental Plants. *Sustainability*, *13*(7), 3707. doi:10.3390u13073707

Hawker, L., & Hendy, R. J. (1963). An Electron Microscope Study of Germination of Conindia of Botrysis Cineria. [Google Scholar]. *Journal of General Microbiology*, *33*(1), 43–46. doi:10.1099/00221287-33-1-43 PMID:14072834

Pourreza, A., Lee, W., Reza, E., Schueller, J., & Raveh, E. (2015). An Optimum Method for Real-Time Infield Detection of Huanglongbing Disease using A Vision Sensor. [Google Scholar]. *Computers and Electronics in Agriculture*, *110*, 221–232. doi:10.1016/j.compag.2014.11.021

Qin, J., Chao, K., & Kim, M. (2011). Investigation of Raman Chemical Imaging for Detection of Lycopene Changes in Tomatoes during Postharvest Ripening. [Google Scholar]. *Journal of Food Engineering*, *107*(3-4), 277–28. doi:10.1016/j.jfoodeng.2011.07.021

Sankaran, S., Mishra, A., Ehsani, R., & Davis, C. (2010). A Review of Advanced Techniques for Detecting Plant Diseases. [Google Scholar]. *Computers and Electronics in Agriculture*, *72*(1), 1–13. doi:10.1016/j.compag.2010.02.007

Maxwell, K., & Johnson, G. (2000). Chlorophyll Fluorescence, a Practical Guide. [Google Scholar]. *Journal of Experimental Botany*, *51*(345), 659–668. doi:10.1093/jexbot/51.345.659 PMID:10938857

. Larenas, M., Guzmán Estrada, R., Contreras J., C. Meriño Gergichevich C. (2015). Determination of the Fluorescence Spectrum of Botrytis cinerea Pers.:Fr. Isolated from Highbush Blueberry (Vaccinium corymbosum L.). *Journal of Soil Sci.band Plant Nutri.*, 15 (4), 938-948. Google Scholar

Bragg, P. L., Govi, G., & Cannell, R. Q. (1983). A Comparison of Methods, Including Angled and Vertical Minirhizotrons for Studying Root Growth and Distribution In a Spring Oat Crop. [Google Scholar]. *Plant and Soil*, *73*(3), 435–440. doi:10.1007/BF02184322

Barraclough, P. B. (1984). The Growth and the Activity of Winter Wheat Roots in the Field: Root Growth of High-Yielding Crops in Relation to Shoot Growth. [Google Scholar]. *J. agr. Sci. (Camb.)*, *103*(2), 439–442. doi:10.1017/S0021859600047419

Kage, H., & Stützel, H. (1999a). A Simple Empirical Model for Predicting Development and Dry Matter Partitioning in Cauliflower (Brassica Oleracea L. Botrytis). [Google Scholar]. *Scientia Horticulturae*, *80*(1-2), 19–38. doi:10.1016/S0304-4238(98)00226-X

Kage, H., & Stützel, H. HUME. (1999b). An Object Oriented Component Library for Generic Modular Modelling of Dynamic Systems in Modelling Cropping Systems. Eds. CSM Donatelli, F. Villalobos and JM Villar. pp 299–300. European Society of Agronomy,Lleida. Google Scholar

González-Fernández, R., Valero-Galván, J., Gómez-Gálvez, F. J., & Jorrín-Novo, J. V. (2015). Unraveling the in vitro secretome of the phytopathogen Botrytis cinerea to understand the interaction with its hosts. *Frontiers in Plant Science*, *6*, 839. doi:10.3389/fpls.2015.00839 PMID:26500673

Kage, H., Kochler, M., & Stützel, H. (2000). Root Growth of Cauliflower (Brassica oleracea L. botrytis) under Unstressed Conditions: Measurement and Modeling. [Google Scholar]. *Plant and Soil*, *223*(1/2), 131–145. doi:10.1023/A:1004866823128

Fernandez, R., & Novo, H. (2010). *Proteomics of Fungal Plant Pathogens: the Case of Botrytis cinerea, Current Research, Techn. and Edu. Topics in App. Microbiology and Microbial Tech., 205-2017*. Google Scholar.

Fernandez-Acero, F., Thomas Colby, T., & Anne Harzen, A. (2009). Proteomic Analysis of the Phytopathogenic Fungus Botrytis cinerea during Cellulose Degradation. [Google Scholar]. *Proteomics*, *9*(10), 2892–2902. doi:10.1002/pmic.200800540 PMID:19415670

Padliya, N. D., & Cooper, B. (2006). Mass Spectrometry-based Proteomics for the Detection of Plant Pathogens. [Google Scholar]. *Proteomics*, *6*(14), 4069–4075. doi:10.1002/pmic.200600146 PMID:16791831

Fernandez-Acero, F. J., Jorge, I., Calvo, E., Vallejo, I., Carbú, M., Camafeita, E., Garrido, C., López, J. A., Jorrin, J., & Cantoral, J. M. (2007). Proteomic Analysis of Phytopathogenic Fungus Botrytis cinerea as a Potential Tool for Identifying Pathogenicity Factors. *Therapeutic Targets and for Basic Res. Archives of Microbiology*, *187*(3), 207–215. doi:10.100700203-006-0188-3 PMID:17124592

Fernandez-Acero, F. J., Carbu´, M., Garrido, C., Vallejo, I., & Cantoral, J. M. (2007). Proteomic Advances in Phytopathogenic Fungi. *Current Proteomics*, *4*(2), 79–88. doi:10.2174/157016407782194620

Dhingra, V., Gupta, M., Andacht, T., & Fu, Z. F. (2005). New Frontiers in Proteomics Research: A Perspective. *International Journal of Pharmaceutics*, *299*(1-2), 1–18. doi:10.1016/j.ijpharm.2005.04.010 PMID:15979831

Myung, K., Hamilton-Kemp, T. R., & Archbold, D. D. (2007). Interaction with and Effects on the Profile of Proteins of Botrytis cinerea by C6 Aldehydes. *Journal of Agricultural and Food Chemistry*, *55*(6), 2182–2188. doi:10.1021/jf0631629 PMID:17323971

Tournu, H., Serneels, J., & Van Dijck, P. (2005). Fungal Pathogens Research: Novel and Improved Molecular Approaches for the Discovery of Antifungal Drug Targets. *Current Drug Targets*, *6*(8), 909–922. doi:10.2174/138945005774912690 PMID:16375674

Bar-Nun, N., L'Hyvernay, A., Donèche, B., & Mayer, M. (2007). Changes in mycelial structure of Botrytis cinerea Induced by Removal of the Glucan Matrix. *Int. J. Vine and Wine Sci.*, *41*(3), 149–153. doi:10.20870/oeno-one.2007.41.3.843

González-Fernández, E., Piña-Rey, A., Fernández-González, M., Aira, M., & Rodríguez-Rajo, F. (2020). Identification and evaluation of the main risk periods of Botrytis cinerea infection on grapevine based on phenology, weather conditions and airborne conidia. *The Journal of Agricultural Science*, *158*(1-2), 88–98. doi:10.1017/S0021859620000362

Williamson, B., Tudzynski, B., Tudzynski, P., & Van Kan, J. A. (2007). Botrytis cinerea: The Cause of Grey Mould Disease. *Molecular Plant Pathology*, *8*(5), 561–580. doi:10.1111/j.1364-3703.2007.00417.x PMID:20507522

Giraud, T., Fortini, D., Levis, C., Lamarque, C., Leroux, P., LoBuglio, K., & Brygoo, Y. (1999). Two Sibling Species of the Botrytis cinerea Complex, Transposa and Vacuma, Are Found in Sympatry on Numerous Host Plants. *Phytopathology*, *89*(10), 967–973. doi:10.1094/PHYTO.1999.89.10.967 PMID:18944743

Diolez, A., Marches, F., Fortini, D., & Brygoo, Y. (1995). Boty, a Long-Terminal-Repeat Retroelement in the Phytopathogenic Fungus Botrytis cinerea. *Applied and Environmental Microbiology*, *61*(1), 103–108. doi:10.1128/aem.61.1.103-108.1995 PMID:7887592

Levis, C., Fortini, D., & Brygoo, Y. (1997). Flipper, a Mobile Fot1-Like Transposable Element in Botrytis cinerea. *Molecular & General Genetics*, *254*(6), 674–680. doi:10.1007004380050465 PMID:9202383

Wahab, H. A. (2015). Characterization of Egyptian Botrytis cinerea Isolates from Different Host Plants. *Advances in Microbiology*, *5*(03), 177–189. doi:10.4236/aim.2015.53017

Finkers, R., Bai, Y., Berg, P., Berloo, R., Meijer-Dekens, F., Have, A., Kan, J., Lindhout, P., & Heusden, A. (2008). Quantitative Resistance to Botrytis cinerea from Solanum Neorickii. *Euphytica*, *159*(1-2), 83–92. doi:10.100710681-007-9460-0

Wukasch, R., & Hofstra, G. (1977). Ozone and Botrytis Interactions in Onion-Leaf Dieback: Open-Top Chamber Studies. Etiology [Google Scholar]. *Phytopathology*, *67*(9), 1080–1084. doi:10.1094/Phyto-67-1080

Baker, R. S., & Wilcox, G. (1961). Effect of Foliage Ozone Injury and Infection of Potato Damage and Stand Reduction on Onion Yield. [Google Scholar]. *Proceedings of the American Society for Horticultural Science*, *78*, 400–405.

Finkers, R., van den Berg, P., van Berloo, R., ten Have, A., van Heusden, A. W., van Kan, J. A. L., & Lindhout, P. (2007). Three QTLs for Botrytis cinerea Resistance in Tomato. [Google Scholar]. *Theoretical and Applied Genetics*, *114*(4), 585–593. doi:10.100700122-006-0458-0 PMID:17136515

Van Heusden, A., Koornneef, M., Voorrips, R. E., Brüggemann, W., Pet, G., Vrielink-van Ginkel, R., Chen, X., & Lindhout, P. (1999). Three QTLs from Lycopersicon peruvianum confer a high level of resistance to Clavibacter michiganensis ssp michiganensis. [Google Scholar]. *Theoretical and Applied Genetics*, *99*(6), 1068–1074. doi:10.1007001220051416

<unknown>Zhao, S., Guo, Y., & Wang, Q. (2020) Antifungal effects of lycorine on Butrytis Cenerea with Possible Mechanism. *Biotechnology Letters,* </unknown> doi:10.21203/rs.3.rs-251600/v1

Ji, D., Chen, T., Ma, D., Liu, J., Xu, Y., & Tian, S. P. (2018). Inhibitory effects of Methyl Thujate on mycelial growth of, botrytis cinerea, and possible mechanisms. *Postharvest Biology and Technology*, *142*, 46–54. doi:10.1016/j.postharvbio.2018.04.003

Nechita, A., Filimon, R. V., Filimon, R. M., Colibaba, L.-C., Gherghel, D., Damian, D., Pașa, R., & Cotea, V. V. (2018). In vitro Antifungal Activity of a New Bioproduct Obtained from Grape Seed Proanthocyanidins on Botrytis cinerea Mycelium and Spores. *Notulae Botanicae Horti Agrobotanici Cluj-Napoca*, *47*(2), 418–425. doi:10.15835/nbha47111367

Lopes, U. P., Zambolim, L., Pereira, C. N., & Gracia, O. (2017). Resistance of Botrytis cinerea to fungicides controlling gray mold on strawberry. *Brazil Bragantia*, *76*(2), 266–272. doi:10.1590/1678-4499.055

Alforja, S. I. R., Rico, P. M. B., Caoili, B. L., & Latina, R. A. (2021). Two Philippine *Photorhabdus luminescens* strains inhibit the in vitro growth of *Lasiodiplodia theobromae, Fusarium oxysporum* f. sp. *lycopersici*, and *Colletotrichum* spp. *Egyptian Journal of Biological Pest Control*, *31*(1), 108. doi:10.118641938-021-00454-x

Muangpat, P., Yooyangket, T., Fukruksa, C., Suwannaroj, M., Yimthin, T., Sitthisak, S., Chantratita, N., Vitta, A., Tobias, N. J., Bode, H. B., & Thanwisai, A. (2017). Screening of the anti-microbial activity against drug resistant bacteria of *Photorhabdus* and *Xenorhabdus* associated with entomopathogenic nematodes from Mae Wong National Park, Thailand. *Frontiers in Microbiology*, *8*, 1142. doi:10.3389/fmicb.2017.01142 PMID:28702004

Gustafson, S. (2019) FAO SOFA report 2019: New insights into food loss and waste. https://www.ifpri.org/blog/fao-sofa-report-2019-new-insights-food-loss-and-waste. Accessed 11 July 2020.

Patzke, H., & Schieber, A. (2018). Growth-inhibitory activity of phenolic compounds applied in an emulsifiable concentrate - ferulic acid as a natural pesticide against Botrytis cinerea. *Food Research International*, *113*, 18–23. doi:10.1016/j.foodres.2018.06.062 PMID:30195511

Xu, D., Yu, G., Xi, P., Kong, X., Wang, Q., Gao, L., & Jiang, Z. (2018b). Synergistic effects of resveratrol and pyrimethanil against Botrytis cinerea on grape. *Molecules (Basel, Switzerland)*, *23*(6), 1455. doi:10.3390/molecules23061455 PMID:29914082

Xu, D., Deng, Y., Han, T., Jiang, L., Xi, P., Wang, Q., Jiang, Z., & Gao, L. (2018). In vitro and in vivo effectiveness of phenolic compounds for the control of postharvest gray mold of table grapes. *Postharvest Biology and Technology*, *139*, 106–114. doi:10.1016/j.postharvbio.2017.08.019

Compilation of References

Abati, A., Skarulis, M. C., Shawker, T., & Solomon, D. (1995). Ultrasound-guided fine-needle aspiration of parathyroid lesions: A morphological and immunocytochemical approach. *Human Pathology*, *26*(3), 338–343.

Acs, Z. J., Arenius, P., Hay, M., & Minniti, M. (2005). *Global Entrepreneurship Monitor. Executive Report 2004*. Global Entrepreneurship Monitor.

Acs, Z. J., Braunerhjelm, P., Audretsch, D. B., & Carlsson, B. (2009). The knowledge spillover theory of entrepreneurship. *Small Business Economics*, *32*(1), 15–30. doi:10.100711187-008-9157-3

Acs, Z. J., Desai, S., & Hessels, J. (2008). Entrepreneurship, economic development and institutions. *Small Business Economics*, *31*(3), 219–234. https://doi.org/10.1007/s11187-008-9135-9

Adnan, M., Hamada, M. S., Hahn, M., Li, G.-Q., & Luo, C.-X. (2019). Fungicide resistance of *Botrytis cinerea* from strawberry to procymidone and zoxamide in Hubei, China. *Phytopathol Res*, *1*(1), 17. doi:10.118642483-019-0024-8

Adongo, J. O. Alice, Njue A.W., Omolo J.O.: Cheplogoi P.K., Dan O. Otaye D.O.(2012). In Vitro Inhibition of Botrytis Cinerea - Causative Agent for Grey Mold by Crude Extracts of Basidiomycetes Fungi. Science Journal of Biotech., Vol. 2012, Article ID sjbt-175, 3 Pages. doi: . Google Scholar doi:10.7237jbt/175

Adrian, M., Sparkes, A., & Thomas, B. (2000). Strategic and regional initiatives in the Agri-food industry in Wales. *British Food Journal*, *102*(4), 274–289. doi:10.1108/00070700010327689

Afandi, E., Kermani, M., & Mammadov, F. (2017). Social capital and entrepreneurial process. *The International Entrepreneurship and Management Journal*, *13*(3), 685–716. https://doi.org/10.1007/s11365-016-0421-8

Aggarwal, V. K., & Jain, S. (2014). Role of women entrepreneurs in 21st century India: Challenges and empowerment. *International Journal of Advanced Research*, *2*(9), 834–840.

Aguayo, R., Riquelme, J., Valenzuela, P. D. T., Hahn, M., & Silva Moreno, E. (2011). Bchex Virulence Gene of Botrytis Cinerea: Characterization and Functional Analysis. [Google Scholar]. *Journal of General Plant Pathology*, *77*(4), 230–238. doi:10.100710327-011-0311-4

Ahl, H. J. (2002). *The making of the female entrepreneur: A discourse analysis of research texts on women's entrepreneurship* (Doctoral dissertation). Internationella Handelshögskolan.

Airriess, C. A., & Clawson, D. L. (1994). Vietnamese market gardens in New Orleans. *Geographical Review*, 16–31.

Ajayi, M. T. (1987). Effects of spacing/number of plants per stand and fertilizer placement on performance of two maize varieties intercropped with cowpea in the forest zone of Nigeria. *Ghana Journal of Agricultural Science*, *20*, 39–46.

Ajzen, I. (1991). The theory of planned behavior. *Organizational Behavior and Human Decision Processes*, *50*(2), 179–211. doi:10.1016/0749-5978(91)90020-T

Ajzen, I. (1991). *The theory of planned behaviour*. Organizational Behaviour and Human Decision.

Aldrich, H. (1999). Promise, Failure, and Redemption: A Life Course Perspective on Teaching as a Career. *The Social Worlds of Higher Education: Handbook for Teaching in a New Century*, 533-42..

Aldrich, H. E., & Martinez, M. A. (2007). Many are called, but few are chosen: An evolutionary perspective for the study of entrepreneurship. In Entrepreneurship (pp. 293–311). Concepts, Theory and Perspective. https://doi.org/10.1007/978-3-540-48543-8_14.

Aldrich, H., & Martinez, M. (2001). *Many are called, but few are chosen: An evolutionary*. Academic Press.

Aldrich, H., & Zimmer, C. (1986). Entrepeneurship through social networks. *The Art and Science of Entrepeneurship*. doi:10.2139/ssrn.1497761

Aldrich, H. E., & Fiol, M. (1994). Fools rush in? The institutional context of industrial creation. *Academy of Management Review*, *19*(4), 645–670. https://doi.org/10.5465/AMR.1994.9412190214

Aldrich, H. E., & Kim, P. H. (2007). Small worlds, infinite possibilities? How social networks affect entrepreneurial team formation and search. *Strategic Entrepreneurship Journal*, *1*(1/2), 147–165. doi:10.1002ej.8

Aldrich, H., Zimmer, C., & Jones, T. (1986). Small business still speaks with the same voice: A replication of 'the voice of small business and the politics of survival'. *The Sociological Review*, *34*(2), 335–356. doi:10.1111/j.1467-954X.1986.tb02705.x

Alessa, A. A., Zaabi, E. A., & Diab, A. M. (2018). Impact of Environmental Factors on Financing Agriculture Entrepreneurs. *International Journal of Entrepreneurship*, *22*(4), 1–13. https://www.abacademies.org/articles/Impact-of-environmental-factors-on-financing-agriculture-entrepreneurs-1939-4675-22-4-228.pdf

Alex, L. (2011). *A Review and Analysis of Policies on Farmers' Entrepreneurship Development*. PELUM.

Alforja, S. I. R., Rico, P. M. B., Caoili, B. L., & Latina, R. A. (2021). Two Philippine *Photorhabdus luminescens* strains inhibit the in vitro growth of *Lasiodiplodia theobromae, Fusarium oxysporum* f. sp. *lycopersici*, and *Colletotrichum* spp. *Egyptian Journal of Biological Pest Control*, *31*(1), 108. doi:10.118641938-021-00454-x

Ali, J. (2016). Performance of small and medium-sized food and agribusiness enterprises: evidence from Indian firms. *International Food and Agribusiness Management Review, 19*, 53-64.

Ali, I., & Tajddini, K., urRehman, K., Ali, J. F., & Ahmed, I. (2010). University student's inclination of governance and its effects on entrepreneurial intentions: An empirical analysis. *International Journal of Trade. Economics and Finance*, *1*(1), 36.

Ali, J. (2016). Performance of small and medium-sized food and agribusiness enterprises: Evidence from indian firms. *The International Food and Agribusiness Management Review*, *19*(4), 53–64. https://doi. org/10.22434/IFAMR2016.0024

Alkon, A., & Mares, T. (2012). Food sovereignty in US food movements: Radical visions and neoliberal constraints. *Agriculture and Human Values, 29*(3), 347-359. doi:10.1007/s10460-012-9356-z

Allen, P., FitzSimmons, M., Goodman, M., & Warner, K. (2003). Shifting plates in the agrifood landscape: The tectonics of alternative agrifood initiatives in California. *Journal of Rural Studies*, *19*(1), 61–75.

Alsagoff, S. A. K., Clonts, H. A., & Jolly, C. M. (1990). An integrated poultry, multi-species aqua culture for Malaysian rice farmers: A mixed integer programming approach. *Agricultural Systems*, *32*(3), 207–231.

Alsos, G. A., Carter, S., & Ljunggren, E. (2011). *The handbook of research on entrepreneurship in agriculture and rural development*. Edward Elgar Publishing.

Alsos, G. A., Hogskolani, J., Internationella, H., & Ihh, E. (2011). *The handbook of research onentrepreneurship in agriculture and rural development*. Edward Elgar Publishing Limited. doi:10.4337/9780857933249

Altieri, M. A., &Nicholls, C. I. (2005). *Agroecology and the search for a truly sustainable agriculture*. United Nations Environmental Programme, Environmental Training Network for Latin America and the Caribbean.

Altieri, M. A. (1989). Agroecology: A new research and development paradigm for world agriculture. *Agriculture, Ecosystems & Environment*, *27*(1-4), 37–46. doi:10.1016/0167-8809(89)90070-4

Amba-Rao, S. C. (2008). Multinational Institution Social Responsibility, Ethics, Interactions and Third World Governments: An Agenda for the 1990s. *Journal of Business Ethics*, *12*(3), 553–572.

Amit, R., Glosten, L., & Muller, E. (1993). Challenges to theory development in entrepreneurship research. *Journal of Management Studies*, *30*(5), 815–834.

Amy, B. (2014). *Understanding food: principles and preparation*. Nelson Education.

Anand Singh, K., & Krishna, K. (1994). Agricultural entrepreneurship: The concept and evidence. *The Journal of Entrepreneurship*, *3*(1), 97–111. doi:10.1177/097135579400300106

Anco, D. J., & Michael, A. E. (2011). Fruit Rots of Blueberry: Alternaria, Anthracnose, and Botrytis, *Agri. and Nat. Res. HYG-3213-11*, 1-2. https://ohioline.osu.edu/factsheet/plpath-fru-44

Andre, R., & Burns, C. (2006). Post-migration food habits of sub-Saharan African migrants in Victoria: A cross-sectional study. *Nutrition & Dietetics: the Journal of the Dietitians Association of Australia*, *63*(2), 91–102. doi:10.1111/j.1747-0080.2006.00055.x

Andrew, D. (2002). *March: Local and regional food.* IGD.

Angela, T., Arfini, F., Belletti, G., & Marescotti, A. (2007). Regional foods and rural development: The role of product qualification. *Journal of Rural Studies*, *23*(1), 12–22. doi:10.1016/j.jrurstud.2006.09.010

Annie, H.-L. (2004). *Introduction.* Taylor & Francis.

Anwar, I., & Saleem, I. (2018). Effect of entrepreneurial education on entrepreneurial intention of Indian students. *International Journal of Research*, *5*(12), 2306–2316.

Anwar, I., Saleem, I., Islam, K. B., Thoudam, P., & Khan, R. (2020). Entrepreneurial intention among female university students: Examining the moderating role of entrepreneurial education. *Journal for International Business and Entrepreneurship Development*, *12*(4), 217–234. doi:10.1504/JIBED.2020.110254

Anwar, I., Thoudam, P., & Saleem, I. (2021). Role of entrepreneurial education in shaping entrepreneurial intention among university students: Testing the hypotheses using mediation and moderation approach. *Journal of Education for Business*, 1–13. doi:10.1080/08832323.2021.1883502

Arafat, M. Y., & Saleem, I. (2017a). *Determinants of entrepreneurial startups in India: A study using GEM data.* Academic Press.

Arafat, M. Y., & Saleem, I. (2017b). Determinants of entrepreneurial startups in India: A study using GEM data. In S. Misra, S. Shukla, & G. Batthini (Eds.), *Twelfth biennial conference on Entrepreneurship. Paper presented at twelfth biennial conference on entrepreneurship, entrepreneurship development Institute of India, Ahmedabad 22–24 February* (pp. 656–665). Delhi: Bookwell.

Arafat, M. Y., Ali, J., Dwivedi, A. K., & Saleem, I. (2020a). Social and Cognitive Aspects of Women Entrepreneurs: Evidence from India. *Vikalpa*, *45*(4), 223–239. doi:10.1177/02560909211015457

Arafat, M. Y., Khan, A. M., Saleem, I., Khan, N. A., & Khan, M. M. (2020b). Intellectual and cognitive aspects of women entrepreneurs in India. *International Journal of Knowledge Management Studies*, *11*(3), 278–297. doi:10.1504/IJKMS.2020.109092

Arafat, M. Y., & Saleem, I. (2017a). Examining start-up intention of Indians through cognitive approach: A study using GEM data. *Journal of Global Entrepreneurship Research*, *7*(1), 13. https://doi.org/10.1186 /s40497-017-0073-3

Arafat, M. Y., & Saleem, I. (2017b). Examining start-up Intention of Indians through cognitive approach: A study using GEM data. *Journal of Global Entrepreneurship Research*, *7*(1), 1–11. doi:10.118640497-017-0073-3

Arafat, M. Y., Saleem, I., & Dwivedi, A. K. (2020). Understanding entrepreneurial intention among Indian youth aspiring for self-employment. *International Journal of Knowledge and Learning*, *13*(3), 185–200. doi:10.1504/IJKL.2020.109882

Arafat, M. Y., Saleem, I., Dwivedi, A. K., & Khan, A. (2018). Determinants of agricultural entrepreneurship: A GEM data-based study. *The International Entrepreneurship and Management Journal*, •••, 1–26.

Arafat, M. Y., Saleem, I., Dwivedi, A. K., & Khan, A. (2020). Determinants of agricultural entrepreneurship: A GEM data based study. *The International Entrepreneurship and Management Journal*, *16*(1), 345–370. doi:10.100711365-018-0536-1

Arenius, P., & Kovalainen, A. (2006). Similarities and differences across the factors associated with women's self-employment preference in the Nordic countries. *International Small Business Journal*, *24*(1), 31–59.

Arenius, P., & Minniti, M. (2005). Perceptual variables and nascent entrepreneurship. *Small Business Economics*, *24*(3), 233–247.

Armstrong, D. (2000). A survey of community gardens in upstate New York: Implications for health promotion and community development. *Health & Place*, *6*(4), 319–327.

Arregle, J. L., Hitt, M. A., Sirmon, D. G., & Very, P. (2007). The development of organizational social capital: Attributes of family firm. *Journal of Management Studies*, *44*(1), 73–95. doi:10.1111/j.1467-6486.2007.00665.x

Ashraf, S., Amanatidis, S., & Samman, S. (2002). The effect of migration on dietary intake, type 2 diabetes and obesity: The Ghanaian Health and Nutrition Analysis in Sydney, Australia (GHANAISA). *Ecology of Food and Nutrition*, *41*(3), 255–270. doi:10.1080/0367-020291909778

Astuti, U. P., Wahyu, W., & Andi, I. (2011). *Mempengaruhi Alih Fungsi Lahan Pangan menjadi Kelapa Sawit di Bengkulu : Kasus Petani di Desa Kungkai Baru*. Academic Press.

Autio, E., Kenney, M., Mustar, P., Siegel, D., & Wright, M. (2014). Entrepreneurial innovation: The importance of context. *Research Policy*, *43*(7), 1097–1108. https://doi.org/10.1016/j.respol.2014.01.015

Avis, R. (2012) Compost teas and extracts: brewin' and bubblin' basics. *Res. Inst. - Permac. Forums Courses Inf. News*. https://permaculturenews.org/2012/07/11/compost-teas-and-extracts-brewin-and-bubblin-basics/

Ayaz, M., Ammad-Uddin, M., Zubair Sharif, Z., Ali Mansour, A., & Aggoune, E.M. (2019). Internet-Of-Things (IoT)-Based Smart Agriculture: Toward Making the Fields Talk. *Special Section on New Technologies for Smart Farming 4.0*, *7*, 129551–129583. doi:.2019.2932609 doi:10.1109/ACCESS

Azeta, J., Bolu, C. A., Alele, F., Daranijo, E. O., Onyeubani, P., & Abioye, A. A. (2019). Application of Mechatronics in Agriculture: A review. *International Conference on Engineering for Sustainable World Journal of Physics: Conference Series*, *1378*, 1-11. 10.1088/1742-6596/1378/3/032006

Bairwa, S. L., & Kushwaha, S. (2012). Agro Industry scenario in India. In Agricultural research and Sustainable development in India. Bharti Publications.

Bairwa, S. L., Kushwaha, S., Meena, L. K., Lakra, K., & Kumar, P. (2014a). Agribusiness Potential of North Eastern States: A SWOT Analysis. In Agribusiness Potentials in India: experience from hill states. EBH Publishers.

Bairwa, S. L., Kalia, A., Meena, L. K., Lakra, K., & Kushwaha, S. (2014b). Agribusiness Management Education: A Review on Employment Opportunities. *International Journal of Scientific and Research Publications*, *4*(2), 1–4.

Bairwa, S. L., Lakra, K., Kushwaha, S., Meena, L. K., & Kumar, P. (2014). Agripreneurship development as a tool to the upliftment of agriculture. *International Journal of Scientific and Research Publications*, *4*(3), 1–4. https://www.academia.edu/download/33331824/ijsrp-p2770_agripreneurship.pdf

Baker, R. S., & Wilcox, G. (1961). Effect of Foliage Ozone Injury and Infection of Potato Damage and Stand Reduction on Onion Yield. [Google Scholar]. *Proceedings of the American Society for Horticultural Science*, *78*, 400–405.

Baker, T., Gedajlovic, E., & Lubatkin, M. (2005). A framework for comparing entrepreneurship processes across nations. *Journal of International Business Studies*, *36*(5), 492–504. https://doi.org/10.1057/palgrave.jibs.8400153

Bali Swain, R., & Wallentin, F. Y. (2012). Factors empowering women in Indian self-help group programs. *International Review of Applied Economics*, *26*(4), 425–444.

Bandura, A. (1977). Self-efficacy: Toward a unifying theory of behavioral change. *Psychological Review*, *84*(2), 191–215. doi:10.1037/0033-295X.84.2.191 PMID:847061

Bandura, A. (1977). Self-efficacy: Toward a unifying theory of behavioural change. *Psychological Review*, *84*(2), 191–215.

Bane, P. (2012). *The permaculture handbook: garden farming for town and country*. New SocietyPublishers.

Banerjee, A., Duflo, E., Glennerster, R., & Kinnan, C. (2015). The miracle of microfinance?Evidence from a randomized evaluation. *American Economic Journal. Applied Economics*, *7*(1), 22–53. doi:10.1257/app.20130533

Barbieri, C., & Mahoney, E. (2009). Why is diversification an attractive farm adjustment strategy? Insights from Texas farmers and ranchers. *Journal of Rural Studies*, *25*(1), 58–66. .jrurstud.2008.06.001. doi:10.1016/j

Barbosa, A. C., do Carmo, A. E., Graf, L., Tomaz, R., de Souza, C. F., Mendes, J., Randi, M. A. F., Buchi, D., & Schadeck, R. J. G. (2006). Morphology and Lipid Body and Vacuole Dynamics during Secondary Conidia Formation in Colletotrichum Acutatum: Laser scanning confocal analysis. *Canadian Journal of Microbiology*, *52*(2), 117–124. doi:10.1139/w05-104 PMID:16541147

Bardhan, K., & Klasen, S. (1999). UNDP's gender-related indices: A critical review. *World Development*, *27*(6), 985–1010. doi:10.1016/S0305-750X(99)00035-2

Barnes, A. (2003). *How to maintain the trust: A study in banking customer*. Academic Press.

Barnes, A. P., Hansson, H., Manevska-Tasevska, G., Shrestha, S. S., & Thomson, S. G. (2015). The influence of diversification on long-term viability of the agricultural sector. *Land Use Policy*, *49*, 404–412.

Barney, J. B., & Hansen, M. H. (1994). Trustworthiness as a source of competitive advantage. *Strategic Management Journal*, *15*(S1), 175–190. doi:10.1002mj.4250150912

Bar-Nun, N., L'Hyvernay, A., Donèche, B., & Mayer, M. (2007). Changes in mycelial structure of Botrytis cinerea Induced by Removal of the Glucan Matrix. *Int. J. Vine and Wine Sci.*, *41*(3), 149–153. doi:10.20870/oeno-one.2007.41.3.843

Baron, R. A. (1998). Cognitive mechanisms in entrepreneurship: Why and when entrepreneurs think differently than other people. *Journal of Business Venturing*, *13*(4), 275–294. https://doi.org/10.1016 /S0883-9026(97)00031-1

Baron, R. A. (2004). The cognitive perspective: A valuable tool for answering entrepreneurship's basic "why" questions. *Journal of Business Venturing*, *19*(2), 221–239.

Baron, R. A. (2007). Behavioural and cognitive factors in entrepreneurship: Entrepreneurs as the active element in new venture creation. *Strategic Entrepreneurship Journal*, *1*(1/2), 167–182.

Baron, R. A., Markman, G. D., & Hirsa, A. (2001). Perceptions of women and men as entrepreneurs: Evidence for differential effects of attributional augmenting. *The Journal of Applied Psychology*, *86*(5), 923–929. doi:10.1037/0021-9010.86.5.923 PMID:11596808

Barraclough, P. B. (1984). The Growth and the Activity of Winter Wheat Roots in the Field: Root Growth of High-Yielding Crops in Relation to Shoot Growth. [Google Scholar]. *J. agr. Sci. (Camb.)*, *103*(2), 439–442. doi:10.1017/S0021859600047419

Barral, Rey Benayas, Meli, & Maceira. (2015). Quantifying the impacts of ecological restoration on biodiversity and ecosystem services in agroecosystems: a global meta-analysis. *Agriculture, Ecosystems and Environment, 202*, 223–231. doi:10.1016/j.agee.2015.01.009

Barron, J. (2017). Community gardening: Cultivating subjectivities, space, and justice. *Local Environment*, *22*(9), 1142–1158.

Bartling, H. (2012). A chicken ain'tnothin'but a bird: Local food production and the politics of land-use change. *Local Environment, 17*(1), 23–34.

Battistella, C., Biotto, G., & De Toni, A. F. (2012). From design driven innovation to meaning strategy. *Management Decision, 50*(4), 718–743. doi:10.1108/00251741211220390

Baughn, C. C., Chua, B. L., & Neupert, K. E. (2006). The normative context for women's participation in entrepreneruship: A multicountry study. *Entrepreneurship Theory and Practice, 30*(5), 687–708. doi:10.1111/j.1540-6520.2006.00142.x

Bekhet, H. A., & Abdullah, A. (2010). Energy use in agriculture sector: Input-output analysis. *International Business Research, 3*(3), 111. doi:10.5539/ibr.v3n3p111

Bell, G. (2005). *The permaculture garden*. Chelsea Green.

Belsky, J. (1993). Household food security, farmt rees, and agroforestry: A comparative study in Indonesia and the Philippines. *Human Organization, 52*(2), 130–141. doi:10.17730/humo.52.2.308kw181875xpt0r

Benos, T., Kalogeras, N., Wetzels, M., De Ruyter, K., & Pennings, J. M. (2018). Harnessing a 'currency matrix'for performance measurement in cooperatives: A multi-phased study. *Sustainability, 10*(12), 4536.

Berg, H. (2002). Rice monoculture and integrated rice-fish farming in the Mekong Delta, Vietnam—Economic and ecological considerations. *Ecological Economics, 41*(1), 95–107. doi:10.1016/S0921-8009(02)00027-7

Bhasin, S. (2012). An appropriate change strategy for lean success. *Management Decision, 50*(3), 439–458. doi:10.1108/00251741211216223

Bhattacharya, S. N. (1986). *Rural economic growth in India in theory and practice*. Metropolitan Book Company. Retrieved from https://www.abebooks.com/9788120000797/Rural-economic-growth-India-theory-812000079X/plp

Bi, K., Scalschi, L., Jaiswal, N., Mengiste, T., Fried, R., Sanz, A. B., Arroyo, J., Zhu, W., Masrati, G., & Sharon, A. (2021). The *Botrytis cinerea* Crh1 transglycosylase is a cytoplasmic effector triggering plant cell death and defense response. *Nature Communications, 12*(1), 2166. doi:10.103841467-021-22436-1 PMID:33846308

Bika, R., Baysal-Gurel, F., & Jennings, C. (2021). *Botrytis cinerea* management in ornamental production: A continuous battle. *Canadian Journal of Plant Pathology, 43*(3), 345–365. doi:10.1080/07060661.2020.1807409

Bird, B., &Schjoedt, L. (2017). Entrepreneurial behavior: Its nature, scope, recent research, and agenda for future research. *Revisiting the Entrepreneurial Mind*, 379-409.

Birwa, S. L., Kushwaha, S., Lark, K., & Mena, L. K. (2014). Agripreneurship Development as a Tool toUpliftment of Agriculture. *International Journal of Scientific and Research Publications, 4*(3).

Blanchflower, D. G. (2004). *Self-employment: More may not be better*. Academic Press.

Blanchflower, D. G., & Oswald, A. J. (1998). *Entrepreneurship and the youth labour market problem: A report for the OECD*. Report to OECD.

Blanco-Canqui, H., & Lal, R. (2010). Soil and water conservation. In *Principles of soilconservation and management* (pp. 1–19). Springer.

Block, D. R., Chávez, N., Allen, E., & Ramirez, D. (2012). Food sovereignty, urban food access, and food activism: Contemplating the connections through examples from Chicago. *Agriculture and Human Values*, *29*(2), 203–215.

Blyn, G. (2016). *Agricultural trends in India, 1891-1947*. University of Pennsylvania Pre.

Boogaard, B. K., Bock, B. B., Oosting, S. J., Wiskerke, J. S., & van der Zijpp, A. J. (2011). Social acceptance of dairy farming: The ambivalence between the two faces of modernity. *Journal of Agricultural & Environmental Ethics*, *24*(3), 259–282. doi:10.100710806-010-9256-4

Borland, C. M. (1975). *Locus of control, need for achievement and entrepreneurship* (Doctoral dissertation). ProQuest Information & Learning.

Bosma, N., &Schutjens, V. (2007). Patterns of promising entrepreneurial activity in European regions. *Tijdschriftvooreconomische en socialegeografie, 98*(5), 675-686.

Bosma, N. (2021). *Global Entrepreneurship Monitor Report, 2020/21*. Author.

Bourdieu, P. (1986). The forms of capital. In J. G. Richardson (Ed.), *Handbook of Theory and Research for the Sociology of Education* (pp. 241–258). Greenwood Press.

Bowen, S., & Mutersbaugh, T. (2014). Local or localized? Exploring the contributions of Franco-Mediterranean agrifood theory to alternative food research. *Agriculture and Human Values*, *31*(2), 201–213.

Bowen, S., & Mutersbaugh, T. (2014). Local or localized?Exploring the contributions of Franco-Mediterranean agrifood theory to alternative food research. *Agriculture and Human Values*, *31*(2), 201–213.

Bowens, N. (2015). *The color of food*. New Society.

Boyd, N. G., & Vozikis, G. S. (1994). The influence of self-efficacy on the development of entrepreneurial intentions and actions. *Entrepreneurship Theory and Practice*, *18*, 63–63.

Bradley, M. (2009). Down with the fences: Battles for the commons in South London. *The Land*, *7*, 34–38.

Bragg, P. L., Govi, G., & Cannell, R. Q. (1983). A Comparison of Methods, Including Angled and Vertical Minirhizotrons for Studying Root Growth and Distribution In a Spring Oat Crop. [Google Scholar]. *Plant and Soil*, *73*(3), 435–440. doi:10.1007/BF02184322

Britz, W., van Ittersum, M., Lansink, A. O., & Heckelei, T. (2012). Tools for integrated assessment in agriculture. State of the art and challenges. *Bio-Based and Applied Economics*, *1*(2), 125–150.

Brixy, U., & Hessels, J. (2010). *Human capital and start-up success of nascent entrepreneurs*. EIM Research Reports H, 201013.

Brixy, U., Sternberg, R., & Stüber, H. (2012). The selectiveness of the entrepreneurial process. *Journal of Small Business Management*, *50*(1), 105–131. https://doi.org/10.1111/j.1540-627X.2011.00346.x

Brockhaus, R. H., & Horwitz, P. S. (1986). The psychology of the entrepreneur. In D. L. Sexton & R. W. Smilor (Eds.), The art and science of entrepreneurship. Ballinger publishing company (pp. 25–48). Academic Press.

Brockhaus, R. H. Sr. (1980). Risk taking propensity of entrepreneurs. *Academy of Management Journal*, *23*(3), 509–520.

Browne, M. W., & Cudeck, R. (1992). Alternative ways of assessing model fit. *Sociological Methods & Research*, *21*(2), 230–258. doi:10.1177/0049124192021002005

Brown, K. H., Carter, A., & Bailkey, M. (2003). *Urban agriculture and community food security in the United States: Farming from the city center to the urban fringe*. Community Food Security Coalition.

Brown, K. H., & Jameton, A. L. (2000). Public health implications of urban agriculture. *Journal of Public Health Policy*, *21*(1), 20–39. doi:10.2307/3343472 PMID:10754796

Brunjes, J., & Revilla, J. (2013). 'Recession push' and 'prosperity pull' entrepreneurship in a rural developing context. *Entrepreneurship and Regional Development*, *25*, 251–271.

Brush, C. G. (1990). Women and enterprise creation: Barriers and opportunities. *Enterprising women: Local initiatives for job creation, 37.*

Brush, C. G. (1992). Research on women business owners: Past trends, a new perspective and future directions. *Entrepreneurship Theory and Practice*, *16*(4), 5–30. doi:10.1177/104225879201600401

Bryk, H., Barbars, D., & Piotr, S. (2004). Inhibitory Effect of Pseudomonas spp. on the Development of Botrytis cinerea and Penicillium expansum. *Plant Protection Science*, *40*(4), 128–134. doi:10.17221/467-PPS

Burge, W. D. (2017). *The relationship between the human and social capital characteristics of nascent entrepreneurs and expected job growth in the United States* (Doctoral dissertation). The University of Southern Mississippi.

Burt, R. S. (2000). The network structure of social capital. *Research in Organizational Behavior*, *22*, 345–423. doi:10.1016/S0191-3085(00)22009-1

Butler, W. H. (2012). Welcoming animals back to the city: Navigating the tensions of urban livestock through municipal ordinances. *Journal of Agriculture, Food Systems, and Community Development*, *2*(2), 193–215.

Cadji, J., &Alkon, A. (2014). One day, the white people are going to want these houses again. *Incomplete streets: Processes, practices and possibilities*, 154-175.

Caron, B. (2003). Food choices among newly married couples: Convergence, conflict, individualism, and projects. *Appetite*, *40*(1), 25–41. doi:10.1016/S0195-6663(02)00147-2 PMID:12631502

Carr, S., & Rollin, A. (2016). *An exploration of Agripreneurship Scope, Actors and Prospects*. Academic Press.

Carroll, A. (2005). Impact of Institution Social Responsibility and Higher Education. *Journal of Business and Society*, *38*(3), 268–295. doi:10.1177/000765039903800303

Carter, N. M., Williams, M., & Reynolds, P. D. (1997). Discontinuance among new firms in retail: The influence of initial resources, strategy, and gender. *Journal of Business Venturing*, *12*(2), 125–145. doi:10.1016/S0883-9026(96)00033-X

Carter, R., & Van Auken, H. (2006). Small firm bankruptcy. *Journal of Small Business Management*, *44*(4), 493–512. doi:10.1111/j.1540-627X.2006.00187.x

Carter, S. (1998). Portfolio entrepreneurship in the farm sector: Indigenous growth in rural areas? *Entrepreneurship and Regional Development*, *10*(1), 17–32.

Carter, S., & Rosa, P. (1998). Indigenous rural firms: Farm enterprises in the UK. *International Small Business Journal*, *16*(4), 15–27.

Cegarra-Navarro, J. G., Sánchez-Vidal, M. E., & Cegarra-Leiva, D. (2011). Balancing exploration and exploitation of knowledge through an unlearning context: An empirical investigation in SMEs. *Management Decision*.

Chandra Shekhra, P. (2003) Third wave in Indian agriculture: introduction to Agriclinics and agribusiness centre scheme. *Manage Ext. Res. Rev.*, 10-20.

Charlotte, W., Tregear, A., & Allinson, J. (2003). In search of the concerned consumer: UK public perceptions of food, farming and buying local. *Journal of Rural Studies*, *19*(2), 233–244. doi:10.1016/S0743-0167(02)00083-9

Chatterjee, S., Gupta, S. D., & Upadhyay, P. (2018). Empowering women and stimulating development at bottom of pyramid through micro-entrepreneurship. *Management Decision*.

Chen, C. C., Greene, P. G., & Crick, A. (1998). Does entrepreneurial self-efficacy distinguish entrepreneurs from managers? *Journal of Business Venturing*, *13*(4), 295–316.

Chilvers, M. I., & Toit, L. J. (2006). Detection and Identification of Botrytis Species Associated with Neck Rot, Scape Blight, and Umbel Blight of Onion, Plant Management Network. *Plant Health Progress*, 1–20.

Chvatalová, Kocmanová, & Dočekalová. (2011). Institution Sustainability Reporting and Measuring Institution Performance. *International Symposium on Environmental Software Systems*, 245-254.

Cinar, E. M., Du, Y., & Hienkel, T. (2018). Chinese entrepreneurship attributes: A comparative GEM data analysis. *Journal of Entrepreneurship in Emerging Economies*, *10*(2), 217–248. https://doi.org/10.1108/JEEE-03-2017-0016

Clark, K. H., & Nicholas, K. A. (2013). Introducing urban food forestry: A multifunctional approach to increase food security and provide ecosystem services. *Landscape Ecology*, *28*(9), 1649–1669. doi:10.100710980-013-9903-z

Claude, F. (1988). Food, self and identity. *Information (Basel)*, *27*(2), 275.

Claude, L.-S. (1963). *Totemism Pa Txt*. Beacon Press.

Claude, L.-S. (1966). Anthropology: Its achievements and future. *Nature*, *209*(5018), 10–13. doi:10.1038/209010a0 PMID:5925326

Cohen, M. J., & Garrett, J. L. (2010). The food price crisis and urban food (in) security. *Environment and Urbanization*, *22*(2), 467–482.

Cohen, N., & Reynolds, K. (2014). Urban agriculture policy making in New York's "New Political Spaces" strategizing for a participatory and representative system. *Journal of Planning Education and Research*, *34*(2), 221–234.

Cohn, A. (2006). *Agroecology and the Struggle for Food Sovereignty in the Americas: A Collaborative Project of the International Institute for Environment and Development (IIED), the IUCN Commissionon Environmental, Economic and Social Policy (CEESP) and the Yale School of Forestry & Environmental Studies (Yale F&ES)*. IIED.

Collado, I. G., Sa'nchez, A. J. M., & Hanson, J. R. (2007). Fungal Terpene Metabolites: Biosynthetic Relationships and the Control of the Phytopathogenic Fungus Botrytis cinerea. [Google Scholar]. *Natural Product Reports*, *24*(4), 674–686. doi:10.1039/b603085h PMID:17653354

Collins, O. F., & Moore, D. G. (1964). The enterprising man. East Lansing: Michigan State University. Coleman, J. S. (1988). Social Capital in the Creation of human capital. *American Journal of Sociology*, *94*, S95–S120. https://doi.org/10.1086/228943

Companioni, N., Rodríguez, A., Carrión, M., Alonso, M., Ojeda, R. M., & Peña, E. (1997). La Agricultura Urbana en Cuba: suparticipaciónen la seguridadalimentaria. *Proceedings III encuentronacional de agriculturaorganica*, 9-13.

Condor, R. (2020). Entrepreneurship in agriculture: A literature review. *International Journal of Entrepreneurship and Small Business*, *40*(4), 516. doi:10.1504/IJESB.2020.109013

Conroy, D. E., Willow, J. P., & Metzler, J. N. (2002). Multidimensional fear of failure measurement: The performance failure appraisal inventory. *Journal of Applied Sport Psychology*, *14*(2), 76–90. doi:10.1080/10413200252907752

Cook, I. (2008). Geographies of food: Mixing. *Progress in Human Geography*, *32*(6), 821–833.

Cooper, A. C., & Dunkelberg, W. C. (1987). Entrepreneurial research: Old questions, new answers and methodological issues. *American Journal of Small Business*, *11*(3), 11–24.

Cotoras, M., Garcı'a, C., & Mendoza, L. (2009). Botrytis Cinerea Isolates Collected from Grapes Present Different Requirements for Conidia Germination. [Google Scholar]. *Mycologia*, *101*(3), 287–295. doi:10.3852/08-012 PMID:19537202

Cowling, M., & Taylor, M. (2001). Entrepreneurial women and men: Two different species? *Small Business Economics*, *16*(3), 167–175. doi:10.1023/A:1011195516912

CPRE. (2003). *Down your way? A CPRE briefing on supermarkets and local food.* London: CPRE. [Available from: https://www.cpre.org.uk/library/results/local-foods

CPRE. (2008). *Help local food survive the supermarket onslaught.* CPRE. Available from: http://support.cpre.org.uk/cpre/supermarket_solutions.asp

Crane, A., Viswanathan, L., & Whitelaw, G. (2013). Sustainability through intervention: A case study of guerrilla gardening in Kingston, Ontario. *Local Environment*, *18*(1), 71–90.

Crawford, M. (2010). *Creating a forest garden.* Green Books.

Cromie, S. (2000). Assessing entrepreneurial inclinations: Some approaches and empirical evidence. *European Journal of Work and Organizational Psychology*, *9*(1), 7–30.

Cronbach, L. J. (1951). Coefficient alpha and the internal structure of tests. *Psychometrika, 16*(3), 297-334.

Crowder, D. (2009). The roles of Institution Reporting, Stakeholders and the Internet: Mapping the New Institution Landscape. *Journal of Urban Studies*, *37*(10), 1837–1848. doi:10.1080/00420980020080451

Crowther, D. (2008). *The Social and Environmental Accounting and its applicability.* The Financial Times.

Cuervo, Á., Ribeiro, D., & Roig, S. (2007). Entrepreneurship: Concepts, Theory and Perspective. Introduction. In Entrepreneurship (pp. 1–20). Springer.

Cueva Beteta, H. (2006). What is missing in measures of women's empowerment? *Journal of Human Development*, *7*(2), 221–241. doi:10.1080/14649880600768553

Curado, C., Henriques, L., & Bontis, N. (2011). Intellectual capital disclosure payback. *Management Decision*, *49*(7), 1080–1098. doi:10.1108/00251741111151154

Dahl, M. S., & Sorenson, O. (2009). The embedded entrepreneur. *European Management Review*, *6*(3), 172–181. doi:10.1057/emr.2009.14

Dahlqvist, J., Davidsson, P., & Wiklund, J. (2000). Initial conditions as predictors of new venture performance: A replication and extension of the Cooper et al. study. *Enterprise and Innovation Management Studies*, *1*(1), 1–17.

Dalgaard, T. H., & Porter, J. R. (2003). Agroecología: escalamiento e interdisciplinariedad. *Agricultura, ecosistemas y medioambiente, 100*(1), 39-51.

Dalgaard, T. H., & Porter, J. R. (2003). Agroecología: escalamiento e interdisciplinariedad. Agricultura. *Ecosistemas y Medio Ambiente, 100*(1), 39-51.

Dalsgaard, J. P. T., & Oficial, R. T. (1997). A quantitative approach for assessing the productive performance and ecological contributions of smallholder farms. *Agricultural Systems*, *55*(4), 503–533. doi:10.1016/S0308-521X(97)00022-X

Daly, H. (1999). The Allocation, Distribution and Scale towards an Economics that is Efficient, Just and Sustainable. *Journal of Ecological Economics*, *6*(3), 185–193. doi:10.1016/0921-8009(92)90024-M

Darmon, N., Darmon, M., Maillot, M., & Drewnowski, A. (2005). A nutrient density standard for vegetables and fruits: Nutrients per calorie and nutrients per unit cost. *Journal of the American Dietetic Association, 105*(12), 1881–1887.

Daugstad, K., Rønningen, K., & Skar, B. (2006). Agriculture as an upholder of cultural heritage? Conceptualizations and value judgements—A Norwegian perspective in international context. *Journal of Rural Studies*, *22*(1), 67–81.

David & Hillman. (1989). An evolutionary continuum of people-plant interaction. *Foraging and Farming: The Evolution of Plant Exploitation, 11.*

David. (2001). *Remembrance of repasts: an anthropology of food and memory*. Academic Press.

Davidsson, P. (1995). *Re: Determinants Of Entrepreneurial Intentions.* Type To Research, W. I. E.

Davidsson, P. (2005). The types and contextual fit of entrepreneurial processes. *International Journal of Entrepreneurship Education*, *2*, 4–407.

Davidsson, P., Delmar, F., & Wiklund, J. (2006). Entrepreneurship as growth; growth as entrepreneurship. *Entrepreneurship and the Growth of Firms*, *1*(2), 21–38.

Davidsson, P., & Honig, B. (2003). The role of social and human capital among nascent entrepreneurs. *Journal of Business Venturing*, *18*, 301–331. https://doi.org/10.1016/S0883-9026(02)00097-6

Dawadi, S., Baysal-Gurel, F., Addesso, K. M., Liyanapathiranage, P., & Simmons, T. (2021). Fire Ant Venom Alkaloids: Possible Control Measure for Soilborne and Foliar Plant Pathogens. *Pathogens (Basel, Switzerland)*, *10*(6), 659. doi:10.3390/pathogens10060659 PMID:34071926

De Massis, A., Kotlar, J., Wright, M., & Kellermanns, F. W. (2018). *Sector-based entrepreneurial capabilities and the promise of sector studies in entrepreneurship*. doi:10.1177/1042258717740548

De Simone, N., Pace, B., Grieco, F., Chimienti, G., Tyibilika, V., Santoro, V., Capozzi, V., Colelli, G., Spano, G., & Russo, P. (2020). Botrytis cinerea and Table Grapes: A Review of the Main Physical, Chemical, and Bio-Based Control Treatments in Post-Harvest. *Foods*, *9*(9), 1138. doi:10.3390/foods9091138 PMID:32824971

De Zeeuw, H., Van Veenhuizen, R., & Dubbeling, M. (2011). The role of urban agriculture in building resilient cities in developing countries. *Thai Journal of Agricultural Science*, *149*(S1), 153–163. doi:10.1017/S0021859610001279

De Zeeuw, H., Van Veenhuizen, R., & Dubbeling, M. (2011). The role of urban agriculture in building resilient cities in developing countries. *The Journal of Agricultural Science*, *149*(S1), 153–163.

Deakins, D., Bensemann, J., & Battisti, M. (2016). Entrepreneurial skill and regulation: Evidence from primary sector rural entrepreneurs. *International Journal of Entrepreneurial Behaviour & Research*, *22*(2), 234–259.

Deavers. (1997). Outsourcing: A corporate competitiveness strategy, not a search for low wages. *Journal of Labour Research, 18*(4), 504.

Debbie, K., Anderson, A., & Marshall, D. (1998). Living together and eating together: Changes in food choice and eating habits during the transition from single to married/cohabiting. *The Sociological Review*, *46*(1), 48–72. doi:10.1111/1467-954X.00089

Deborah, H., & Levy, S. (1991). Auto driving: A photo elicitation technique. *The Journal of Consumer Research*, *18*(3), 257. doi:10.1086/209258

Deborah, L. (1994). Food, memory and meaning: The symbolic and social nature of food events. *The Sociological Review*, *42*(4), 664–685. doi:10.1111/j.1467-954X.1994.tb00105.x

Deepthi, V. (2016). *A critical study on entrepreneurial behaviour of agripreneurs in Andhra Pradesh* (Doctoral dissertation). Acharya NG Ranga Agricultural University.

DEFRA. (2002). The strategy for sustainable farming and food: facing the future. London: DEFRA.

DEFRA. (2003). Local food – A snapshot of the sector. Report of the working group on local food. London: DEFRA.

DeJong, K., & Sarma, J. (1993). Generation Gaps Revisited. In *Foundations of Genetic Algorithms 2*. Morgan-Kaufmann Publishers.

Demeke, M. & Meschi, M. (2010). Technology adoption in the presence of constraints: the case of fertilizer demand in Ethiopia. *Review of Development Economics, 7*, 58–70.

Depositario, D. P. T., Aquino, N. A., & Feliciano, K. C. (2011). Entrepreneurial skill development needs of potential agri-based technopreneurs. Journal of ISSAAS.

DeRosa, M., McElwee, G., & Smith, R. (2019). Farm diversification strategies in response to rural policy: A case from rural Italy. *Land Use Policy*, *81*, 291–301. https://doi.org/10.1016/j.landusepol.2018.11.006

Despommier, D. (2010). *The vertical farm: feeding the world in the 21st century*. Macmillan.

DeTienne, D. R., & Chandler, G. N. (2007). The role of gender in opportunity identification. *Entrepreneurship Theory and Practice*, *31*(3), 365–386. doi:10.1111/j.1540-6520.2007.00178.x

Devendra, C. (2002). Crop–animal systems in Asia: Implications for research. *Agricultural Systems*, *71*(1-2), 169–177. doi:10.1016/S0308-521X(01)00042-7

Devi, M. (2015). A study on the influencing factors for a literate youth to take up agricultural entrepreneurship. *International Journal of Management and Commerce Innovations*, *3*(1), 692–700.

Dey, M. M., Paraguas, F. J., Kambewa, P., & Pemsl, D. E. (2010). The impact of integrated aquaculture–agriculture on small-scale farms in Southern Malawi. *Agricultural Economics*, *41*(1), 67–79. doi:10.1111/j.1574-0862.2009.00426.x

Dhingra, V., Gupta, M., Andacht, T., & Fu, Z. F. (2005). New Frontiers in Proteomics Research: A Perspective. *International Journal of Pharmaceutics*, *299*(1-2), 1–18. doi:10.1016/j.ijpharm.2005.04.010 PMID:15979831

Dinur, A. R. (2011). Common and un-common sense in managerial decision making under task uncertainty. *Management Decision*.

Diolez, A., Marches, F., Fortini, D., & Brygoo, Y. (1995). Boty, a Long-Terminal-Repeat Retroelement in the Phytopathogenic Fungus Botrytis cinerea. *Applied and Environmental Microbiology*, *61*(1), 103–108. doi:10.1128/aem.61.1.103-108.1995 PMID:7887592

Directorate General of Estate Crops. (2010). *Tree Crop Estate Statistic of Indonesia: 2009-2011, Palm Oil. Directorate General of Estate Crops*. Ministry of Agriculture.

Directorate General of Estate Crops. (2016). *Tree Crop Estate Statistic of Indonesia: 2015-2027, Palm Oil. Directorate General of Estate Crops*. Ministry of Agriculture.

Directorate General of Estate Crops. (2019). *Tree Crop Estate Statistic of Indonesia: 2018-2020, Palm Oil. Directorate General of Estate Crops*. Ministry of Agriculture.

Dixit, A. K., Rawat, I., Sharma, S., Mann, S., & Kumar, R. (2016). Entrepreneurship through EDP and Licensing of Post-Harvest Technology: An Impact Assessment. *Indian Journal of Economics and Development*, *12*(4), 679–686. doi:10.5958/2322-0430.2016.00191.8

Doepke, M., & Tertilt, M. (2011). *Does female empowerment promote economic development? IZA Discussion Papers 5637. Institute for the Study of Labor*. IZA.

Dollinger, M. J. (2003). Entrepreneurship – Strategies and Resources. Pearson International Edition.

Doll, W. J., & Torkzadeh, G. (1987). The relationship of MIS steering committees to size of firm and formalization of MIS planning. *Communications of the ACM*, *30*(11), 972–978. doi:10.1145/32206.32213

Douglas, A., & Kelly, K. (2000). Settlement patterns and the origins of African Jamaican society: Seville plantation, St. Ann's Bay, Jamaica. *Ethnohistory (Columbus, Ohio)*, *47*(2), 369–397. doi:10.1215/00141801-47-2-369

Drucker, P. (1985). *Innovation and Entrepreneurship*. Harper and Row. www.worlconnectors.nl

Duany, A. (2011). *Garden cities: theory & practice of agrarian urbanism*. Duany Plater-Zyberk& Company.

Dubé, L., Pingali, P., & Webb, P. (2012). Paths of convergence for agriculture, health, and wealth. *Proceedings of the National Academy of Sciences of the United States of America*, *109*(31), 12294–12301. doi:10.1073/pnas.0912951109 PMID:22826252

Duflo, E. (2011). *Women empowerment and economic development.* NBER Working Paper No. 17702.

Dung, T. Q., Bonney, L. B., Adhikari, R. P., & Miles, M. P. (2020). Entrepreneurial orientation, knowledge acquisition and collaborative performance in agri-food value-chains in emerging markets. *Supply Chain Management*, *25*(5), 521–533. doi:10.1108/SCM-09-2019-0327

Duru, M., & Therond, O. (2015). Designing agroecological transitions; A review. *Agronomy for Sustainable Development*, *35*(4), 1237–1257.

Duval-Couetil, N., Reed-Rhoads, T., & Haghighi, S. (2012). Engineering students and entrepreneurship education: Involvement, attitudes and outcomes. *International Journal of Engineering Education*, *28*(2), 425.

Duží, B., Tóth, A., Bihuňová, M., & Stojanov, R. (2014). Challenges of urban agriculture. Highlights on Czech and Slovak experience. In Current Challenges of Central Europe: Society and Environment (pp. 82–107). Charles University.

Duží, B., Tóth, A., Bihuňová, M., & Stojanov, R. (2014): Challenges of urban agriculture. Highlights on Czech and Slovak experience. In Current Challenges of Central Europe: Society and Environment (pp. 82–107). Charles University.

Dyllick, T., & Hockerts, K. (2002). Beyond the Business Case for Institution Sustainability. *Journal of Business Strategy and the Environment*, *11*(2), 130–141. doi:10.1002/bse.323

Economic Survey. (2020). *State of the Economy*. https://www.indiabudget.gov.in/economicsurvey

Ekelund, J., Johansson, E., Järvelin, M. R., & Lichtermann, D. (2005). Self-employment and risk aversion—Evidence from psychological test data. *Labour Economics*, *12*(5), 649–659. doi:10.1016/j.labeco.2004.02.009

Elfving, J., Brännback, M., & Carsrud, A. (2009). Toward a contextual model of entrepreneurial intentions. In *Understanding the entrepreneurial mind* (pp. 23–33). Springer. doi:10.1007/978-1-4419-0443-0_2

Elliot, S. R. (2005). Sustainability: An Economic Perspective, Resources Conservations and Recycling. *Journal of Economics*, *44*(1), 263–277.

Emerson, R. M. (1972). Exchange theory, part II: Exchange relations and networks. *Sociological Theories in Progress, 2*, 58-87.

Emerson, R. M. (1972). Exchange theory, part II: Exchange relations and networks. *Sociological Theories in Progress*, *2*, 58–87.

Eric, S. (2012). *Fast food nation: The dark side of the all-American meal*. Houghton Mifflin Harcourt.

Escandon-Barbosa, D., Urbano-Pulido, D., & Hurtado-Ayala, A. (2019). Exploring the Relationship between Formal and Informal Institutions, Social Capital, and Entrepreneurial Activity in Developing and Developed Countries. *Sustainability*, *11*(2), 550. doi:10.3390u11020550

Esiobu, N. S., & Ibe, G. O. G. (2015). Analysis of entrepreneurship development in agriculture among arable crop farmers in Imo State, Nigeria. *International Journal of African and Asian Studies*, *7*(3), 92–99.

Evans, D. S., & Jovanovic, B. (1989). An estimated model of entrepreneurial choice under liquidity constraints. *Journal of Political Economy*, *97*(4), 808–827. doi:10.1086/261629

Evenhuis, A., & Wanten, P. J. (2006) Effect of polythene Tunnels and Cultivars on Grey Mould Caused by Botrytis cinerea in Organically Grown Strawberries. *Agri. Consp. Scient.* 71(4), 111-114. https://hrcak.srce .hr/ 7873

Fabio, P. (2014). Food, identity, and cultural reproduction in immigrant communities. *Social Research*, *81*(2), 415.

Fairlie, S. (2009). A short history of enclosure in Britain. *TheLand*, *7*, 16–31.

Fanfani, D. (2006). Ilgoverno del territorio e del paesaggioruralenellospazio "terzo" periurbano. Il parco agricolo come strumento di politiche e di progetto. *Ri-Vista. Researchforlandscapearchitecture*, *6*(2), 54–69.

Fanfani, D. (2013). Local development and "agri-urban" domain: Agricultural park as promotion of an "active ruralship". *Planum: The Journal of Urbanism*, *27*(2), 38–47.

FAO, Food and Agricultural Organization of the United Nations. (2016). In *FAO Statistical Yearbook 2012, Part 1—The setting*. Available at www.fao.org/docrep/015/i2490e/i2490e00.htm

FAO. (2007). *Profitability and Sustainability of Urban and Peri-urban Agriculture*. FAO.

FAO. (2008). *Climate change and food security: a framework document.* Food and Agro ecology Organization of the United Nations. http://www.fao.org/forestry/15538-079b31d45081fe9c3dbc6ff34de4807e4.pdf

FAO. (2011). *Food, agro ecology and cities: Challenges of food and nutrition security, agro ecology and ecosystem management in an urbanizing world.* http://www.fao.org/fileadmin/templates/FCIT/PDF/FoodAgriCities_Oct2011.pdf

FAO/WB. (2008). *Urban agriculture for sustainable poverty alleviation and food security.* Food and Agriculture Organization of the United Nations. http://www.fao.org/fileadmin/templates/FCIT/PDF/UPA_-WBpaper-Final_October_2008.pdf

FARMA. (2008). *21 Reasons to support local foods.* FARMA. Available from: http://www.farmauk.org/21reasons.htm

Favre-Godal, Q., Schwob, P., Lecoultre, N., Hofstetter, V., Gourguillon, L., Riffault-Valois, L., Lordel-Madeleine, S., Gindro, K., & Choisy, P.Favre—Godal. (2021). Plant-microbe features of *Dendrobium fimbriatum* (Orchidaceae) fungal community. *Symbiosis.* Advance online publication. doi:10.100713199-021-00786-0

Fayolle, A. (2005). Evaluation of entrepreneurship education: Behaviour performing or intention increasing? *International Journal of Entrepreneurship and Small Business, 2*(1), 89–98. doi:10.1504/IJESB.2005.006072

Fayolle, A., & Gailly, B. (2015). The impact of entrepreneurship education on entrepreneurial attitudes and intention: Hysteresis and persistence. *Journal of Small Business Management, 53*(1), 75–93. doi:10.1111/jsbm.12065

Fedora, G. (2009). The cultural meaning of food and its polyvalent role in the construction of identity among Senegalese migrants in Italy. In *Food and Migration Workshop.* School of Oriental and African Studies, University of London. Retrieved from https://www.soas.ac.uk/migrationdiaspora/seminarsevents/food_migration_abstracts/file49

Feintrenie, L., Chong, W. K., & Levang, P. (2010). Why do farmers prefer oil palm? Lessons learnt from Bungo district, Indonesia. *Small-scale Forestry, 9*(3), 379–396. doi:10.100711842-010-9122-2

Fernandez-Acero, F. J., Carbu´, M., Garrido, C., Vallejo, I., & Cantoral, J. M. (2007). Proteomic Advances in Phytopathogenic Fungi. *Current Proteomics, 4*(2), 79–88. doi:10.2174/157016407782194620

Fernandez-Acero, F. J., Jorge, I., Calvo, E., Vallejo, I., Carbú, M., Camafeita, E., Garrido, C., López, J. A., Jorrin, J., & Cantoral, J. M. (2007). Proteomic Analysis of Phytopathogenic Fungus Botrytis cinerea as a Potential Tool for Identifying Pathogenicity Factors. *Therapeutic Targets and for Basic Res. Archives of Microbiology, 187*(3), 207–215. doi:10.100700203-006-0188-3 PMID:17124592

Fernandez-Acero, F., Thomas Colby, T., & Anne Harzen, A. (2009). Proteomic Analysis of the Phytopathogenic Fungus Botrytis cinerea during Cellulose Degradation. [Google Scholar]. *Proteomics, 9*(10), 2892–2902. doi:10.1002/pmic.200800540 PMID:19415670

Fernandez, R., & Novo, H. (2010). *Proteomics of Fungal Plant Pathogens: the Case of Botrytis cinerea, Current Research, Techn. and Edu. Topics in App. Microbiology and Microbial Tech., 205-2017.* Google Scholar.

Finkers, R., Bai, Y., Berg, P., Berloo, R., Meijer-Dekens, F., Have, A., Kan, J., Lindhout, P., & Heusden, A. (2008). Quantitative Resistance to Botrytis cinerea from Solanum Neorickii. *Euphytica, 159*(1-2), 83–92. doi:10.100710681-007-9460-0

Finkers, R., van den Berg, P., van Berloo, R., ten Have, A., van Heusden, A. W., van Kan, J. A. L., & Lindhout, P. (2007). Three QTLs for Botrytis cinerea Resistance in Tomato. [Google Scholar]. *Theoretical and Applied Genetics, 114*(4), 585–593. doi:10.100700122-006-0458-0 PMID:17136515

Fishbein, M., & Ajzen, I. (1975). *Belief, Attitude, Intention and Behaviour: An Introduction to Theory and Research.* Addison-Wesley.

Fitz-Koch, S., & Nordqvist, M. (2017). The Reciprocal Relationship of Innovation Capabilities and Socioemotional Wealth in a Family Firm. *Journal of Small Business Management, 55*(4), 547–570. doi:10.1111/jsbm.12343

Fitz-Koch, S., Nordqvist, M., Carter, S., & Hunter, E. (2018). Entrepreneurship in the agricultural sector: A literature review and future research opportunities. *Entrepreneurship Theory and Practice, 42*(1), 129–166. doi:10.1177/1042258717732958

Fitzsimmons, J. R., & Douglas, E. J. (2011). Interaction between feasibility and desirability in the formation of entrepreneurial intentions. *Journal of Business Venturing, 26*(4), 431–440. doi:10.1016/j.jbusvent.2010.01.001

Flyvbjerg, B. (2002). Bringing power to planning research: One researcher's praxis story. *Journal of Planning Education and Research, 21*(4), 353–366.

Franke, N., &Lüthje, C. (2004). Entrepreneurial intentions of business students—A benchmarking study. *International Journal of Innovation and Technology Management, 1*(3), 269-288.

Frei, M., & Becker, K. (2005, May). Integrated rice-fish culture: Coupled production saves resources. *Natural Resources Forum, 29*(2), 135–143. doi:10.1111/j.1477-8947.2005.00122.x

Freytag, A., & Thurik, R. (2010). Introducing entrepreneurship and culture. In *Entrepreneurship and culture* (pp. 1–8). Springer. doi:10.1007/978-3-540-87910-7_1

Fuentes Fuentes, M. M., Ruiz Arroyo, M., Bojica, A. M., & Fernández Pérez, V. (2010). Prior knowledge and social networks in the exploitation of entrepreneurial opportunities. *The International Entrepreneurship and Management Journal, 6*(4), 481–501. https://doi.org/10.1007/s11365-010-0136-1

Fukuyama, F. (1999). *Social capital and civil society*. Paper prepared for delivery at the International Monetary Fund Conference on Second Generation Reforms.

Fukuyama, F. (2001). Social capital, civil society and development. *Third World Quarterly*, *22*(1), 7–20. doi:10.1080/713701144

Galindo, M. Á., & Méndez-Picazo, M. T. (2013). Innovation, entrepreneurship and economic growth. *Management Decision*, *51*(3), 501–514. doi:10.1108/00251741311309625

Garbach, K., Milder, J. C., DeClerck, F. A., Montenegro de Wit, M., Driscoll, L., & Gemmill-Herren, B. (2017). Examining multi-functionality for crop yield and ecosystem services in fives ystems of agroecological intensification. *International Journal of Agricultural Sustainability*, *15*(1), 11–28. doi:10.1080/14735903.2016.1174810

Garg, S., & Agarwal, P. (2017). Problems and prospects of woman entrepreneurship–a review of literature. *IOSR Journal of Business and Management*, *19*(1), 55–60. doi:10.9790/487X-1901065560

Garikipati, S. (2012). Microcredit and women's empowerment: Through the lens of time-use data from rural India. *Development and Change*, *43*(3), 719–750. doi:10.1111/j.1467-7660.2012.01780.x

Garrity, D. P. (2004). Agroforestry and the achievement of the Millennium Development Goals. *Agroforestry Systems*, *61*(1), 5–17. doi:10.1023/B:AGFO.0000028986.37502.7c

Gartner, W. B. (1985). A conceptual framework for describing the phenomenon of new venture creation. *Academy of Management Review*, *10*(4), 696–706. doi:10.5465/amr.1985.4279094

Gartner, W. B. (1989). Some Suggestions for Research on Entrepreneurial Traits and Characteristics. *Entrepreneurship Theory and Practice*, *14*(1), 27–38.

Gartner, W. B. (1990). What are we talking about when we talk about entrepreneurship? *Journal of Business Venturing*, *5*(1), 15–28. doi:10.1016/0883-9026(90)90023-M

Gatewood, E. J., Shaver, K. G., Powers, J. B., & Gartner, W. B. (2002). Entrepreneurial expectancy, task effort, and performance. *Entrepreneurship Theory and Practice*, *27*(2), 187–206. doi:10.1111/1540-8520.00006

Gatto, M., Wollni, M., Asnawi, R., & Qaim, M. (2017). Oil palm boom, contract farming, and rural economic development: Village-level evidence from Indonesia. *World Development*, *95*, 127–140. doi:10.1016/j.worlddev.2017.02.013

Geoffrey, A., Sánchez, J. G., Vignali, G., & Chaillot, A. (2003). Latent consumers' attitude to farmers' markets in North West England. *British Food Journal*, *105*(8), 487–497. doi:10.1108/00070700310497264

Giacomin, O., Janssen, F., Pruett, M., Shinnar, R. S., Llopis, F., & Toney, B. (2011). Entrepreneurial intentions, motivations and barriers: Differences among American, Asian and European students. *The International Entrepreneurship and Management Journal*, *7*(2), 219–238. doi:10.100711365-010-0155-y

Gibb, N., & Wittman, H. (2013). Parallel alternatives: Chinese-Canadian farmers and the Metro Vancouver local food movement. *Local Environment*, *18*(1), 1–19.

Gina, A. (2014). Food and identity: Food studies, cultural, and personal identity. *Journal of International Business and Cultural Studies*, *8*, 1.

Ginige, A., & Sivagnanasundaram, J. (2019). Enhancing Agricultural Sustainability through Crowdsensing: A Smart Computing Approach. *Journal of Advanced Agricultural Technologies*, *6*(3), 161–165. doi:10.18178/joaat.6.3.161-165

Giraud, T., Fortini, D., Levis, C., Lamarque, C., Leroux, P., LoBuglio, K., & Brygoo, Y. (1999). Two Sibling Species of the Botrytis cinerea Complex, Transposa and Vacuma, Are Found in Sympatry on Numerous Host Plants. *Phytopathology*, *89*(10), 967–973. doi:10.1094/PHYTO.1999.89.10.967 PMID:18944743

Gliessman, S. (2015). *Agroecology: A growing field.* Academic Press.

Gliessman, S. R. (2015). *Agroecology: the ecology of sustainable food systems* (3rd ed.). Academic Press.

Gliessman, S. (2013). Agroecology: Growing the roots of resistance. *Agroecology and Sustainable Food Systems*, *37*(1), 19–31.

Gliessman, S. R. (2009). The framework for conversion. In S. R. Gliessman & M. Rosemeyer (Eds.), *The conversion to sustainable agriculture: principles, processes, and practices* (pp. 3–16). CRC Press., doi:10.1201/9781420003598-c1

Global Agrisystem. (2010). *Evaluation Study of Agriclinics and agribusiness centre Scheme.* Global Agrisystem Private Limited.

Goetz, A. M., & Gupta, R. S. (1996). Who takes the credit? Gender, power, and control over loan use in rural credit programs in Bangladesh. *World Development*, *24*(1), 45–63. doi:10.1016/0305-750X(95)00124-U

Golden, S. (2013). *Urban agro ecology impacts: Social, health, and economic: An annotated bibliography.* https://asi.ucdavis.edu/programs/sarep/publications/food-and-society/uaannotatedbiblio-2013.pdf

Gold, M. A. (1994). Agroforestry systems for temperate climates: lessons from Roman Italy. *Forest & Conservation History*, *38*(3), 118–126.

Gómez-Araujo, E., & Lafuente, , I. E., Vaillant, Y., & Gómez Núñez, L. M. (2015). The differential impact of selfconfidence, reference models and the fear of failure in young entrepreneurs. *Innovar (Universidad Nacional de Colombia)*, *25*(57), 157–174.

Gomiero, T., Giampietro, M., Bukkens, S. G., & Paoletti, M. G. (1999). Environmental and socio economic constraints to the development of freshwater fish aquaculture in China. *Critical Reviews in Plant Sciences*, *18*(3), 359–371. doi:10.1080/07352689991309261

González-Fernández, E., Piña-Rey, A., Fernández-González, M., Aira, M., & Rodríguez-Rajo, F. (2020). Identification and evaluation of the main risk periods of Botrytis cinerea infection on grapevine based on phenology, weather conditions and airborne conidia. *The Journal of Agricultural Science, 158*(1-2), 88–98. doi:10.1017/S0021859620000362

González-Fernández, R., Valero-Galván, J., Gómez-Gálvez, F. J., & Jorrín-Novo, J. V. (2015). Unraveling the in vitro secretome of the phytopathogen Botrytis cinerea to understand the interaction with its hosts. *Frontiers in Plant Science, 6*, 839. doi:10.3389/fpls.2015.00839 PMID:26500673

Goodman, D. (2004). Rural Europe redux? Reflections on alternative agro-food networks and paradigm change. *Sociologia Ruralis, 44*(1), 3–16.

Gore, M. S. (1994). *Indian education: Structure and process*. Rawat Publications.

Gottlieb, R., & Joshi, A. (2010). *Food justice*. MIT Press.

Government of India. (2016). Agricultural statistics at a Glance 2016. Ministry of Agriculture and Farmers Welfare, Department of Agriculture.

Government of India. (2019). *Swarnajayanti Gram SwarozgarYojna*. Retrieved from https://www.india.gov.in/swarnjayanti-gram-swarozgar-yojana

Government South Sumatra. (2015). Decree of the Governor of South Sumatra, Number: 838 / KPTS / Disnakertrans / 2015, concerning the minimum wage of South Sumatra Province in 2016

Goyal, S. Z., & Obeng, F. K. (2004). Public-Private Sector partnerships for improved agriculture services delivery: How do we make them work? *Journal of Social Development in Africa, 19*(2), 49–67.

Grabke, A. (2014). *Fungicide Resistance in Botrytis cinerea from Strawberry - Molecular Mechanisms and Management. In Agronomy and Crop Sciences Commons, Biology Commons, and the Plant Path*. Comm.

Grande, J. (2011). New venture creation in the farm sector–critical resources and capabilities. *Journal of Rural Studies, 27*(2), 220–233.

Grande, J. (2011). New venture creation in the farm sector–Critical resources and capabilities. *Journal of Rural Studies, 27*(2), 220–233. doi:10.1016/j.jrurstud.2011.02.003

Grande, J., Madsen, E. L., & Borch, O. J. (2011). The relationship between resources, entrepreneurial orientation and performance in farm-based ventures. *Entrepreneurship and Regional Development, 23*(34), 89–111. https://doi.org/10.1080/08985620903183710

Granovetter, M. S. (1992). Problems of explanation in economic sociology. In N. Nohria & R. G. Eccles (Eds.), *Networks and organizations: Structure, form and action* (pp. 25–56). Harvard Business School Press.

Gray, C. (2002). Entrepreneurship, Resistance to change and Growth in Small Firms. *Journal of Small Business and Enterprise Development, 9*(1), 61–72. doi:10.1108/14626000210419491

Great, B. (2002). *Farming & Food: A Sustainable Future. Report of the Policy Commission on the Future of Farming and Food.* Cabinet Office.

Greene, W. H. (2002). *Econometric analysis* (5th ed.). Prentice Hall.

Greve, A., & Salaff, J. W. (2003). Social networks and entrepreneurship. *Entrepreneurship Theory and Practice, 28*(1), 1–22. doi:10.1111/1540-8520.00029

Grootaert, C., & Van Bastelaer, T. (2001). Understanding and measuring social capital: A synthesis of findings and recommendations from the social capital initiative. *Social Capital Initiative Working Paper, 24*(24), 1–31. doi:10.1227/00006123-197907010-00058

Guitart, D., Pickering, C., & Byrne, J. (2012). Past results and future directions in urban community gardens research. *Urban Forestry & Urban Greening, 11*(4), 364–373. doi:10.1016/j.ufug.2012.06.007

Gulati, A., Sharma, P., Samantara, A., & Terway, P. (2018). *Agriculture extension system in India: Review of current status, trends and the way forward.* Academic Press.

Gupta, V. K., & Bhawe, N. M. (2007). The influence of proactive personality and stereotype threat on women's entrepreneurial intentions. *Journal of Leadership & Organizational Studies, 13*(4), 73–85. doi:10.1177/10717919070130040901

Gustafson, S. (2019) FAO SOFA report 2019: New insights into food loss and waste. https://www.ifpri.org/blog/fao-sofa-report-2019-new-insights-food-loss-and-waste. Accessed 11 July 2020.

Guthman, J. (2008). Bringing good food to others: Investigating the subjects of alternative food practice. *Cultural Geographies, 15*(4), 431–447.

Guzman, J., & Santos, F. J. (2001). The booster function and the entrepreneurial quality: An application to the province of Seville. *Entrepreneurship and Regional Development, 13*(3), 211–228.

Hadimani, R. N. (1985). *Dynamics of industrial entrepreneurship.* APH Publishing.

Hagen, E. E. (1960). The entrepreneur as rebel against traditional society. *Human Organization, 19*(4), 185–187. doi:10.17730/humo.19.4.tl57064355861441

Hagey, A., Rice, S., & Flournoy, R. (2012). Growing urban agriculture: Equitable strategies and policies for improving access to healthy food and revitalizing communities. Report by PolicyLink, Oakland, CA.

Hahn. (2008). Consumption, Identities and Agency in Africa: Introduction. *Consumption in Africa: Anthropological Approaches,* 9.

Hair, J. F., Anderson, R. E., Babin, B. J., & Black, W. C. (2010). *Multivariate data analysis: A global perspective* (Vol. 7). Academic Press.

Hair, J. F., Black, W. C., Babin, B. J., Anderson, R. E., & Tatham, R. L. (1998). Multivariate data analysis. Prentice Hall.

Hair, J. F., Celsi, M., Ortinau, D. J., & Bush, R. P. (2010). *Essentials of marketing research* (Vol. 2). McGraw-Hill/Irwin.

Hall, J., Matos, S., Sheehan, L., & Silvestre, B. (2012). Entrepreneurship and innovation at the base of the pyramid: A recipe for inclusive growth or social exclusion? *Journal of Management Studies*, *49*(4), 785–812. doi:10.1111/j.1467-6486.2012.01044.x

Hansen, M. A. (2016). *Botrytis Blight of Peony. Communications and Marketing, College of Agriculture and Life Sciences, Virginia Tech.*

Hansson, H., Ferguson, R., Olofsson, C., & Rantamäki-Lahtinen, L. (2013). Farmers' motives for diversifying their farm business -the influence of family. *Journal of Rural Studies*, *32*, 240–250. https://doi. org/10.1016/j.jrurstud.2013.07.002

Hardi, P., & Zdan, T. (1997). Assessing sustainable development: principles in practice. Principles in Practice.

Harjanti, D., & Noerchoidah, N. (2017). The Effect of Social Capital and Knowledge Sharing on Innovation Capability. *Jurnal Manajemen dan Kewirausahaan, 19*, 72–78.

Harris, E. (2009). Neoliberal subjectivities or a politics of the possible?Reading for difference in alternative food networks. *Area*, *41*(1), 55–63.

Hart, S. L. (1997). Beyond Greening: Strategies for a Sustainable World. *Harvard Business Review*, *75*(1), 66–76.

Hart, S. L., & Milstein, M. B. (2003). Creating Sustainable Value. *The Academy of Management Executive*, *17*(2), 56–67. doi:10.5465/ame.2003.10025194

Hashemi, S. M., Schuler, S. R., & Riley, A. P. (1996). Rural credit programs and women's empowerment in Bangladesh. *World Development*, *24*(4), 635–653. doi:10.1016/0305-750X(95)00159-A

Hassan, A., Anwar, I., Saleem, I., Islam, K. B., & Hussain, S. A. (2021). Individual entrepreneurial orientation, entrepreneurship education and entrepreneurial intention: The mediating role of entrepreneurial motivations. *Industry and Higher Education.*

Hassan, A., Saleem, I., Anwar, I., & Hussain, S. A. (2020). Entrepreneurial intention of Indian university students: The role of opportunity recognition and entrepreneurship education. *Education + Training*, *62*(7/8), 843–861. doi:10.1108/ET-02-2020-0033

Hassink, J., Grin, J., & Hulsink, W. (2012). Multifunctional Agriculture Meets Health Care: Applying the Multi-Level Transition Sciences Perspective to Care Farming in the Netherlands. *Sociologia Ruralis*, *53*(2), 223–245. doi:10.1111/j.1467-9523.2012.00579.x

Hassink, J., Hulsink, W., & Grin, J. (2016). Entrepreneurship in agriculture and healthcare: Different entry strategies of care farmers. *Journal of Rural Studies*, *43*, 27–39. doi:10.1016/j.jrurstud.2015.11.013

Hawker, L., & Hendy, R. J. (1963). An Electron Microscope Study of Germination of Conindia of Botrysis Cineria. [Google Scholar]. *Journal of General Microbiology, 33*(1), 43–46. doi:10.1099/00221287-33-1-43 PMID:14072834

Hedberg & Malmborg. (2003). The Global Education and Institution sustainability reporting in Swedish companies. *Institution Social Responsibility and Environmental Management, 24*(2), 153-164.

Helen, V., & Raine, K. (2008). Consuming food and constructing identities among Arabic and South Asian immigrant women. *Food, Culture, & Society, 11*(3), 355–373. doi:10.2752/175174408X347900

Hemenway, T. (2009). *Gaia's garden: a guide to home-scale permaculture*. Chelsea Green Publishing.

Heredero, J. M. (1979). Chapter. In T. V. Rao & T. K. Moulik (Eds.), Agricultural Entrepreneurship: Identification and selection of small scale entrepreneur (pp. 151–159). IIM.

Hernández-Carrión, C., Camarero-Izquierdo, C., & Gutiérrez-Cillán, J. (2016). Entrepreneurs' Social Capital and the Economic Performance of Small Businesses: The Moderating Role of Competitive Intensity and Entrepreneurs' Experience. *Strategic Entrepreneurship Journal, 11*(1), 61–89. doi:10.1002ej.1228

Hern, M. (2016). *What a City is for: Remaking the Politics of Displacement*. MIT Press.

Hessels, J., Grilo, I., Thurik, R., & van der Zwan, P. (2011). Entrepreneurial exit and entrepreneurial engagement. *Journal of Evolutionary Economics, 21*(3), 447–471.

Hiroki, S., Garnevska, E., & McLaren, S. (2016). Consumer perceptions about local food in New Zealand, and the role of lifecycle-based environmental sustainability. *Journal of Agricultural & Environmental Ethics, 29*(3), 479–505. doi:10.100710806-016-9616-9

Hislop, R. S. (2014). *Reaping equity: A survey of Food Justice Organizations in the USA*. University of California.

Hodgson, K. (2012). *Planning for food access and community-based food systems: A national scan and evaluation of local comprehensive and sustainability plans*. Report by American Planning Association.

Holdsworth, B. (2005). *Continuous Productive Urban Landscapes: designing urban agriculture for sustainable cities*. Academic Press.

Holland, J. (1975). *Adaptation in Natural and Artificial Systems*. University of Michigan Press.

Holloway, L., Kneafsey, M., Venn, L., Cox, R., Dowler, E., & Tuomainen, H. (2007). Possible food economies: A methodological framework for exploring food production–consumption relationships. *Sociologia Ruralis, 47*(1), 1–19. doi:10.1111/j.1467-9523.2007.00427.x

Holmgren, D. (2004). *Permaculture: principles and pathways beyond sustainability*. Holmgren Design, Hepburn.

Holt-Giménez, E. (2008). Out of AGRA: The green revolution returns to Africa. *Development*, *51*(4), 464–471.

Hondagneu-Sotelo, P. (2014). *Paradise transplanted: Migration and the making of California gardens*. University of California Press.

Honjo, Y. (2015). Why are entrepreneurship levels so low in Japan? *Japan and the World Economy*, *36*, 88101. https://doi.org/10.1016/j.japwor.2015.08.002

Horst, M., McClintock, N., & Hoey, L. (2017). The intersection of planning, urban agriculture, and food justice: A review of the literature. *Journal of the American Planning Association*, *83*(3), 277–295.

Hosen, M., Ahmed, A. M. R., & Islam, M. (2010). Physiological Variability and in Vitro Antifungal Activity against Botrytis Cinerea Causing Botrytis Gray Mold of Chickpea (Cicer arietinum L.). *Spanish Journal of Agricultural Research*, *8*(3), 750–756. doi:10.5424jar/2010083-1274

Huarng, K. H., & Yu, T. H. K. (2011). Entrepreneurship, process innovation and value creation by a non-profit SME. *Management Decision*, *49*(2), 284–296. doi:10.1108/00251741111109160

Hu, L. T., & Bentler, P. M. (1999). Cutoff criteria for fit indexes in covariance structure analysis: Conventional criteria versus new alternatives. *Structural Equation Modeling*, *6*(1), 1–55. doi:10.1080/10705519909540118

Hulsink, W., & Dons, H. (2008). *Venturing and Clustering in Agri-Food and High-Technology Hot Spots*. Wageningen UR Frontis Series. doi:10.1007/978-1-4020-8768-4_1

Hunjo, H. (2015). A Critical Discourse Evaluation of Decolonisation and Democratisation: Issues in Africa as Exemplified in Soyinka's Non-fictional Texts. *African Literature and the Future*, 49.

Hvitsand, C. (2016). Community supported agriculture (CSA) as a transformational act—Distinct values and multiple motivations among farmers and consumers. *Agroecology and Sustainable Food Systems*, *40*(4), 333–351. doi:10.1080/21683565.2015.1136720

Iain, W. (2012). Ntsambu, the foul smell of home: Food, commensality and identity in the Comoros and in the diaspora. *Food & Foodways*, *20*(3-4), 187–210. doi:10.1080/07409710.2012.715962

Ibn, K., & Rosenthal, F. (1967). *The Muqaddimah: An introduction to history*. Princeton University Press.

IGD. (2005). *The local and regional food opportunity*. IGD. Available from: http://www.defra.gov.uk/foodfarm/food/industry/regional/pdf/localregfoodopps.pdf

IGD. (2010). *Local food gains ground with shoppers*. Available from IGD: https://www.igd.com/index.asp?id=1&fid=1&sid=8&tid=16&cid=1427

IPR. (2015). *Cooperative planning for periurban park at Vltava and Berounka River*. Available at: http://en.iprpraha/clanek/1409/cooperative-planning-for-periurban-park-at-vltava-and-berounka-river

Ishaya, S., & Abaje, I. (2008). Indigenous people's perception of climate change and adaptation strategies in Jema's local government area of Kaduna State, Nigeria. *Journal of Geography and Regional Planning*, *1*, 138–143.

Izquierdo, E., & Buelens, M. (2011). Competing models of entrepreneurial intentions: The influence of entrepreneurial self-efficacy and attitudes. *International Journal of Entrepreneurship and Small Business*, *13*(1), 75–91. doi:10.1504/IJESB.2011.040417

Jacke, D., & Toensmeier, E. (2005). Edible forest gardens, volume II: ecological design and practice for temperate-climate permaculture. Chelsea Green Publishing.

Jamu, D. M., & Piedrahita, R. H. (2002). An organic matter and nitrogen dynamics model for the ecological analysis of integrated aquaculture/agriculture systems: II. Model evaluation and application. *Environmental Modelling & Software*, *17*(6), 583–592. doi:10.1016/S1364-8152(02)00017-8

Jaramillo, C. (2016). *Urban agriculture leaders ask for citywide commitments to garden preservation and creation*. Academic Press.

Jean, P. (1990). Ethnic identity in adolescents and adults: Review of research. *Psychological Bulletin*, *108*(3), 499–514. doi:10.1037/0033-2909.108.3.499 PMID:2270238

Jelsma, I., Slingerland, M., Giller, K. E., & Bijman, J. (2017). Collective action in a smallholder oil palm production system in Indonesia: The key to sustainable and inclusive smallholder palm oil? *Journal of Rural Studies*, *54*, 198–210. doi:10.1016/j.jrurstud.2017.06.005

Jennings, D. E., & Zeithaml, C. P. (1983). Locus of Control: A Review and Directions for Entrepreneurial Research. In Academy of Management Proceedings (No. 1, pp. 417–421). Briarcliff Manor: Academy of Management.

Ji, D., Chen, T., Ma, D., Liu, J., Xu, Y., & Tian, S. P. (2018). Inhibitory effects of Methyl Thujate on mycelial growth of, botrytis cinerea, and possible mechanisms. *Postharvest Biology and Technology*, *142*, 46–54. doi:10.1016/j.postharvbio.2018.04.003

John, R., & Kispotta, W. (2016). Constraints Faced by the Women Entrepreneurs in Agribusiness in District of Allahabad, U.P. *IOSR Journal of Business and Management, 18*(1), 13-17.

John, K. (2010). Food consumption trends and drivers. *Philosophical Transactions of the Royal Society of London. Series B, Biological Sciences*, *365*(1554), 2793–2807. doi:10.1098/rstb.2010.0149 PMID:20713385

John, K., Holdsworth, D., & Mather, D. (2007). Determinants of trust in imported food products: Perceptions of European gatekeepers. *British Food Journal*, *109*(10), 792–804. doi:10.1108/00070700710821331

Jon, H. (2006). Food and memory. *Annual Review of Anthropology*, *35*(1), 361–378. doi:10.1146/annurev.anthro.35.081705.123220

Jose, S. (2014). *Innovation in agriculture: A key process for sustainable development.* Inter-American Institute for Cooperation on Agriculture (IICA). Institutional position paper, 1-20.

Juerg, S., & Hall, R. (1981). *Criteria of food acceptance: how man chooses what he eats.* Papers presented at an interdisciplinary symposium, Einsiedeln.

Julia, P. (2002). Like baby minnows we came with the current": social memory among Angolan refugees in Meheba settlement, Zambia. *Annual Meeting of the Association of Social Anthropologists of Great Britain and the Common wealth.*

Kage, H., & Stützel, H. HUME. (1999b). An Object Oriented Component Library for Generic Modular Modelling of Dynamic Systems in Modelling Cropping Systems. Eds. CSM Donatelli, F. Villalobos and JM Villar. pp 299–300. European Society of Agronomy,Lleida. Google Scholar

Kage, H., Kochler, M., & Stützel, H. (2000). Root Growth of Cauliflower (Brassica oleracea L. botrytis) under Unstressed Conditions: Measurement and Modeling. [Google Scholar]. *Plant and Soil*, *223*(1/2), 131–145. doi:10.1023/A:1004866823128

Kage, H., & Stützel, H. (1999a). A Simple Empirical Model for Predicting Development and Dry Matter Partitioning in Cauliflower (Brassica Oleracea L. Botrytis). [Google Scholar]. *Scientia Horticulturae*, *80*(1-2), 19–38. doi:10.1016/S0304-4238(98)00226-X

Kalleberg, A. L., & Leicht, K. T. (1991). Gender and organizational performance: Determinants of small business survival and success. *Academy of Management Journal*, *34*(1), 136–161. https://doi.org/10.2307 /256305

Kalra, R. K., Anil, B., Tonts, M., & Siddique, K. H. M. (2013). Self-help groups in Indian agriculture: A case study of farmer groups in Punjab, Northern India. *Agroecology and Sustainable Food Systems*, *37*(5), 509–530. doi:10.1080/10440046.2012.719853

Kar, B., Subudhi, R. N., & Padhy, R. (2017). Impact of self-efficacy and contextual variables on entrepreneurial intention. *Pertanika Journal of Social Science & Humanities*, *25*(3), 1121–1138.

Keinath, A. P., Fravel, D. R., & Papavizas, G. C. (1991). Potential of Gliocladium Roseum for Biocontrol of Verticillium Dahliae. [Google Scholar]. *Hytopathology*, *81*(6), 644–648. doi:10.1094/Phyto-81-644

Kelegama, S., & Hirimuthugodage, D. (2019). Food Security Policy Framework in Sri Lanka: Market reforms, diversification and safety nets. In Regional Cooperation for Sustainable Food Security in South Asia (pp. 227-247). Routledge India.

Kelley, D., Singer, S., & Herrington, M. (2016). Global entrepreneurship monitor. Global Report, Global Entrepreneurship Research Association, London Business School.

Khan, A. M., Arafat, M. Y., Raushan, M. A., Khan, M. S., Dwivedi, A. K., Khan, M. M., & Saleem, I. (2019b). Role of intellectual capital in augmenting the start-up intentions of Indians—An analysis using GEM data. *Journal of Global Entrepreneurship Research*, *9*(1), 1–14. doi:10.118640497-019-0146-6

Khan, A. M., Arafat, M. Y., Raushan, M. A., & Saleem, I. (2020). Examining the relevance of intellectual capital in improving the entrepreneurial propensity among Indians. *International Journal of Knowledge Management*, *16*(1), 123–140. doi:10.4018/IJKM.2020010106

Khan, A. M., Arafat, M. Y., Raushan, M. A., Saleem, I., Khan, N. A., & Khan, M. M. (2019a). Does intellectual capital affect the venture creation decision in India? *Journal of Innovation and Entrepreneurship*, *8*(1), 1–15. doi:10.118613731-019-0106-y

Kickul, J., Wilson, F., Marlino, D., & Barbosa, S. D. (2008). Are misalignments of perceptions and self-efficacy causing gender gaps in entrepreneurial intentions among our nation's teens? *Journal of Small Business and Enterprise Development*, *15*(2), 321–335. doi:10.1108/14626000810871709

Kihlstrom, R. E., & Laffont, J. J. (1979). A general equilibrium entrepreneurial theory of firm formation based on risk aversion. *Journal of Political Economy*, *87*(4), 719–748.

Kim, J., Shin, J., Gumilang, A., Chung, K., Choi, K., & Kyoung Su Kim, K. (2016). Effectiveness of Different Classes of Fungicides on Botrytis cinerea Causing Gray Mold on Fruit and Vegetables. *Plant Path. Jour.*, *32*(6), 570–574. doi:10.5423/PPJ.NT.05.2016.0114 PMID:27904464

Kim, M. S., & Hunter, J. E. (1993). Relationships among attitudes, behavioral intentions, and behavior: A meta-analysis of past research, part 2. *Communication Research*, *20*(3), 331–364. doi:10.1177/009365093020003001

Kim, N., & Shim, C. (2018). Social capital, knowledge sharing and innovation of small- and medium-sized enterprises in a tourism cluster. *International Journal of Contemporary Hospitality Management*, *30*(6), 2417–2437. doi:10.1108/IJCHM-07-2016-0392

Kirby, D. A. (2004). Entrepreneurship education: Can business schools meet the challenge? *Education + Training*, *46*(8/9), 510–519. doi:10.1108/00400910410569632

Kirwan, J., Ilbery, B., Maye, D., & Carey, J. (2013). Grassroots social innovations and food localisation: An investigation of the Local Food programme in England. *Global Environmental Change*, *23*(5), 830–837. doi:10.1016/j.gloenvcha.2012.12.004

Kirzner, I. M. (1979). *Competition & Entrepreneurship*. University of Chicago Press.

Kirzner, I. M. (1979). *Perception, opportunity, and profit: Studies in the theory of entrepreneurship*. University of Chicago Press.

Klyver, K., & Hindle, K. (2007). The role of social networks at different stages of business formation. *Small Enterprise Research*, *15*(1), 22–38.

Kneafsey, M., Venn, L., Schmutz, U., Balázs, B., Trenchard, L., Eyden-Wood, T., ... Blackett, M. (2013). Short food supply chains and local food systems in the EU. A state of play of their socio-economic characteristics. *JRC Scientific and Policy Reports, 123*, 129.

Koellinger, P., Minniti, M., & Schade, C. (2007). "I think I can, I think I can": Overconfidence and entrepreneurial behavior. *Journal of Economic Psychology, 28*(4), 502–527. doi:10.1016/j.joep.2006.11.002

Koellinger, P., Minniti, M., & Schade, C. (2013). Gender differences in entrepreneurial propensity. *Oxford Bulletin of Economics and Statistics, 75*(2), 213–234.

Kolvereid, L. (1996). Prediction of Employment Status Choice Intentions. *Entrepreneurship Theory and Practice, 21*(1), 47–58.

Kolvereid, L. (1996). Prediction of employment status choice intentions. *Entrepreneurship Theory and Practice, 21*(1), 47–58. doi:10.1177/104225879602100104

Konijnendijk, C. C. (2003). A decade of urban forestry in Europe. *Forest Policy and Economics, 5*(2), 173–186. doi:10.1016/S1389-9341(03)00023-6

Kourilsky, M. L., & Walstad, W. B. (1998). Entrepreneurship and female youth: Knowledge, attitudes, gender differences, and educational practices. *Journal of Business Venturing, 13*(1), 77–88. doi:10.1016/S0883-9026(97)00032-3

Krueger, J. R. F. N., Reilly, M. D., & Carsrud, A. L. (2000). Competing models of entrepreneurial intentions. *Journal of Business Venturing, 15*(5–6), 411–432. doi:10.1016/S0883-9026(98)00033-0

Krueger, N. F. (2003). The cognitive psychology of entrepreneurship. In *Handbook of entrepreneurship research* (Vol. 1, pp. 105–140). doi:10.1007/0-387-24519-7_6

Krueger, N. (1993). The impact of prior entrepreneurial exposure on perceptions of new venture feasibility and desirability. *Entrepreneurship Theory and Practice, 18*(1), 5–21. doi:10.1177/104225879301800101

Krueger, N. F. (2007). The cognitive infrastructure of opportunity emergence. In *Entrepreneurship* (pp. 185–206). Springer.

Krueger, N. F., & Brazeal, D. V. (1994). Entrepreneurial potential and potential entrepreneurs. *Entrepreneurship Theory and Practice, 18*, 91–91.

Krueger, N. F., & Carsrud, A. L. (1993). Entrepreneurial intentions: Applying the theory of planned behaviour. *Entrepreneurship and Regional Development, 5*(4), 315–330.

Kuebler, M., & Rugh, J. S. (2013). New evidence on racial and ethnic disparities in homeownership in the United States from 2001 to 2010. *Social Science Research, 42*(5), 1357–1374.

Kumar, B. M. (2007). Agroforestry: The new old paradigm for Asian food security. *Journal of Tropical Agriculture, 44*, 1–14.

Kumar, P. (2017). Agro-Industry, Agricultural Marketing, Entrepreneurship, Agri-Business, Trade and Innovations. *Indian Journal of Agricultural Economics*, *72*(1), 76–80.

Lado-Sestayo, R., & Vivel-Bua, M. (2018). Women and Entrepreneurship: An Analysis of the Determinants in the Tourism Sector. In Nascent Entrepreneurship and Successful New Venture Creation (pp. 203-214). IGI Global.

Lakhiar, I. A., Jianmin, G., Syed, T. N., Chandio, F. A., Buttar, N. A., & Qureshi, W. A. (2018). Monitoring and Control Systems in Agriculture Using Intelligent Sensor Techniques: A Review of the Aeroponic System. *Journal of Sensors*, 1–18. doi:10.1155/2018/8672769

Lancaster, B., & Marshall, J. (2008). *Water-harvesting earthworks*. Rainsource.

Landis, D. A. (2017). Designingagriculturallandscapesforbiodiversity-basedecosystemservices. *Basic and Applied Ecology*, *18*, 1–12. doi:10.1016/j.baae.2016.07.005

Langowitz, N., & Minniti, M. (2005). Gender differences and nascent entrepreneurship. *Documento de trabajo*.

Langowitz, N. S., Minniti, M., & Arenius, P. (2005). *Global entrepreneurship monitor: 2004 report on women and entrepreneurship*. University of Illinois at Urbana-Champaign's Academy for Entrepreneurial Leadership Historical Research Reference in Entrepreneurship.

Langowitz, N., & Minniti, M. (2007). The entrepreneurial propensity of women. *Entrepreneurship Theory and Practice*, *31*(3), 341–364.

Lang, T., & Barling, D. (2012). Food security and food sustainability: Reformulating the debate. *The Geographical Journal*, *178*(4), 313–326.

Lang, T., & Barling, D. (2012). Food security and food sustainability: Reformulating the debate. *TheGeographicalJournal*, *178*(4), 313–326.

Lans, T., Seuneke, P., & Klerkx, L. (2013). Agricultural entrepreneurship. In E. G. Carayannis (Ed.), Encyclopedia of creativity, invention, innovation and entrepreneurship (pp. 44–49). Springer. https://doi.org/10.1007/978-1-4614-3858-8_496.

Lans, T., Blok, V., & Wesselink, R. (2014). Learning apart and together: Towards an integrated competence framework for sustainable entrepreneurship in higher education. *Journal of Cleaner Production*, *62*, 37–47. doi:10.1016/j.jclepro.2013.03.036

Lans, T., Seuneke, P., & Klerkx, L. (2017). Agricultural entrepreneurship. In *Encyclopedia of creativity, invention, innovation and entrepreneurship* (pp. 1–7). Springer.

Lans, T., Van Galen, M. A., Verstegen, J. A. A. M., Biemans, H. J. A., & Mulder, M. (2014). Searching for entrepreneurs among small business owner-managers in agriculture. *NJAS Wageningen Journal of Life Sciences*, *68*, 41–51.

Larson, A., & Starr, J. A. (1993). A network model of organization formation. *Entrepreneurship Theory and Practice*, *17*(2), 5–16.

Latha, M., & Chandrakumar, G. (2012). A study on agricultural women Self Help Groups (SHGs) members' micro credit analysis in Trichy district, Tamil Nadu. Academic Press.

Laussel, D., & Le Breton, M. (1995). A general equilibrium theory of firm formation based on individual unobservable skills. *European Economic Review*, *39*(7), 1303–1319. (94)00092-E doi:10.1016/0014-2921

Lawson, L. J. (2005). *City bountiful: A century of community gardening in America*. Univ of California Press.

Learned, K. E. (1992). What happened before the organization? A model of organization formation. *Entrepreneurship Theory and Practice*, *17*(1), 39–48. doi:10.1177/104225879201700105

Lee, S. H., & Wong, P. K. (2004). An exploratory study of technopreneurial intentions: A career anchor perspective. *Journal of Business Venturing*, *19*(1), 7–28. doi:10.1016/S0883-9026(02)00112-X

Levis, C., Fortini, D., & Brygoo, Y. (1997). Flipper, a Mobile Fot1-Like Transposable Element in Botrytis cinerea. *Molecular & General Genetics*, *254*(6), 674–680. doi:10.1007004380050465 PMID:9202383

Levkoe, C. Z. (2011). Towards a transformative food politics. *Local Environment*, *16*(7), 687–705.

Liao, J., & Welsch, H. (2005). Roles of social capital in venture creation: Key dimensions and research implications. *Journal of Small Business Management*, *43*, 345–362. X.2005.00141.x doi:10.1111/j.1540-627

Liao, J., & Welsch, H. (2005). Roles of social capital in venture creation: Key dimensions and research implications. *Journal of Small Business Management*, *43*(4), 345–362. doi:10.1111/j.1540-627X.2005.00141.x

Light, I., & Dana, L. P. (2013). Boundaries of social Capital in Entrepreneurship. *Entrepreneurship Theory and Practice*, *37*(3), 603–624. https://doi.org/10.1111/etap.12016

Li, H., Chen, Y., Zhang, Z., Li, B., Qin, G., & Tian, S. (2018). Pathogenic mechanisms and control strategies of *Botrytis cinerea* causing post-harvest decay in fruits and vegetables. *Food Quality and Safety*, *2*(3), 111–119. doi:10.1093/fqsafe/fyy016

Liñán, F., & Chen, Y. W. (2009). Development and cross-cultural application of a specific instrument to measure entrepreneurial intentions. *Entrepreneurship Theory and Practice*, *33*(3), 593–617.

Liñán, F., & Chen, Y. W. (2009). Development and cross–cultural application of a specific instrument to measure entrepreneurial intentions. *Entrepreneurship Theory and Practice*, *33*(3), 593–617. doi:10.1111/j.1540-6520.2009.00318.x

Liñán, F., Santos, F. J., & Fernández, J. (2011). The influence of perceptions on potential entrepreneurs. *The International Entrepreneurship and Management Journal*, *7*(3), 373.

Lindemann-Matthies, P., Briegel, R., Schüpbach, B., & Junge, X. (2010). Aesthetic preference for a Swiss alpine landscape: The impact of different agricultural land-use with different biodiversity. *Landscape and Urban Planning*, *98*(2), 99–109. doi:10.1016/j.landurbplan.2010.07.015

Lin, N. (2003). *Social capital, a theory of social structure and action*. Cambridge University Press.

Lin, N., Ensel, W. M., & Vaughn, J. C. (1981). Social resources and strength of ties: Structural factors in occupational status attainment. *American Sociological Review*, *46*, 393–405.

Liu, T. C., & Wu, L. W. (2007). Customer retention and cross-buying in the banking industry: An integration of service attributes, satisfaction and trust. *Journal of Financial Services Marketing*, *12*(2), 132–145. doi:10.1057/palgrave.fsm.4760067

London, T., & Hart, S. L. (2010). *Next generation business strategies for the base of the pyramid: New approaches for building mutual value*. FT Press.

Loomis, J. L. (1959). Communication, the development of trust, and cooperative behavior. *Human Relations*, *12*(4), 305–315. doi:10.1177/001872675901200402

Lopes, U. P., Zambolim, L., Pereira, C. N., & Gracia, O. (2017). Resistance of Botrytis cinerea to fungicides controlling gray mold on strawberry. *Brazil Bragantia*, *76*(2), 266–272. doi:10.1590/1678-4499.055

Lortie, J., Castrogiovanni, G. J., & Cox, K. C. (2017). Gender, social salience, and social performance: How women pursue and perform in social ventures. *Entrepreneurship and Regional Development*, *29*(1-2), 155–173. doi:10.1080/08985626.2016.1255433

Lovell, S. T. (2010). Multifunctional urban agriculture for sustainable land use planning in the United States. *Sustainability*, *2*(8), 2499–2522. doi:10.3390u2082499

Lucas, J., & Knights, I. (1987). Spores on Leaves: Endogenous and Exogenous Control of Development. In: Pegg GF, Ayres PG(Eds) Fungal Infection of Plants. Symposium of the British Mycological Society. Cambridge University Press, Cambridge, 45–59. Google Scholar

Lucy, T. (2001). Resisting and reversing language shift: Heritage-language resilience among U.S. native biliterates. *Harvard Educational Review*, *71*(4), 676–706. doi:10.17763/haer.71.4.ku752mj536413336

Lüthje, C., & Franke, N. (2003). The 'making' of an entrepreneur: Testing a model of entrepreneurial intent among engineering students at MIT. *Research Management*, *33*(2), 135–147.

MacLeod, C. J., & Moller, H. (2006). Intensification and diversification of New Zealand agriculture since 1960: An evaluation of current indicators of land use change. *Agriculture, Ecosystems & Environment*, *115*(1-4), 201–218. doi:10.1016/j.agee.2006.01.003

Mahua, Podder, & Samanta. (2019). Factors that influence Sustainable Education with respect to Innovation and Statistical Science. *International Journal of Recent Technology and Engineering*, *7*(5S2).

Mall, R. K., Singh, R., Gupta, A., Srinivasan, G., & Rathore, L. S. (2006). Impact of climate change on Indian agriculture: A review. *Climatic Change*, *78*(2), 445–478. doi:10.100710584-005-9042-x

Mancuso, J. R. (1975). The entrepreneur's quiz. In C. M. Baumack & J. R. Mancuso (Eds.), *Entrepreneurship and venture management*. Prentice Hall.

Manjula, M. (2021). The Smallholder in the Agriculture Market Reforms in India. *Economic and Political Weekly*, *56*(15), 23.

Marcuse, P. (2009). From critical urban theory to the right to the city. *City*, *13*(2-3), 185–197.

Maria, F. (2006). Slow Food's presidia: what do small producers do with big retailers? In *Between the Local and the Global* (pp. 203–240). Emerald Group Publishing Limited.

Maria, K., & Venkatesh, A. (2007). Food for thought: A study of food consumption in postmodern US culture. *Journal of Consumer Behaviour*, *6*(6), 419–435. doi:10.1002/cb.232

Marion, N. (2013). *Food politics: How the food industry influences nutrition and health* (Vol. 3). Univ of California Press.

Mark, S., & Wolfe, R. (1999). Consumption as self-presentation: A socioanalytic interpretation of Mrs. Cage. *Journal of Marketing*, *63*(4), 135–138. doi:10.1177/002224299906300411

Markussen, S., & Røed, K. (2017). The gender gap in entrepreneurship–the role of peer effects. *Journal of Economic Behavior & Organization*, *134*, 356–373.

Marques, C. S., Ferreira, J. J., Gomes, D. N., & Rodrigues, R. G. (2012). Entrepreneurship education: How psychological, demographic and behavioural factors predict the entrepreneurial intention. *Education + Training*, *54*(8/9), 657–672. doi:10.1108/00400911211274819

Marsden, T., & Sonnino, R. (2012). Human health and wellbeing and the sustainability of urban–regional food systems. *Current Opinion in Environmental Sustainability*, *4*(4), 427–430. doi:10.1016/j.cosust.2012.09.004

Mars, M. M., & Schau, H. J. (2016). Institutional entrepreneurship and the negotiation and blending of multiple logics in the Southern Arizona local food system. *Agriculture and Human Values*, *34*(2), 407–422. doi:10.100710460-016-9722-3

Mars, R. (2005). *The basics of permaculture design*. Chelsea Green Publishing.

Martin, H., Boone, J., & Haley, S. (2010). Local food marketing as a development opportunity for small UK agri-food businesses. *International Journal on Food System Dynamics*, *1*(3), 194.

Martin-Sanchez, V., Contín-Pilart, I., & Larraza-Kintana, M. (2017). The influence of entrepreneurs' social referents on start-up size. *The International Entrepreneurship and Management Journal*, *14*, 173–194. https://doi.org/10.1007/s11365-017-0454-7

Maula, M., Autio, E., & Arenius, P. (2003). *What drives micro-angel investments? A large sample study of the factors explaining micro-angel investments*. Unpublished working paper.

Maula, M., Autio, E., & Arenius, P. (2005). What drives micro-angel investments? *Small Business Economics*, *25*(5), 459–475.

Mauro, F., & Hardison, P. D. (2000). Traditional knowledge of indigenous and local communities: International debate and policy initiatives. *Ecological Applications*, *10*(5), 1263–1269. doi:10.1890/1051-0761(2000)010[1263:TKOIAL]2.0.CO;2

Maxwell, K., & Johnson, G. (2000). Chlorophyll Fluorescence, a Practical Guide. [Google Scholar]. *Journal of Experimental Botany*, *51*(345), 659–668. doi:10.1093/jexbot/51.345.659 PMID:10938857

Maye, D. (2013). Moving alternative food networks beyond the niche. *International Journal of Sociology of Agriculture and Food*, *20*(3), 383–389.

McAllister, D. J. (1995). Affect-and cognition-based trust as foundations for interpersonal cooperation in organizations. *Academy of Management Journal*, *38*(1), 24–59.

McClelland, D. C. (1967). *Achieving society* (Vol. 92051). Simon and Schuster.

McClintock, N. (2010). Why farm the city? Theorizing urban agriculture through a lens of metabolic rift. *Cambridge Journal of Regions, Economy and Society*, *3*(2), 191–207.

McDougall, P. P. (1989). International versus domestic entrepreneurship: New venture strategic behavior and industry structure. *Journal of Business Venturing*, *4*(6), 387–400. https://doi.org/10.1016/0883-9026(89)90009-8

McElwee, G. (2006). Farmers as entrepreneurs: Developing competitive skills. *Journal of Developmental Entrepreneurship*, *11*(03), 187–206.

McElwee, G. (2008). A taxonomy of entrepreneurial farmers. *International Journal of Entrepreneurship and Small Business*, *6*(3), 465–478.

McIntyre. (Eds.). (2009). *The International Assessment of Agricultural Knowledge, Science and Technology for Development: Summary for Decision-Makers Report of the Global Report*. Island Press.

McKee, L. (1995). The Earth is their witness. *The Sciences*, *35*(2), 36–41. doi:10.1002/j.2326-1951.1995.tb03183.x

McKelvie, A., & Wiklund, J. (2010). Advancing firm growth research: A focus on growth mode instead of growth rate. *Entrepreneurship Theory and Practice*, *34*(2), 261–288. https://doi.org/10.1111/j.15406520.2010.00375.x

McMullen, J. S., & Shepherd, D. A. (2006). Entrepreneurial action and the role of uncertainty in the theory of the entrepreneur. *Academy of Management Review*, *31*, 132–152. https://doi.org/10.5465 /amr.2006.19379628

Mehmood, A., & Parra, C. (2013). Social innovation in an unsustainable world. In *The international handbook on social innovation: Collective action, social learning and transdisciplinary research* (pp. 53–66). Edward Elgar Publishing.

Melanie, W., & Arnould, E. (1991). "We gather together": Consumption rituals of Thanksgiving Day. *The Journal of Consumer Research*, *18*(1), 13. doi:10.1086/209237

Méndez, V. E., Bacon, C. M., & Cohen, R. (2013). Agroecology as a transdisciplinary, participatory, and action-oriented approach. *Agroecology and Sustainable Food Systems*, *37*(1), 3–18.

Menzies, T. V., & Tatroff, H. (2006). The propensity of male vs. female students to take courses and degree concentrations in entrepreneurship1. *Journal of Small Business and Entrepreneurship*, *19*(2), 203–223. doi:10.1080/08276331.2006.10593367

Michael, S. C., & Pearce, J. A. II. (2009). The need for innovation as a rationale for government involvement in entrepreneurship. *Entrepreneurship and Regional Development*, *21*(3), 285–302. doi:10.1080/08985620802279999

Milburn, L. A. (2010). Greening Cities, Growing Communities: Learning From Seattle's Urban Community Gardens by Jeffrey Hou, Julie M. Johnson, and Laura J. Lawson. 2009. Seattle: University of Washington Press and the Landscape Architecture Foundation. 232 pages, 140 illustrations, 130 in color. $40.00, paperback. ISBN-13: 978-0-295989-28-0. *Landscape Journal*, *29*(2), 238–239. doi:10.3368/lj.29.2.238

Milestad, R., Ahnström, J., & Björklund, J. (2010). Essential multiple functions of farms in rural communities and landscapes. *Renewable Agriculture and Food Systems*, *26*(2), 137–148. doi:10.1017/S1742170510000529

Mills, J. (2012). Exploring the social benefits of agri-environment schemes in England. *Journal of Rural Studies*, *28*(4), 612–621. doi:10.1016/j.jrurstud.2012.08.001

Ministry of Rural Development - Govt. of India. (2019). *DeendayalAntyodayaYojana-National Rural Livelihoods Mission (DAY-NRLM)*. Retrieved from https://aajeevika.gov.in/en/content/social-inclusion-development

Minniti, M., & Nardone, C. (2007). Being in someone else's shoes: The role of gender in nascent entrepreneurship. *Small Business Economics*, *28*(2), 223–238.

Mintel. (2008). *Locally sourced foods*. Mintel. Available from: https://academic.mintel.com/

Mintel. (2010). *Consumer Food Labelling*. London: Mintel.

Mintz, S. W., & Du Bois, C. M. (2002). The anthropology of food and eating. *Annual Review of Anthropology*, *31*(1), 99–119. doi:10.1146/annurev.anthro.32.032702.131011

Mitchell, R. K., Busenitz, L., Lant, T., McDougall, P. P., Morse, E. A., & Smith, J. B. (2002). Toward a theory of entrepreneurial cognition: Rethinking the people side of entrepreneurship research. *Entrepreneurship Theory and Practice*, *27*(2), 93–104.

Mitchell, R. K., Busenitz, L., Lant, T., McDougall, P. P., Morse, E. A., & Smith, J. B. (2004). The distinctive and inclusive domain of entrepreneurial cognition research. *Entrepreneurship Theory and Practice*, *28*(6), 505–518.

Mitchell, T. D., & Jones, P. D. (2005). An improved method of constructing a database of monthly climate observations and associated high-resolution grids. *Journal of Climatology*, *25*(6), 693–712. doi:10.1002/joc.1181

Mittal, R. K. (1999). *Entrepreneurial Management: Marketing Strategy and Product Portfolio.* Fine Art Press.

Moisio, R., Arnould, E., & Price, L. (2004). Between mothers and markets: Constructing family identity through homemade food. *Journal of Consumer Culture*, *4*(3), 361–384. doi:10.1177/1469540504046523

Mok, H. F., Williamson, V. G., Grove, J. R., Burry, K., Barker, S. F., & Hamilton, A. J. (2014). Strawberry fields forever? Urban agriculture in developed countries: A review. *Agronomy for Sustainable Development*, *34*(1), 21–43.

Mollison, B. (1988). Permaculture: A Designers' Manual. Tagari Publications.

Mollison, B., &Slay, R. M. (1988). *Permaculture: a designers' manual* (Vol. 1). Tyalgum: Tagari Publications.

Mollison, B. C., & Holmgren, D. (1978). *Permaculture 1: A perennial agriculture system for human settlements*. University of Tasmania.

Mollison, B., Slay, R. M., Girard, J. L., & Girard, J. L. (1991). *Introduction to permaculture.* Tagari Publications.

Morales-Gualdrón, S. T., & Roig, S. (2005). The new venture decision: An analysis based on the GEM project database. *The International Entrepreneurship and Management Journal*, *1*(4), 479–499.

Moraru, R. A., Ungureanu, G., Bodescu, D., & Donosă, D. (2016). Motivations and challenges for entrepreneurs in agritourism. *Agronomy Series of Scientific Research/Lucrari Stiintifice Seria Agronomie, 59*(1).

Morgan, K. (2009). *Feeding the city: The challenge of urban food planning*. Academic Press.

Morton, J. F. (2007). The impact of climate change on smallholder and subsistence agriculture. *Proceedings of the National Academy of Sciences of the United States of America, 104*(50), 680–685. doi:10.1073/pnas.0701855104 PMID:18077400

Mouekouba, L., Zhang, Z., & Erinle Kehinde Olajide, E. (2013). Biological Control of Botrytis Cinerea in Tomato Leaves. *2013 International Conf. on Agri. and Biotech. IPCBEE, 60*, 64-68. 10.7763/IPCBEE

Mougeot, L. J. (2006). *Growing better cities: Urban agriculture for sustainable development*. IDRC.

Mougeot, L. J. (Ed.). (2005). *Agropolis: The social, political, and environmental dimensions of urban agriculture*. IDRC.

Mousa, F. T., & Wales, W. (2012). Founder effectiveness in leveraging entrepreneurial orientation. *Management Decision, 50*(2), 305–324. doi:10.1108/00251741211203588

Muangpat, P., Yooyangket, T., Fukruksa, C., Suwannaroj, M., Yimthin, T., Sitthisak, S., Chantratita, N., Vitta, A., Tobias, N. J., Bode, H. B., & Thanwisai, A. (2017). Screening of the anti-microbial activity against drug resistant bacteria of *Photorhabdus* and *Xenorhabdus* associated with entomopathogenic nematodes from Mae Wong National Park, Thailand. *Frontiers in Microbiology, 8*, 1142. doi:10.3389/fmicb.2017.01142 PMID:28702004

Mukherji, N., & Morales, A. (2010). *Zoning for urban agro ecology: Zoning practice*. American Planning Association.

Mulligan, M., & Hill, S. (2001). *Ecological pioneers: a social history of Australian ecologica lthought and action*. Cambridge UniversityPress.

Mundler, P., & Laughrea, S. (2016). The contributions of short food supply chains to territorial development: A study of three Quebec territories. *Journal of Rural Studies, 45*, 218–229. doi:10.1016/j.jrurstud.2016.04.001

Mustafa, K., & Welsh, J. (2001). *Food, foodways and immigrant experience*. Centre for Studies in Food Security.

Mutasowifin, A. (2002). Penerapan Balanced scorecard sebagai tolok ukur penilaian pada badan usaha berbentuk koperasi [The application of the Balanced scorecard as a measure of assessment for cooperative business entities]. *Paramadina University Journal, 1*(3), 245–26.

Myung, K., Hamilton-Kemp, T. R., & Archbold, D. D. (2007). Interaction with and Effects on the Profile of Proteins of Botrytis cinerea by C6 Aldehydes. *Journal of Agricultural and Food Chemistry, 55*(6), 2182–2188. doi:10.1021/jf0631629 PMID:17323971

NABARD. (2019). *Report on Status of Micro Finance in India 2019-20*. Available at: www.nabard.org/ auth/writereaddata/tender /1307174808Status%20of%20Microfinance /1307174808Status%20of%20Microfinance%20in%20India%2020 16-17.pdf

Nabi, G., & Liñán, F. (2013). Considering business start-up in recession time: The role of risk perception and economic context in shaping the entrepreneurial intent. *International Journal of Entrepreneurial Behaviour & Research, 19*(6), 633–655.

Nagalakshmi, T., & Sudhakar, A. (2013). Agri-preneurs: A Case Study of Dharmapuri Farmers. *International Journal of Scientific Research, 2*(8).

Nagler, P., & Naudé, W. (2014). *Non-Farm Enterprises in Rural Africa: New Empirical Evidence*. Policy Research Working Papers. doi:10.1596/1813-9450-7066

Nahapiet, J., & Ghoshal, S. (1998). Social capital, intellectual capital, and the organizational advantage. *Academy of Management Review, 23*(2), 242–266.

Nair, P. K. R. (1993). *An introduction to agroforestry*. Kluwer Academic Publishers. doi:10.1007/978-94-011-1608-4

Nandhini, S., Bhrathi, S., Goud, D. D., & Krishna, K. P. (2019). Smart Agriculture IOT with Cloud Computing, Fog Computing and Edge Computing. *International Journal of Engineering and Advanced Technology*, *9*(2), 3578–3583. doi:10.35940/ijeat.B2600.129219

Nasip, S., Amirul, S. R., Sondoh, S. L. Jr, & Tanakinjal, G. H. (2017). Psychological characteristics and entrepreneurial intention: A study among university students in North Borneo, Malaysia. *Education + Training*, *59*(7/8), 825–840. doi:10.1108/ET-10-2015-0092

Navis, C., & Ozbek, O. V. (2016). The right people in the wrong places: The paradox of entrepreneurial entry and successful opportunity realization. *Academy of Management Review*, *41*(1), 109–129. https://doi.org/10.5465/amr.2013.0175

Neira, I., Calvo, N., Fernández, L., & Portela, M. (2017). Entrepreneur: Do social capital and culture matter? *The International Entrepreneurship and Management Journal*, *13*(2), 665–683.

Neuner, K., Kelly, S., & Raja, S. (2011). *Planning to eat? Innovative local government plans and policies to build healthy food systems in the United States*. Academic Press.

Newman, A., North-Samardzic, A., Bedarkar, M., & Brahmankar, Y. (2021). *Entrepreneurship in India*. Routledge.

NewsDesk. (2021, July 18). Goa is Committed to Encouraging Women Entrepreneurs: CM. *Goa Chronicle*. Retrieved from https://goachronicle.com/goa-is-committed-to-encouraging-women-entrepreneurs-cm/

Nguyen, C. (2017). Entrepreneurial intention of international business students in Viet Nam: A survey of the country joining the Trans-Pacific Partnership. *Journal of Innovation and Entrepreneurship*, *6*(1), 1–13. doi:10.118613731-017-0066-z

Nieto, M., & González-Álvarez, N. (2016). Social capital effects on the discovery and exploitation of entrepreneurial opportunities. *The International Entrepreneurship and Management Journal*, *12*(2), 507530. https://doi.org/10.1007/s11365-014-0353-0

Nilsson, T. (1997). *Kundageranden i små tillverkande företag* (PhD diss.). Luleå University of Technology, Luleå.

Nkuranga, T., & Wilcox, K. (2013). *Cooperative performance index field results and analysis. In USAID-enabling market integration through rural group empowerment*. EMIRGE.

Noguera, M., Alvarez, C., & Urbano, D. (2013). Socio-cultural factors and female entrepreneurship. *The International Entrepreneurship and Management Journal*, *9*(2), 183–197.

Nordahl, D. (2009). *Public Produce: The New Urban Agriculture*. Island Press.

Norgaard, R. B. (1984). Traditional agricultural knowledge: Past performance, future prospects, and institutional implications. *American Journal of Agricultural Economics*, *66*(5), 874–878. doi:10.2307/1241018

Nowak, D. J. (2006). Institutionalizing urban forestry as a "biotechnology" to improve environmental quality. *Urban Forestry & Urban Greening*, *5*(2), 93–100. doi:10.1016/j.ufug.2006.04.002

Nowak, D. J., & Dwyer, J. F. (2007). Benefits and costs of urban forest ecosystems. In J. E. Kuser (Ed.), *Urban and community forestry in the Northeast* (pp. 25–46). Springer. doi:10.1007/978-1-4020-4289-8_2

OECD/ICRIER. (2018). Agricultural policies in India. In *OECD Food and Agricultural Reviews* (p. 296). OECD Publishing Paris.

Omoruyi, E. M. M., Olamide, K. S., Gomolemo, G., & Donath, O. A. (2017). Entrepreneurship and economic growth: Does entrepreneurship bolster economic expansion in Africa. *Journal of Socialomics*, *6*(4), 1–11.

Onchiri, M. A., Maurice, S., & Patrick, M. (2015). Social economic factors and access to finance: The case of women managed agri-based small and medium enterprises Nyamira County, Kenya. *International Journal of Marketing and Technology*, *5*(9), 56–74.

Ostgaard, T. A., & Birley, S. (1994). Personal networks and firm competitive strategy—A strategic or coincidental match? *Journal of Business Venturing*, *9*(4), 281–305. doi:10.1016/0883-9026(94)90009-4

Östman, Ö., Ekbom, B., & Bengtsson, J. (2003). Yield increase attributable to aphid predation by ground-living polyphagous natural enemies in spring barley in Sweden. *Ecological Economics*, *45*(1), 149–158.

Padliya, N. D., & Cooper, B. (2006). Mass Spectrometry-based Proteomics for the Detection of Plant Pathogens. [Google Scholar]. *Proteomics*, *6*(14), 4069–4075. doi:10.1002/pmic.200600146 PMID:16791831

Pamela, K. (2012). *Food and Culture.* Cengage Learning.

Pandey, G. (2013). Agripreneurship Education and Development: Need of the Day. *Asian Resonance*, *2*(4), 155–157.

Pant, J., Demaine, H., & Edwards, P. (2005). Bio-resource flow in integrated agriculture–aquaculture systems in a tropical monsoonal climate: A case study in Northeast Thailand. *Agricultural Systems*, *83*(2), 203–219. doi:10.1016/j.agsy.2004.04.001

Parham, S. (2015). The productive periphery: food space and urbanism on the edge. *Localizing urban food strategies. Farming cities and performing rurality*, 118-130.

Partalidou M and Anthopoulou T. (2016) Urban Allotment Gardens During Precarious Times: From Motives to Lived Experiences. *SociologiaRuralis*.

Pathak, G. S., & Wang, M. (2012). Entrepreneurial self-efficacy and entrepreneurial intention in India: A comparative study of professional students. *International Journal of Psychology*, 47.

Patzke, H., & Schieber, A. (2018). Growth-inhibitory activity of phenolic compounds applied in an emulsifiable concentrate - ferulic acid as a natural pesticide against Botrytis cinerea. *Food Research International, 113*, 18–23. doi:10.1016/j.foodres.2018.06.062 PMID:30195511

Paula, A. (2013). Ethnic food: the other in ourselves. In D. Sanderson & M. Crouch (Eds.), *Food: Expressions and Impressions* (pp. 45–56). Inter-Disciplinary Press.

Paulson, A. L., & Townsend, R. (2004). Entrepreneurship and financial constraints in Thailand. *Journal of Corporate Finance, 10*(2), 229–262. doi:10.1016/S0929-1199(03)00056-7

Perfecto, I., & Vandermeer, J. (2010). The agro ecological matrix as alternative to the land-sparing/agriculture intensification model. *Proceedings of the National Academy of Sciences of the United States of America, 107*(13), 5786–5791. doi:10.1073/pnas.0905455107 PMID:20339080

Periodic Labour Force Survey. (2017). *Annual Report, PLFS, 2017-18*. www.mospi.nic.in/publication/annual-report-plfs-2017-18

Peter, J., Comfort, D., & Hillier, D. (2004). A case study of local food and its routes to market in the UK. *British Food Journal, 106*(4), 328–335. doi:10.1108/00070700410529582

Petrasch, S., Knapp, S. J., van Kan, J. A. L., & Blanco-Ulate, B. (2019). Grey mould of strawberry, a devastating disease caused by the ubiquitous necrotrophic fungal pathogen Botrytis cinerea. *Molecular Plant Pathology, 20*(6), 877–892. doi:10.1111/mpp.12794 PMID:30945788

Petrova, K. (2012). Part-time entrepreneurship and financial constraints: Evidence from the Panel Study of Entrepreneurial Dynamics. *Small Business Economics, 39*(2), 473–493. doi:10.100711187-010-9310-7

Pia, C. (2001). Mac and Gravy. In *Pilaf, Pozole and Pad Thai*. University of Massachusetts Press.

Pimbert, M. P., Thompson, J., Vorley, W. T., Fox, T., Kanji, N., & Tacoli, C. (2001). Global restructuring, agri-food systems and livelihoods. Gatekeeper Series 100. London: International Institute for Environment and Development.

Pindado, E., & Sánchez, M. (2017). Researching the entrepreneurial behaviour of new and existing ventures in European agriculture. *Small Business Economics, 49*(2), 421–444. https://doi.org/10.1007/s11187-0179837-y

Pindado, E., Sánchez, M., Verstegen, J. A., & Lans, T. (2018). Searching for the entrepreneurs among new entrants in European Agriculture: The role of human and social capital. *Land Use Policy, 77*, 19–30.

Pinderhughes, R. (2007). *Green collar jobs: An analysis of the capacity of green businesses to provide high quality jobs for men and women with barriers to employment*. The City of Berkeley Office of Energy and Sustainable Development. Retrieved from https://community-wealth.org/sites/clone.community-wealth.org/files/downloads/paper-pinderhughes-schecter.pdf

Piotti, G. (2007). [Book Review] Osti, Giorgio: Nuoviasceti: Consumatori, imprese e istituzioni di fronteallacrisiambientale (Bologna: Il Mulino, 2006). Economic Sociology: The European Electronic Newsletter, 21-23.

Pirolo, L., & Presutti, M. (2010). The impact of social capital on the start-ups' performance growth. *Journal of Small Business Management*, *48*(2), 197–227. doi:10.1111/j.1540-627X.2010.00292.x

Pitt, M. M., Khandker, S. R., & Cartwright, J. (2006). Empowering women with micro finance: Evidence from Bangladesh. *Economic Development and Cultural Change*, *54*(4), 791–831.

Podder & Samanta. (2019). Impact of Climate Changes on Rural Livelihoods and Re-Orienting the Situation through Human Resources towards a Sustainable Society. In *International Conference on Sustainable Computing in Science, Technology & Management (SUSCOM-2019)*. Elsevier.

Podder, S. K. (2019). Impact of Business Analytics for Smart Education System and Management Functions. In *Proc. of International Conference on Advancements in Computing & Management (ICACM-2019)*. Springer

Ponisio, L., M'Gonigle, L. K., Mace, K. C., Palomino, J., de Valpine, P., & Kremen, C. (2014). Diversification practices reduce organic to conventional yield gap. *Proceedings. Biological Sciences*, *282*(1799), 20141396. doi:10.1098/rspb.2014.1396 PMID:25621333

Porter, J., Costanza, R., Sandhu, H., Sigsgaard, L., & Wratten, S. (2009). The value of producing food, energy, and ecosystem services within an agroecosystem. *AMBIO: A Journal of the Human Environment, 38*(4), 186–193. doi:10.1579/0044-7447-38.4.186

Portes, A. (1998). Social capital: Its origins and applications in modern sociology. *Annual Review of Sociology*, *24*(1), 1–24.

Pothukuchi, K. (2009). Community and regional food planning: Building institutional support in the United States. *International Planning Studies*, *14*(4), 349–367.

Pourreza, A., Lee, W., Reza, E., Schueller, J., & Raveh, E. (2015). An Optimum Method for Real-Time Infield Detection of Huanglongbing Disease using A Vision Sensor. [Google Scholar]. *Computers and Electronics in Agriculture*, *110*, 221–232. doi:10.1016/j.compag.2014.11.021

Power, A. G. (2010). Ecosystem services and agriculture: Tradeoffs and synergies. *Philosophical Transactions of the Royal Society of London. Series B, Biological Sciences*, *365*(1554), 2959–2971. doi:10.1098/rstb.2010.0143 PMID:20713396

Pradeep, Sangeetha, & Singhb. (2016). Relationship between Food Production and Consumption Diversity in India–Empirical Evidences from Cross Section Analysis. *Agricultural Economics Research Review, 29*, 139.

Pramudya, E. P., Hospes, O., & Termeer, C. J. A. M. (2017). Governing the palm-oil sector through finance: The changing roles of the Indonesian State. *Bulletin of Indonesian Economic Studies*, *53*(1), 57–82.

Pratella, G. C., & Mari, M. (1993). Effectiveness of Trichoderma, Gliocladium, and Paecilomyces in Postharvest Fruit Protection. *Postharvest Biol.* [Google Scholar]. *Tech., 3*, 49–56.

Public Health Law and Policy. (2009). *Establishing land use protections for community gardens*. Author.

Pursell, C. (1993). The rise and fall of the appropriate technology movement in the United States, 1965-1985. *Technology and Culture, 34*(3), 629–637.

Putnam, R. D. (1995). Bowling alone: America's declining social capital. *Journal of Democracy, 6*(1), 65–78. doi:10.1353/jod.1995.0002

Putnam, R. D. (2001). *Bowling Alone: The Collapse and Revival of American Community*. Touchstone.

Putnam, R. D. (2001). Social Capital: Measurement and Consequences. *Isuma: Canadian Journal of Policy Research, 2*, 41–51.

Qin, J., Chao, K., & Kim, M. (2011). Investigation of Raman Chemical Imaging for Detection of Lycopene Changes in Tomatoes during Postharvest Ripening. [Google Scholar]. *Journal of Food Engineering, 107*(3-4), 277–28. doi:10.1016/j.jfoodeng.2011.07.021

Rachel, D., Fearne, A., & Healing, V. (2005). Reconnection in the UK food chain: Bridging the communication gap between food producers and consumers. *British Food Journal, 107*(1), 17–33. doi:10.1108/00070700510573177

Rajalahti, R., Woelcke, J., & Pehu, E. (2005). *Developing research systems to support the changing agricultural sector* (Vol. 14). Agriculture & Rural Development Department, World Bank.

Ramos-Rodríguez, A. R., Medina-Garrido, J. A., & Ruiz-Navarro, J. (2012). Determinants of hotels and restaurants entrepreneurship: A study using GEM data. *International Journal of Hospitality Management, 31*(2), 579–587. https://doi.org/10.1016/j.ijhm.2011.08.003

Ranaweera, C., & Prabhu, J. (2003). The influence of satisfaction, trust and switching barriers on customer retention in a continuous purchasing setting. *International Journal of Service Industry Management.*

Rapidel, B., Ripoche, A., Allinne, C., Metay, A., Deheuvels, O., Lamanda, N., Blazy, J.-M., Valdés-Gómez, H., & Gary, C. (2015). Analysis of ecosystem services trade-offs to design agroecosystems with perennial crops. *Agronomy for Sustainable Development, 34*(4), 1373–1390. doi:10.100713593-015-0317-y

Raposo, M., & Do Paço, A. (2011). Entrepreneurship education: Relationship between education and entrepreneurial activity. *Psicothema, 23*(3), 453–457. PMID:21774900

Redwood, M. (2008). *Agriculture in Urban Planning: Generating Livelihoods and Food Security*. Earthscan.

Regina, F. (2001). A Imigração Portuguesa e as Associações Como Forma de Manutenção da Identidade Lusitana-Sul do Brazil. *Scrita Nova, 94*, 27. http://www.ub.edu/geocrit/sn-94-27/htm

Reglinski, T., Elmer, P., Taylor, H., Wood, P., & Hoyte, S. (2010). Inhibition of Botrytis cinerea Growth and Suppression of Botrytis Bunch Rot in Grapes using Chitosan. *Plant Pathology, 59*(5), 882–890. doi:10.1111/j.1365-3059.2010.02312.x

Reijntjes, C. B., Haverkort, B., & Waters-Bayer, A. (1992). *Farming for the future. Anintroduction to low-external input and sustainable agriculture*. Macmillian Press.

Renting, H., Marseden, T. K., & Banks, J. (2003). Understanding alternative food networks: Exploring the role of short food supply chains in rural development. *Environment & Planning A, 35*(3), 393–411. doi:10.1068/a3510

ReportU. N. I. D. O. (2017). *Women and Youth*. Available at: https://www.unido.org/fileadmin/user_media_upgrade/What_we_do/Topics/Women_and_Youth/Brochure_low_resolution.pdf

Reynolds, P. D. (1995). Who starts new firms? Linear additive versus interaction based models. *Frontiers of Entrepreneurship Research*, 32-46.

Reynolds, K. (2015). Disparity despite diversity: Social injustice in New York City's urban agriculture system. *Antipode, 47*(1), 240–259.

Reynolds, K., & Cohen, N. (2016). *Beyond the kale: Urban agriculture and social justice activism in New York City* (Vol. 28). University of Georgia Press.

Reynolds, P. D., Bygrave, W. D., Autio, E., & Arenius, P. (2004). *GEM 2003 global report*. Babson College.

Reynolds, P., Bosma, N., Autio, E., Hunt, S., De Bono, N., Servais, I., & … . (2005). Global entrepreneurship monitor: Data collection design and implementation 1998–2003. *Small Business Economics, 24*(3), 205231.

Reynolds, P., Storey, D. J., & Westhead, P. (1994). Cross-national comparisons of the variation in new firm formation rates. *Regional Studies, 28*(4), 443–456.

Richards, S. T., & Bulkley, S. (2007). Agricultural Entrepreneurs: The First and the Forgotten? SSRN *Electronic Journal*. doi:10.2139/ssrn.1020697

Richardson, K. (2003). Vegemite, Soldiers and Rosy Cheeks. *Gastronomica, 3*(4), 60–62. doi:10.1525/gfc.2003.3.4.60

Ridha, R. N., &Wahyu, B. P. (2017). Entrepreneurship intention in agricultural sector of young generation in Indonesia. *Asia Pacific Journal of Innovation and Entrepreneurship*.

Rist, L., Feintrenie, L., & Levang, P. (2010). The livelihood impacts of oil palm: Smallholders in Indonesia. *Biodiversity and Conservation, 19*(4), 1009–1024.

Robert, B., Brown, A., Ansfield, M., & Paschall, G. (2002). "You will eat all of that!": A retrospective analysis of forced consumption episodes. *Appetite*, *38*(3), 211–219. doi:10.1006/appe.2001.0482 PMID:12071687

Robichaud, Y., Zinger, J. T., & LeBrasseur, R. (2007). Gender differences within early stage and established small enterprises: An exploratory study. *The International Entrepreneurship and Management Journal*, *3*(3), 323–343. https://doi.org/10.1007/s11365-007-0039-y

Robin, F. (2003). Food and eating: an anthropological perspective. Social Issues Research Centre.

Robinson, P. B., Stimpson, D. V., Huefner, J., & Hunt, H. K. (1991). An attitude approach to the prediction of entrepreneurship. *Entrepreneurship Theory and Practice*, *15*(4), 13–31.

Rooi, C. V., & Holz, G. (2003). Fungicide Efficacy Against Botrytis cinerea at Different Positions on Grape Shoots. *Afr. J. Enol. Vitic*, *24*(1), 11–15.

Rosset, P. M. (1997a). Cuba: Ethics, biological control, and crisis. *Agriculture and Human Values*, *14*(3), 291–302. doi:10.1023/A:1007433501248

Rosset, P. M. (1997b). Alternative agriculture and crisis in Cuba. *Technology in Society*, *16*(2), 19–15.

Roth, M., Frixen, M., Tobisch, C., & Scholle, T. (2015). Finding spaces for urban food production–matching spatial and stakeholder analysis with urban agriculture approaches in the urban renewal area of Dortmund-Hörde. Academic Press.

Roy, R., Akhtar, F., & Das, N. (2017). Entrepreneurial intention among science & technology students in India: Extending the theory of planned behavior. *The International Entrepreneurship and Management Journal*, *13*(4), 1013–1041. doi:10.100711365-017-0434-y

Roy, R., Akhtar, F., & Das, N. (2017). Entrepreneurial intention among science & technology students in India: Extending the theory of planned behaviour. *The International Entrepreneurship and Management Journal*, 1–29.

Rukera, S.T., Mutanga, O., & Micha, J-C. (2012). Optimization of an integrated rabbit-fish-rice system for sustainable production in Rwanda. *Rwanda J, 24*.

Runquist, S., Zhang, N., & Taylor, R. K. (2001). Development of a field-level geographic information system. *Computers and Electronics in Agriculture*, *31*(2), 201–209. doi:10.1016/S0168-1699(00)00155-1

Sabrina, R., Weber Roland W. S., Rieger Daniel, R., Peter, D., & Hahn Matthias, H. (2017). Spread of Botrytis cinerea Strains with Multiple Fungicide Resistance in German Horticulture. *Frontiers in Microbiology*, *7*(2075). doi:10.3389/fmicb.2016.02075

Sachez, P. A. (1995). Science in Agroforestry. *Agroforestry Systems*, *30*(1-2), 5–55. doi:10.1007/BF00708912

Safransky, S. (2014). Greening the urban frontier: Race, property, and resettlement in Detroit. *Geoforum*, *56*, 237–248.

Saiz-Rubio, V., & Rovira-Más, F. (2020). From Smart Farming towards Agriculture 5.0: A Review on Crop Data Management. *Agronomy (Basel)*, *10*(207), 1–21. doi:10.3390/agronomy10020207

Saldivar-Tanaka, L., & Krasny, M. E. (2004). Culturing community development, neighborhood open space, and civic agriculture: The case of Latino community gardens in New York City. *Agriculture and Human Values*, *21*(4), 399–412.

Saleem, I., Arafat, M. Y., Balhareth, H. H., Hussain, A., & Dwivedi, A. k. (2021). Cognitive aspects of disadvantaged entrepreneurs: Evidence from India. *Community Development (Columbus, Ohio)*. Advance online publication. doi:10.1080/15575330.2021.197

Sandeepa, M., & Sarala, K. S. (2020). Opportunities and challenges of agriprenuership in India–Some reviews. *PalArch's Journal of Archaeology of Egypt/Egyptology*, *17*(7), 9541–9554.

Sandhu, H. S., Wratten, S. D., & Cullen, R. (2010). Organic agriculture and ecosystem services. *Environmental Science & Policy*, *13*(1), 1–7. doi:10.1016/j.envsci.2009.11.002

Sanford, A. W. (2011). Ethics, narrative, and agriculture: Transforming agricultural practice through ecological imagination. *Journal of Agricultural & Environmental Ethics*, *24*(3), 283–303. doi:10.100710806-010-9246-6

Sankaran, S., Mishra, A., Ehsani, R., & Davis, C. (2010). A Review of Advanced Techniques for Detecting Plant Diseases. [Google Scholar]. *Computers and Electronics in Agriculture*, *72*(1), 1–13. doi:10.1016/j.compag.2010.02.007

Santiago-Roman, J. (2013). *Entrepreneurism in Puerto Rico: Job creation and entrepreneurial development*. Northcentral University.

Santos, F. J. (2004). Convergencia, desarrollo y empresarialidadenelproceso de globalizacióneconómica. *Revista de Economía Mundial*, *10*(11), 171–202.

Sarma, Y. V. S., & Swamy, A. K. (2012). Introspection of women entrepreneurs: A garrette's approach. *Sumedha Journal of Management*, *1*(4), 4–12.

Sasikala, T. (2015). *Socio economic status of women entrepreneurs in Salem district of Tamilnadu*. Academic Press.

Scherer, R. F., Brodzinski, J. D., & Wiebe, F. A. (1990). Entrepreneur career selection and gender: A socialization approach. *Journal of Small Business Management*, *28*(2), 37.

Schiller, B. R., & Crewson, P. E. (1997). Entrepreneurial origins: A longitudinal inquiry. *Economic Inquiry*, *35*(3), 523–531. doi:10.1111/j.1465-7295.1997.tb02029.x

Schlaegel, C., & Koenig, M. (2014). Determinants of entrepreneurial intent: A meta-analytic test and integration of competing models. *Entrepreneurship Theory and Practice*, *38*(2), 291–332.

Schlaegel, C., & Koenig, M. (2014). Determinants of entrepreneurial intent: A meta–analytic test and integration of competing models. *Entrepreneurship Theory and Practice*, *38*(2), 291–332. doi:10.1111/etap.12087

Schmelzkopf, K. (2002). Incommensurability, land use, and the right to space: Community gardens in New York City1. *Urban Geography*, *23*(4), 323–343.

Schmidhuber, J., & Shetty, P. (2003). *The nutrition transition to 2030: why developing countries are likely to bear the major burden*. http://www.fao.org/fileadmin/templates/esa/Global_persepctives/Long_term_papers/JSPStransition.pdf

Schmidhuber, J., & Tubiello, F. N. (2007). Global food security under climate change. *Proceedings of the National Academy of Sciences of the United States of America*, *104*(50), 19703–19708. doi:10.1073/pnas.0701976104 PMID:18077404

Schraufnagel, J., & Hudelson, B. (2004). Gray Mold (Botrytis Blight). *Wisconsin Garden Facts*, 1. wihort.uwex.edu

Schumpeter Joseph, A. (1934). *The Theory of Economic Development, Cambridge*. Harvard University Press.

Schumpeter, J. (1934). *The theory of economic development*. Harvard University Press.

Sebillotte, M. (2000). Territoires: De l'espace physique au construit social. Les enjeuxpourdemainet les apports de la recherche. *OCL Oléagineux Corps Gras Lipides*, *7*(6), 474–479. doi:10.1051/ocl.2000.0474

Segal, G., Borgia, D., & Schoenfeld, J. (2002). Using social cognitive career theory to predict self-employment goals. *New England Journal of Entrepreneurship*, *5*(2), 47–56.

Segal, G., Borgia, D., & Schoenfeld, J. (2005). The motivation to become an entrepreneur. *International Journal of Entrepreneurial Behaviour & Research*, *11*(1), 42–57. doi:10.1108/13552550510580834

Selvaraj, K. N. (1993). Impact of government expenditure on agriculture and performance of agricultural sector in India. *Bangladesh Journal of Agricultural Economics*, *16*(2), 1–13.

Senger, I., Borges, J. A. R., & Machado, J. A. D. (2017). Using the theory of planned behavior to understand the intention of small farmers in diversifying their agricultural production. *Journal of Rural Studies*, *49*, 32–40. doi:10.1016/j.jrurstud.2016.10.006

Senger, I., Borges, J. A. R., & Machado, J. A. D. (2017). Using the theory of planned behaviour to understand the intention of small farmers in diversifying their agricultural production. *Journal of Rural Studies*, *49*, 32–40.

Setini, M., Yasa, N. N. K., Gede Supartha, I. W., Ketut Giantari, I. G. A., & Rajiani, I. (2020). The Passway of Women Entrepreneurship: Starting from Social Capital with Open Innovation, through to Knowledge Sharing and Innovative Performance. *Journal of Open Innovation*, *6*(2), 25. doi:10.3390/joitmc6020025

Seufert, V., Ramankutty, N., & Foley, J. A. (2012). Comparing the yields of organic and conventional agriculture. *Nature*, *485*(7397), 229–232. doi:10.1038/nature11069 PMID:22535250

Seuneke, P., Lans, T., & Wiskerke, J. S. (2013). Moving beyond entrepreneurial skills: Key factors driving entrepreneurial learning in multifunctional agriculture. *Journal of Rural Studies*, *32*, 208–219.

Shah, V. (2020, November 3). Boosting entrepreneurship among women – A golden opportunity to bring the economy back on track. *The Times of India*. Retrieved from https://timesofindia.indiatimes.com/blogs/voices/boosting-entrepreneurship-among-women-a-golden-opportunity-to-bring-the-economy-back-on-track/

Shah, M., Rao, R., & Shankar, P. V. (2007). Rural credit in 20th century India: Overview of history and perspectives. *Economic and Political Weekly*, •••, 1351–1364.

Shane, S. A. (2003). *A general theory of entrepreneurship: The individual-opportunity nexus*. Edward Elgar Publishing. doi:10.4337/9781781007990

Shane, S., Locke, E. A., & Collins, C. J. (2003). Entrepreneurial motivation. *Human Resource Management Review*, *13*, 257–279. https://doi.org/10.1016/S1053-4822(03)00017-2

Shane, S., & Venkataraman, S. (2000). The promise of entrepreneurship as a field of research. *Academy of Management Review*, *25*(1), 217–226.

Shapero, A., & Sokol, L. (1982). The Social Dimensions of Entrepreneurship, in The Encyclopaedia of Entrepreneurship. In *Encyclopedia of Entrepreneurship* (pp. 72–90). doi:10.1093/oxfordhb/9780199546992.003.0019

Shapero, A. (1975). *The displaced, uncomfortable entrepreneur*. University of Illinois at Urbana-Champaign's Academy for Entrepreneurial Leadership Historical Research Reference in Entrepreneurship.

Shapero, A., & Sokol, L. (1982). *The social dimensions of entrepreneurship*. University of Illinois at Urbana-Champaign's Academy for Entrepreneurial Leadership Historical Research Reference in Entrepreneurship.

Sharma, S., & Sharma, S. (2018). *Agri-entrepreneurial innovations for rural prosperity and sustainable development*. Academic Press.

Shaver, K. G., & Scott, L. R. (1992). Person, process, choice: The psychology of new venture creation. *Entrepreneurship Theory and Practice*, *16*(2), 23–46.

Shehrawat, P. S. (2006). Agro Processing Industries-A Challengeing Entrepreneurship For Rural Development. *Journal of Asia Entrepreneurship and Sustainability*, *2*(3), 111.

Shepherd, D. A., & Douglas, E. J. (1999). Attracting Equity Investors. *Sage (Atlanta, Ga.)*.

Sherry, R., & Stubberud, H. A. (2013). Partnerships for innovation among European small businesses. *Journal of International Business Research*, *12*(2), 97–108.

Shinnar, R. S., Giacomin, O., & Janssen, F. (2012). Entrepreneurial perceptions and intentions: The role of gender and culture. *Entrepreneurship Theory and Practice*, *36*(3), 465–493.

Shirley, L. (1986). *Rice and wheat: the meaning of food in Bangladesh*. RS Khare.

Shirur, M., Shivalingegowda, N. S., Chandregowda, M. J., Manjunath, V., & Rana, R. K. (2019). Critical dimensions of entrepreneurship and entrepreneurial behaviour among mushroom growers: Investigation through Principal Component Analysis. *Indian Journal of Agricultural Research*. doi:10.18805/IJARe.A-5261

Shukla, K. H., & Dwivedi, N. U. (2015). Sustainable development in agricultural sector in India. *The Business & Management Review*, *5*(4), 220.

Sidney, L. (1981). Intepreting consumer mythology: A structural approach to consumer behavior. *Journal of Marketing*, 49.

Sidney, L., & Rook, D. (1999). *Brands, Consumers, Symbols and Research: Sidney J Levy on Marketing*. Sage Publications.

Sidney, M. (2003). Remembrance of Repasts: An Anthropology of Food and Memory. *American Ethnologist*, *30*(3), 474–475. doi:10.1525/ae.2003.30.3.474

Sieger, P., Fueglistaller, U., & Zellweger, T. (2016). *Student entrepreneurship 2016: Insights from 50 countries*. Academic Press.

Sierra, J. E., Medina, B., & Vesga, J. C. (2018). Management system in intelligent agriculture based on Internet of Things. *Revista Espacios*, *39*(08), 20.

Simmons, G. J., Durkin, M. G., McGowan, P., & Armstrong, G. A. (2007). Determinants of internet adoption by SME agri-food companies. *Journal of Small Business and Enterprise Development*, *14*(4), 620–640. doi:10.1108/14626000710832730

Simon-Rojo, M., Recanes, X., Cllau, S., Duzi, B., Eiter, S., Hernández-Jiménez, V., ... Vejre, H. (2016). *From urban food gardening to urban farming*. Academic Press.

Singh, N. P. (1969). n/ACH among Successful-Unsuccessful and Traditional-Progressive Agricultural Entrepreneurs of Delhi. *The Journal of Social Psychology*, *79*(2), 271–272. doi:10.1080/00224545.1969.9922421

Sivanesan, R. (2014). Impact of Brand image and advertisement on consumer buying behavior–Comparative study on rural and urban consumers. *International Journal of Research in Management & Business Studies*, *1*(2), 73–80.

Smallbone, D., & Welter, F. (2001). The distinctiveness of entrepreneurship in transition economies. *Small Business Economics*, *16*(4), 249–262. doi:10.1023/A:1011159216578

Smith, J. R. (1929). *Tree crops: a permanent agriculture*. New York Brace & Co.

Sofer, M., & Saada, M. T. (2017). Entrepreneurship of women in the rural space in Israel: Catalysts and obstacles to enterprise development. *Sociologia Ruralis*, *57*, 769–790. doi:10.1111oru.12125

Soleil, S. (2012) The Biochar Miracle. In *Permac. Res. Inst. - Permac.* Forums Courses Inf. News. http://permaculturenews.org/2012/02/03/the-biochar-miracle/

Sonnino, R. (2013). Local foodscapes: Place and power in the agri-food system. *Acta Agriculturae Scandinavica, Section B–Soil & Plant Science, 63*(sup1), 2-7.

Sonnino, R. (2009). Feeding the city: Towards a new research and planning agenda. *International Planning Studies, 14*(4), 425–435. doi:10.1080/13563471003642795

Sonnino, R., & Marsden, T. (2006). Beyond the divide: Rethinking relationships between alternative and conventional food networks in Europe. *Journal of Economic Geography*, *6*(2), 181–199.

Soomro, B. A., & Shah, N. (2015). Developing attitudes and intentions among potential entrepreneurs. *Journal of Enterprise Information Management*, *28*(2), 304–322. doi:10.1108/JEIM-07-2014-0070

Srivastava, U. K. (1989). Agro-processing industries: Potential, constraints and tasks ahead. *Indian Journal of Agricultural Economics*, *44*(3), 242–256.

Staeheli, L. A., Mitchell, D., & Gibson, K. (2002). Conflicting rights to the city in New York's community gardens. *GeoJournal*, *58*(2), 197–205.

Stenholm, P., Acs, Z. J., & Wuebker, R. (2013). Exploring country-level institutional arrangements on the rate and type of entrepreneurial activity. *Journal of Business Venturing*, *28*(1), 176–193. doi:10.1016/j.jbusvent.2011.11.002

Stevenson, H. H., & Jarillo, C. J. (1990). A paradigm of entrepreneurship: Entrepreneurial management. *Strategic Management Journal*, *11*.

Storey, D. (1994). *Understanding The Small Business Sector*. Routledge.

Summers, D. F. (2000). *The formation of entrepreneurial intentions*. Garland Publishing.

Sundar, S. (2019). India Economic Survey 2018: Farmers Gain As Agriculture Mechanisation Speeds Up, But More R&D Needed. *Financial Express*. Https://Www.Financialexpress.Com/Budget/India-Economic-Survey-2018-For-Farmers-Agriculture-Gdp-Msp/1034266/

Suplab, K. & Podder, A. B. K. (2018). Comparison of effectiveness of employee engagement through permanent employees or outsourced employees. *International Journal of Academic Research and Development, 3*(2), 1406-1408.

Suplab, K., & Podder, A. B. K. (2018). Contribution of HR sub-functions outsourcing in the improvement of quality and innovation of Education. *Primax International Journal of Commerce And Management Research, 3*(1), 11-14.

Susan, S. (2011). The real and symbolic importance of food in war: Hunger pains and big men's bellies in Sierra Leone. *Africa Today*, *58*(2), 42.

Sutton, J. C. (1994). Biological Control of Strawberry Diseases. [Google Scholar]. *Adv. Strawberry Res.*, *13*, 1–12.

Sutton, J. C., Li, D., Peng, G., Yu, H., Zhang, P., & Valdebenito-Sanhueza, R. (1997). Gliocladium Roseum-A Versitile Adversary of Butrysis Ceneria in Crops. [Google Scholar]. *Plant Disease*, *81*(4), 316–328. doi:10.1094/PDIS.1997.81.4.316 PMID:30861808

Sutton, J. C., & Peng, G. (1993). Manipulation and Vectoring of Biocontrol Organisms to Manage Foliage and Fruit Diseases in Cropping Systems. [Google Scholar]. *Annual Review of Phytopathology*, *31*(1), 473–493. doi:10.1146/annurev.py.31.090193.002353

Svatoňová, T., Herák, D., & Kabutey, A. (2015). Financial profitability and sensitivity analysis of palm oil plantation in Indonesia. *Acta Universitatis Agriculturae et Silviculturae Mendilianae Brunensis*, *63*(4), 1365–1373.

Svendsen, G. L. H., & Svendsen, G. T. (2004). *The creation and destruction of social capital: Entrepreneurship, co-operative movements, and institutions*. Edward Elgar Publishing.

Swain, R. B., & Wallentin, F. Y. (2009). Does microfinance empower women? Evidence from self-help groups in India. *International Review of Applied Economics*, *23*(5), 541–556.

Syafa'at, N., Friyatno, S., & Mardianto, S,. & Suryadi. (2004). Kinerja Nilai Tambah dan Produksi Sector Pertanian 2000-2003 [Value Added and Production Performance in the Agricultural Sector 2000-2003]. *Agricultural Policy Analysis*, *2*(1), 1–16.

Talpaz, H., & Tsur, Y. (1982). Optimising aquaculture management of a single-species fish population. *Agricultural Systems*, *9*(2), 127–142. doi:10.1016/0308-521X(82)90027-0

Tamilarasi, R. (2013). Constraints of Women Entrepreneurs in Salem District. *Asia Pacific Journal of Marketing & Management Review*, *2*(8).

Tanumihardjo, S. A., Anderson, C., Kaufer-Horwitz, M., Bode, L., Emenaker, N. J., Haqq, A. M., ... Stadler, D. D. (2007). Poverty, obesity, and malnutrition: An international perspective recognizing the paradox. *Journal of the American Dietetic Association*, *107*(11), 1966–1972.

Taylor, A. M. (1996). *International capital mobility in history: the saving-investment relationship*. Academic Press.

Taylor, J. R., & Lovell, S. T. (2014). Urban home food gardens in the Global North: Research traditions and future directions. *Agriculture and Human Values*, *31*(2), 285–305.

Teckchandani, A. (2014). Do membership associations affect entrepreneurship? The effect of type, composition, and engagement. *Nonprofit and Voluntary Sector Quarterly*, *43*(2_suppl), 84S–104S. doi:10.1177/0899764013502580

Tezuka, H. (1997). Success as the source of failure? Competition and cooperation in the Japanese economy. *Sloan Management Review*, *38*(2), 83.

Thompson, E., Jr., Harper, A. M., & Kraus, S. (2008). *Think Globally—Eat Locally: San Francisco Foodshed Assessment, American Farmland Trust*. Accessed June 23, 2009 at: https://www.farmland.org/programs/states/ca/Feature%20Stories/San-Francisco-Foodshed-Report.asp

Tkachev, A., & Kolvereid, L. (1999). Self-employment intentions among Russian students. *Entrepreneurship and Regional Development, 11*(3), 269–280. doi:10.1080/089856299283209

Todd, N. J. (2012). *A safe and sustainable world: The promise of ecological design*. Island Press.

Torghani, C. (2014). Critical geography of urban agriculture. *Progress in Human Geography, 38*(4), 551–567.

Tornaghi, C. (2011). *Steal this veg*. The Red Pepper June.

Tornaghi, C. (2012). *Edible public space. Experimenting with a socio-environmentally just urbanism*. Territorio.

Tournu, H., Serneels, J., & Van Dijck, P. (2005). Fungal Pathogens Research: Novel and Improved Molecular Approaches for the Discovery of Antifungal Drug Targets. *Current Drug Targets, 6*(8), 909–922. doi:10.2174/138945005774912690 PMID:16375674

Travaline, K., & Hunold, C. (2010). Urban agriculture and ecological citizenship in Philadelphia. *Local Environment, 15*(6), 581–590.

Traversari, S., Cacini, S., Galieni, A., Nesi, B., Nicastro, N., & Pane, C. (2021). Precision Agriculture Digital Technologies for Sustainable Fungal Disease Management of Ornamental Plants. *Sustainability, 13*(7), 3707. doi:10.3390u13073707

Trnka, M., Olesen, J. E., Kersebaum, K. C., Skjelvåg, A. O., Eitzinger, J., Seguin, B., & … . (2011). Agroclimatic conditions in Europe under climate change. *Global Change Biology, 17*(7), 2298–2318.

Tsai, H., Nie, W., Blancon, J. C., Stoumpos, C. C., Asadpour, R., Harutyunyan, B., Neukirch, A. J., Verduzco, R., Crochet, J. J., Tretiak, S., Pedesseau, L., Even, J., Alam, M. A., Gupta, G., Lou, J., Ajayan, P. M., Bedzyk, M. J., Kanatzidis, M. G., & Mohite, A. D. (2016). High-efficiency two-dimensional Ruddlesden–Popper perovskite solar cells. *Nature, 536*(7616), 312–316. doi:10.1038/nature18306 PMID:27383783

Tsai, K.-H., Chang, H.-C., & Peng, C.-Y. (2016). Refining the linkage between perceived capability and entrepreneurial intention: Roles of perceived opportunity, fear of failure, and gender. *The International Entrepreneurship and Management Journal, 12*(4), 1127–1145.

Tushman, M. L., & Romanelli, E. (1985). Organizational evolution: A metamorphosis model of convergence and reorientation. *Research in Organizational Behavior*.

Uche, C., & Familusi, L. C. (2018). The Adoption of agripreneurship as a mitigating measure to unemployment in Nigeria: A topical review. *Global Journal of Management and Business Research: G Interdisciplinary, 18*(2), 24 – 31.

Unger, S., & Wooten, H. (2006). *A food systems assessment for Oakland, CA: Toward a sustainable food plan*. Mayor's Office of Sustainability.

UN-Habitat. (2010). The State of African Cities 2010: Governance, inequalities and urban land markets. Nairobi: United Nations Human Settlements Programme (UN-Habitat).

Urbano, D., & Alvarez, C. (2014). Institutional dimensions and entrepreneurial activity: An international study. *Small Business Economics*, *42*(4), 703–716.

Vaillant, Y., & Lafuente, E. (2007). Do different institutional frameworks condition the influence of local fear of failure and entrepreneurial examples over entrepreneurial activity? *Entrepreneurship and Regional Development*, *19*(4), 313–337.

Valentine, G. (1999). Eating in: Home, consumption and identity. *The Sociological Review*, *47*(3), 491–524. doi:10.1111/1467-954X.00182

Van der Kuip, I., & Verheul, I. (2003). *Early development of entrepreneurial qualities: the role of initial education* (No. N200311). EIM Business and Policy Research.

van der Schans, J. W., & Wiskerke, J. S. (2012). Urban agriculture in developed economies. In *Sustainable food planning: Evolving theory and practice* (pp. 245–258). Wageningen Academic Publishers.

Van der Schans, J. W., & Wiskerke, J. S. C. (2012). Urbanagri-culture in developed economies. In A. Viljoen & J. S. C. Wiskerke (Eds.), *Sustainable Food Planning: Evol-ving Theory and Practice* (pp. 245–258). Wageningen Academic Publishers.

Van der Veen, M. (2003). When is food a luxury? *World Archaeology*, *34*(3), 405–427. doi:10.1080/0043824021000026422

Van Gelderen, M., Brand, M., Van Praag, M., Bodewes, W., Poutsma, E., & Van Gils, A. (2008). Explaining entrepreneurial intentions by means of the theory of planned behaviour. *Career Development International*, *13*(6), 538–559. doi:10.1108/13620430810901688

Van Gelderen, M., & Jansen, P. (2006). Autonomy as a start-up motive. *Journal of Small Business and Enterprise Development*, *13*(1), 23–32. doi:10.1108/14626000610645289

Van Heusden, A., Koornneef, M., Voorrips, R. E., Brüggemann, W., Pet, G., Vrielink-van Ginkel, R., Chen, X., & Lindhout, P. (1999). Three QTLs from Lycopersicon peruvianum confer a high level of resistance to Clavibacter michiganensis ssp michiganensis. [Google Scholar]. *Theoretical and Applied Genetics*, *99*(6), 1068–1074. doi:10.1007001220051416

Van Leeuwen, E., Nijkamp, P., & de Noronha Vaz, T. (2010). The multifunctional use of urban greenspace. *International Journal of Agricultural Sustainability*, *8*(1-2), 20–25.

Vandermeer, J., & Perfecto, I. (2007). The agricultural matrix and a future paradigm forconservation. *Conservation Biology*, *21*(1), 274–277. doi:10.1111/j.1523-1739.2006.00582.x PMID:17298536

Veciana, J. M. (2007). Entrepreneurship as a scientific research programme. In A. Cuervo, D. Ribeiro, & S. Roig (Eds.), *Entrepreneurship. Concepts, Theory and Perspective* (pp. 23–71). Springer.

Venkatarajan, S., Sharma, J., Burman, R., & Lenin, V. (2013). Food Security Vs Nutritional Security—Need for Multi-sectoral Convergence. *International Journal of Agriculture and Food Science Technology*, *4*(6), 621.

Verma. (2018). Opportunities in agri-preneurship in India: Need, challenges and future prospects. *Rashtriya Krishi*, *13*(1), 69–72.

Verma, R. K., Sahoo, A. K., & Rakshit, S. (2017). Opportunities in agri-preneurship in India: Need challenges and future prospects. *Rashtriya Krishi*, *13*(1), 71–72.

Vermeulen, S., & Goad, N. (2006). *Towards better practice in smallholder palm oil production*. Iied.

Vesala, H. T., & Vesala, K. M. (2010). Entrepreneurs and producers: Identities of Finnish farmers in 2001 and 2006. *Journal of Rural Studies*, *26*(1), 21–30. https://doi.org/10.1016/j.jrurstud.2009.06.001

Vesper, K. H. (1990). *New venture strategies*. University of Illinois at Urbana-Champaign's Academy for entrepreneurial leadership historical research reference in entrepreneurship.

Vickers, N. J. (2017). Animal communication: Wheni'mcallingyou, willyouanswertoo? *Current Biology*, *27*(14), R713–R715. doi:10.1016/j.cub.2017.05.064 PMID:28743020

Vijayan, B., & Shivkumar, P. S. (2020). Pro-Agripreneurial Factors for the Formation of Agri-startups in India. *Asian Journal of Agricultural Extension. Economia e Sociologia*, 1–8.

Viljoen, A., & Wiskerke, J. S. C. (Eds.). (2012). *Sustainable Food Planning: Evolving Theory and Practice* (pp. 349–364). Wageningen Academic Publishers. doi:10.3920/978-90-8686-187-3

Villa, M. A. B., Jaramillo, I. D. T., & Alzate, B. A. (2017). EmprendimientoCorporativo e Innovación: Una Revisión y FuturasLíneas de Investigación. *Espacios*, *38*, 1–18.

Vitiello, D., & Brinkley, C. (2014). The hidden history of food system planning. *Journal of Planning History*, *13*(2), 91–112.

Vossenberg, S. (2013). Women entrepreneurship promotion in developing countries: What explains the gender gap in entrepreneurship and how to close it. *Maastricht School of Management Working Paper Series, 8*, 1–27.

Vyas, M. (2019). 11 Million Jobs Lost In 2018 - Cmie. *Cmie*. Https://Www.Cmie.Com/Kommon/Bin/Sr.Php?Kall=Warticle&Dt=2019-01-0809:28:37&Msec=666

Wagener, S., Gorgievski, M., & Rijsdijk, S. (2010). Businessman or host? Individual differences between entrepreneurs and small business owners in the hospitality industry. *Service Industries Journal*, *30*(9), 1513–1527.

Wagner, J. (2007). What a difference a Y makes-female and male nascent entrepreneur in Germany. *Small Business Economics*, *28*(1), 1–21.

Wagner, J. (2007). What a difference a Y makes-female and male nascent entrepreneurs in Germany. *Small Business Economics*, *28*(1), 1–21. doi:10.100711187-005-0259-x

Wagner, J., & Sternberg, R. (2004). Start-up activities, individual characteristics, and the regional milieu: Lessons for entrepreneurship support policies from German micro data. *The Annals of Regional Science*, *38*(2), 219–240. doi:10.100700168-004-0193-x

Walker, S. (2016). Urban agriculture and the sustainability fix in Vancouver and Detroit. *Urban Geography*, *37*(2), 163–182.

Wang, J. H., Chang, C. C., Yao, S. N., & Liang, C. (2016). The contribution of self-efficacy to the relationship between personality traits and entrepreneurial intention. *Higher Education*, *72*(2), 209–224. doi:10.100710734-015-9946-y

Wang, N., Zhang, N., & Wang, M. (2006). Wireless sensors in agriculture and food industry - Recent development and future perspective. *Computers and Electronics in Agriculture*, *50*(1), 1–14. doi:10.1016/j.compag.2005.09.003

Warren, B., & Scranton, P. (2014). *Food nations: selling taste in consumer societies*. Routledge.

Wästfelt, A., & Zhang, Q. (2016). Reclaiming localization and revitalising agriculture: A case study of peri-urban agricultural change in Gotheburg, Sweden. *Journal of Rural Studies*, *47*, 172–185. doi:10.1016/j.jrurstud.2016.07.013

Watkins, J. (2008). Regional groups instrumental in encouraging local markets. *Farm Business,* 16.

Weber, R., & Entrop, A. (2011). Multiple Fungicide Resistance in Botrytis: A Growing Problem in German Soft-Fruit Production. In Fungicides - Beneficial and Harmful Aspects. Academic Press.

Welter, F. (2011). Contextualizing Entrepreneurship—Conceptual Challenges And Ways Forward. *Entrepreneurship Theory and Practice*, *35*(1), 165–184. doi:10.1111/j.1540-6520.2010.00427.x

Welter, F. (2011). Contextualizing entrepreneurship—Conceptual challenges and ways forward. *Entrepreneurship Theory and Practice*, *35*(1), 165–184. https://doi.org/10.1111/j.15406520.2010.00427.x

Wezel A, Bellon S, Doré T, Francis C, Vallod D, David C (2009) Agroecology as a science, a movement and a practice. A review. *Agron Sustain Dev, 29*, 503–515. doi:10.1051/agro/2009004

Wezel, A., Casagrande, M., Celette, F., Vian, J. V., Ferrer, A., & Peigné, A. (2014). Agroecological practices for sustainable agriculture. A review. *Agronomy for Sustainable Development*, *34*(1), 1–20. doi:10.100713593-013-0180-7

Wezel, A., & Soldat, V. (2009). A quantitative and qualitative historical analysis of the scientific discipline agroecology. *International Journal of Agricultural Sustainability*, *7*(1), 3–18. doi:10.3763/ijas.2009.0400

Wezel, A., & Soldat, V. (2009). A quantitative and qualitative historical analysis of the scientific discipline of agroecology. *International Journal of Agricultural Sustainability*, *7*(1), 3–18.

White, B. (2012). Agriculture and the generation problem: Rural youth, employment and the future of farming. *IDS Bulletin*, *43*(6), 9–19.

Whitehead, M. (2013). Neoliberal urban environmentalism and the adaptive city: Towards a critical urban theory and climate change. *Urban Studies (Edinburgh, Scotland)*, *50*(7), 1348–1367.

Wilkinson, J. (2009). The Globalization of Agribusiness and Developing World Food Systems. *Monthly Review (New York, N.Y.)*, *61*(4), 29. doi:10.14452/MR-061-04-2009-08_4

Williamson, B., Tudzynski, B., Tudzynski, P., & Van Kan, J. A. (2007). Botrytis cinerea: The Cause of Grey Mould Disease. *Molecular Plant Pathology*, *8*(5), 561–580. doi:10.1111/j.1364-3703.2007.00417.x PMID:20507522

Wilson, R. A., Barker, M. J., & Brigandt, I. (2007). When traditional essentialism fails: Biological natural kinds. *Philosophical Topics*, *35*(1/2), 189–215. doi:10.5840/philtopics2007351/29

World Bank. (2020). *Employment in agriculture (% of total employment) (modeled ILO estimate)*. https://data.worldbank.org/indicator/SL.AGR.EMPL.ZS

Wukasch, R., & Hofstra, G. (1977). Ozone and Botrytis Interactions in Onion-Leaf Dieback: Open-Top Chamber Studies. Etiology [Google Scholar]. *Phytopathology*, *67*(9), 1080–1084. doi:10.1094/Phyto-67-1080

Xia, B., & Sun, D. (2002). Applications of computational fluid dynamics (CFD) in the food industry: A review. *Computers and Electronics in Agriculture*, *34*(1-3), 5–24. doi:10.1016/S0168-1699(01)00177-6

Xu, D., Deng, Y., Han, T., Jiang, L., Xi, P., Wang, Q., Jiang, Z., & Gao, L. (2018). In vitro and in vivo effectiveness of phenolic compounds for the control of postharvest gray mold of table grapes. *Postharvest Biology and Technology*, *139*, 106–114. doi:10.1016/j.postharvbio.2017.08.019

Xu, D., Yu, G., Xi, P., Kong, X., Wang, Q., Gao, L., & Jiang, Z. (2018b). Synergistic effects of resveratrol and pyrimethanil against Botrytis cinerea on grape. *Molecules (Basel, Switzerland)*, *23*(6), 1455. doi:10.3390/molecules23061455 PMID:29914082

Yang, D. T., & An, M. Y. (2002). Human capital, entrepreneurship, and farm household earnings. *Journal of Development Economics*, *68*(1), 65–88. doi:10.1016/S0304-3878(02)00006-8

Yang, M. M., Li, T., & Wang, Y. (2020). What explains the degree of internationalization of early-stage entrepreneurial firms? A multilevel study on the joint effects of entrepreneurial self-efficacy, opportunity-motivated entrepreneurship, and home-country institutions. *Journal of World Business*, *55*(6), 101114. doi:10.1016/j.jwb.2020.101114

Yaseen, A., Saleem, M. A., Zahra, S., & Israr, M. (2018). Precursory effects on entrepreneurial behaviour in the agri-food industry. *Journal of Entrepreneurship in Emerging Economies*.

Yaseen, A., Saleem, M. A., Zahra, S., & Israr, M. (2018). Precursory effects on entrepreneurial behaviour in the agri-food industry. *Journal of Entrepreneurship in Emerging Economies*, *10*(1), 2–22.

Yeomans, P. A. (1954). *The Keyline plan*. PA Yeomans.

Yunus, M. (2004). Grameen Bank, microcredit and millennium development goals. *Economic and Political Weekly*, •••, 4077–4080.

Zahra, S. A. (1999). 'Governance, Ownership, and Corporate Entrepreneurship: The Moderating Impact of Industry Technological Opportunities', Academy of Management Journal, 39 (6), December, 1713-35. *International Library of Critical Writings in Economics*, *106*, 325–347.

Zahra, S. A., & Wright, M. (2011). Entrepreneurship's next act. *The Academy of Management Perspectives*, *25*(4), 67–83. https://doi.org/10.5465/amp.2010.0149

Zahra, S. A., Wright, M., & Abdelgawad, S. G. (2014). Contextualization and the advancement of entrepreneurship research. *International Small Business Journal*, *32*(5), 479–500. https://doi.org/10.1177 /0266242613519807

Zamberi Ahmad, S., Roland Xavier, S., & Rahim Abu Bakar, A. (2014). Examining entrepreneurial intention through cognitive approach using Malaysia GEM data. *Journal of Organizational Change Management*, *27*(3), 449–464.

Zampetakis, L. A., Anagnosti, A., & Rozakis, S. (2013, September). *Understanding entrepreneurial intentions of students in agriculture and related sciences.* Poster session presented at the meeting of the EEAE 2014 Congress" Agri-Food and Rural Innovations for Healthier Societies," Ljubljana, Slovenia.

Zampetakis, L. A., Kafetsios, K., Bouranta, N., Dewett, T., & Moustakis, V. S. (2009). On the relationship between emotional intelligence and entrepreneurial attitudes and intentions. *International Journal of Entrepreneurial Behaviour & Research*, *15*(6), 595–618. doi:10.1108/13552550910995452

Zapata, M. (2017). *Appendix 1: Equity lens for the Portland State University diversity, equity, and inclusion action plan.* Portland State University.

Zapkau, F. B., Schwens, C., Steinmetz, H., & Kabst, R. (2015). Disentangling the effect of prior entrepreneurial exposure on entrepreneurial intention. *Journal of Business Research*, *68*(3), 639–653. doi:10.1016/j.jbusres.2014.08.007

Zaridis. (2016). The Relation Between Entrepreneur's Characteristics and Competitive Advantage. A Roadmap to SME's Success. *International Journal of Emerging Research in Management & Technology.*

Zezza, A., & Tasciotti, L. (2010). Urban agriculture, poverty, and food security: Empirical evidence from a sample of developing countries. *Food Policy*, *35*(4), 265–273.

Zhang, G., Choi, C., Bartzanas, T., Lee, I.-B, & Kacira, M. (2018). Computational Fluid Dynamics (CFD) research and application in Agricultural and Biological Engineering. *Computers and Electronics in Agriculture*, *149*, 1-2. .2018.04.007 doi:10.1016/j.compag

Zhao, H., Seibert, S. E., & Hills, G. E. (2005). The mediating role of self-efficacy in the development of entrepreneurial intentions. *The Journal of Applied Psychology*, *90*(6), 1265–1272. doi:10.1037/0021-9010.90.6.1265 PMID:16316279

Zhu, N. Y., Liu, X., Liu, Z. Q., Hu, K., Wang, Y. K., Tan, J. L., Huang, M., Zhu, Q., Ji, X., Jiang, Y., & Guo, Y. (2018). Deep learning for smart agriculture: Concepts, tools, applications, and opportunities. *International Journal of Agricultural and Biological Engineering*, *11*(4), 32–44. doi:10.25165/j.ijabe.20181104.4475

About the Contributors

Mohd Yasir Arafat has completed PhD in Commerce (Entrepreneurship Area) in 2019 and Master of Commerce from Aligarh Muslim University (AMU), Aligarh, India. Currently, he is working as Assistant Professor at Centre for Distance Education, AMU. He has been Ministry of Minority Affairs Scholar, Senior Research Fellow, Maulana Azad National Fellow, Centre for Research and Entrepreneurship Education (CREED) Fellow at Entrepreneurship Development of India (EDII). He is recipient of Best Paper Award and Best Scholar Award. His research interests include electric vehicles entrepreneurship, agricultural entrepreneurship, renewable energy entrepreneurship and entrepreneurship education. He has published papers in learned journals like International Entrepreneurship and Management Journal (Springer), Journal Global Entrepreneurship Research (Springer), International Journal of Intelligent Enterprise (InderSciences), International Journal of Knowledge Management (IGI Global), and Academy of Management Global Proceedings. He is also Lead Editor of a Book "Driving Factors for Venture Creation and Success in Agricultural Entrepreneurship" to be published by IGI Global.

Adil Khan is an assistant professor at School of Management, OP Jindal University, Raigarh. He teaches marketing and Entrepreneurship subjects. He conducts research in area of marketing and entrepreneurial behaviour.

* * *

Hashem Abdullah Al Nemer is an Associate Professor of Economics and Finance at college of business, University of Jeddah, Saudi Arabia. He is a member of several importance committees at the university (Innovation and Entrepreneurship, Academics Promotion, Smart Food Truck High board member, University High Consultant boards, Scientific and research, Community Serving, etc). Dr. Hashem, was the Founder and General Director for Center of Innovation and Entrepreneurship at University of Jeddah for five years in row. As founding General Director of the center Dr. Hashem he was planned and implemented the center. As Director he initiated various successful programs to promote entrepreneurship and innovation among Saudi nationals. He was also the Vice Dean of College of Business, during his tenure as Vice Dean he initiated and established various Master program including MSc in Entrepreneurship and Innovation.

Lloyd James Baiyegunhi is currently an Associate Professor in Agricultural Economics, School of Agricultural, Earth and Environmental Sciences, University of KwaZulu-Natal (UKZN), South Africa. Before joining the University in January 2009 as a Lecturer, he had begun his career at University Fort Hare, Alice South Africa where he taught from May 2006 - December 2008. He has authored/co-authored over 35 papers in ISI/DoHET accredited journals. Professor Baiyegunhi is registered as a Professional Natural Scientist by the South African Council for Natural Scientific Professions. He is also an NRF C-rated researcher, which confirms that he is an established researcher with a sustained recent record of productivity in the field and with a body of quality work, the core of which has coherence and attests to ongoing engagement with the field. This comes with some international recognition for the quality and impact of his recent research outputs.

Swati Bhatt is a senior faculty in Marketing Area of Department of Management Studies, SRM Institute of Science & Technology, Delhi-NCR Campus. Her research interest are Sustainability, Services Marketing, CSR. She has a number of publications to her credit.

Ermal Bino is a lecturer of management courses under the College of Business at the University of Buraimi, Sultanate of Oman.

Lejdel Brahim was born at EL-Oued (Algeria) in 1982. He obtained his magister degree in 2009 from University of Ouragla (Algeria). He is actually PHD in computer science since September, 2015. He obtained a PHD degree from university of Biskra. He works actually in the university of EL-Oued. He is working on the GIS domain and artificial intelligence. He is member of the Smart Computer Science Laboratory. He is reviewer of some indexed journals, as Visual language and computing, international journal of geographical information science. He is author of some research papers published in refereed journals and international conferences.

Ilya Bystrov has PhD in economics & management. During his academic career he contributed to and published more than 22 research articles in journals, book chapters and articles in conference proceedings. His recent research interests are in the field of sustainability, economic development, and expatriate adjustment. His working experience includes leadership positions in logistics, consulting, and education, as well as self-employment. Last 10 years of his international teaching career he has spent in the GCC region teaching mostly management-related courses. His combined teaching and industry experience exceeds 20 years.

Ferdinand Epoc is an Associate Professor under the College of Business at the University of Buraimi, Sultanate of Oman.

Veer P. Gangwar is working as a Professor at Mittal School of Businesss, Lovely Professional University, India.

Abukasim Idrisi is a Research Scholar working in the field of international economics in the Department of Economics at Aligarh Muslim University.

Pallavi Jain has a very lumin academic and professional career She has obtained her Ph.D. (Chemistry) from Banasthali Vidyapeeth, Rajasthan. Dr. Jain's arch interests include Inorganic Chemistry, Green chemistry, Analytical Chemistry. She has published numerous research paper in journals of repute.

Mohd Asif Khan is an Associate Professor in the Department of Commerce, Aligarh Muslim University and currently serving as Director, Center for Professional & Technical Studies, AMU. Dr. Khan has served as a member of University's Executive Council for two consecutive tenures. He also served as the University Proctor. As director, CPTS, Dr. Khan introduced several employment generating and entrepreneurial skills based courses for the weaker sections of the community. His research interest includes Entrepreneurship Development, New Enterprise Creation, etc.

Mohd Saeem Khan received his Master of Commerce (2010) and Ph.D, Commerce (2016) from department of commerce, Aligarh Muslim University (AMU), Aligarh, India. Currently he is working as an Assistant Professor in the Department of Commerce, AMU, Aligarh. His research interest include sector specific entrepreneurship and new venture creation.

Shad Ahmad Khan is working as Assistant Professor at College of Business, University of Buraimi.

Sunil Kulkarni is working as an Associate Professor in Chemical Engineering Department of Gharda Institute of Technology, Khed, Maharashtra, India. He has 20 years of experience in teaching and research. He is associated with many international journals as reviewer and editor. He has authored 11 books on various areas in chemical and biochemical engineering. His areas of interest include adsorption, biological wastewater treatment and advances in biotechnology.

Ngadi is a researcher at the Research Center for Population, the Indonesian Institute of Sciences (LIPI) since 2005. He is a Ph.D. candidate in Human Geography, the University of Tokyo, Japan. He obtained master's degree in Population and Employment Studies at the University of Indonesia (2005). His research interests relate to population and labor mobility, employment and human resources, demographic change, and rural development. He has published more than 30 articles both as books and journals in national and international publications.

Femi Emmanuel Oni is an Agricultural Scientist researcher with a core interest in an interdisciplinary field. He has a Bachelor of Agriculture from the University of Ilorin Nigeria and a master's degree in Development Studies from the University of KwaZulu-Natal. His research interest is food and all interrelated subject which include agriculture, anthropology, education, economics, environment, health and nutrition, justice, communication, and sociology. He has also author articles in journals and books as well as attended conferences.

Anjila Saleem is an undergraduate in commerce from Aligarh Muslim University, Aligarh. She is currently pursuing post-graduation from the same university. Her area of interest for research lies in the field of entrepreneurship.

Sharat Sharma is currently working as a senior faculty in Department of Management Studies, SRM Institute of Science & Technology, Delhi-NCR Campus. His research interest are anchored within Finance, Sustainability, General Management and Decision Sciences.

Mohd Taqi has completed UGC Dr. S. Radhakrishnan Post Doctoral Fellowship (Commerce/Management) in 2018 and Ph.D. (Commerce) in 2014 from Department of Commerce, AMU, Aligarh. He has qualified UGC-NET in Commerce in 2013. He has completed M.Com. from CCS Univeristy, Meerut in 2009. His area of interest is Accounting and Finance. Dr. Taqi has visited France, Thailand, Bangladesh and Nepal for the presentation of his research paper in various international conferences. He has published more than fourteen (14) research papers in reputed peer reviewed international journals. He has also participated and presented research papers in more than fifteen (15) national as well as international conferences. Dr. Taqi is working as Assistant Professor of Commerce at Centre for Distance and Online Education, AMU, Aligarh. He is engaged in teaching of Undergraduate and Post Graduate students.

Amaan Ullah is working as a Research Scholar in the field of Green and Agricultural entrepreneurship and Water Management in the Department of Commerce at Aligarh Muslim University, Aligarh.

José G. Vargas-Hernández, M.B.A., Ph.D., Member of the National System of Researchers of Mexico and a research professor at University Center for Economic and Managerial Sciences, University of Guadalajara. Professor Vargas-Hernández has a Ph. D. in Public Administration and a Ph.D. in Organizational Economics. He has undertaken studies in Organisational Behaviour and has a Master of Business Administration, published four books and more than 200 papers in international journals and reviews (some translated to English, French, German, Portuguese, Farsi, Chinese, etc.) and more than 300 essays in national journals and reviews. He has obtained several international Awards and recognition.

Index

E

F

G

I

L

M

N

O

P

R

S

T

U

V

W

www.ingramcontent.com/pod-product-compliance
Lightning Source LLC
LaVergne TN
LVHW081348300926
849161LV00008B/2097